Springer Series in Information Sciences   27

Editor: Thomas S. Huang

Springer Series in Information Sciences

Editors: Thomas S. Huang Teuvo Kohonen Manfred R. Schroeder
Managing Editor: H. K. V. Lotsch

Zhengyou Zhang Olivier Faugeras

3D Dynamic Scene Analysis

A Stereo Based Approach

With 181 Figures

Springer-Verlag Berlin Heidelberg GmbH

Dr. Zhengyou Zhang
Dr. Olivier Faugeras

INRIA Sophia Antipolis, 2004 Route des Lucioles, BP 93,
F-06902 Sophia Antipolis Cedex, France

Series Editors:

Professor Thomas S. Huang

Department of Electrical Engineering and Coordinated Science Laboratory,
University of Illinois, Urbana, IL 61801, USA

Professor Teuvo Kohonen

Laboratory of Computer and Information Sciences, Helsinki University of Technology,
SF-02150 Espoo 15, Finland

Professor Dr. Manfred R. Schroeder

Drittes Physikalisches Institut, Universität Göttingen, Bürgerstrasse 42–44,
W-3400 Göttingen, Fed. Rep. of Germany

Managing Editor: Helmut K. V. Lotsch

Springer-Verlag, Tiergartenstrasse 17,
W-6900 Heidelberg, Fed. Rep. of Germany

ISBN 978-3-642-63485-7   ISBN 978-3-642-58148-9 (eBook)
DOI 10.1007/978-3-642-58148-9

Typesetting: Camera ready by authors

54/3140-5 4 3 2 1 0

Preface

The problem of analyzing sequences of images to extract three-dimensional motion and structure has been at the heart of the research in computer vision for many years. It is very important since its success or failure will determine whether or not vision can be used as a sensory process in *reactive* systems.

The considerable research interest in this field has been motivated at least by the following two points:

1. The redundancy of information contained in time-varying images can overcome several difficulties encountered in interpreting a single image.
2. There are a lot of important applications including automatic vehicle driving, traffic control, aerial surveillance, medical inspection and global model construction.

However, there are many new problems which should be solved: how to efficiently process the abundant information contained in time-varying images, how to model the change between images, how to model the uncertainty inherently associated with the imaging system and how to solve inverse problems which are generally ill-posed. There are of course many possibilities for attacking these problems and many more remain to be explored. We discuss a few of them in this book based on work carried out during the last five years in the Computer Vision and Robotics Group at INRIA (Institut National de Recherche en Informatique et en Automatique).

In fact, image sequence analysis is a rather vague term and can be used with several meanings. Our definition is that, given one or several sequences of images from one (*monocular sequences*) or several cameras whose relative positions are known (*stereo sequences*) acquired while the cameras are moving in an unknown environment containing a number of mobile rigid objects, we must determine the various relative motions (cameras and objects) and the structure of the scene. There has been a tremendous amount of work on the analysis of monocular sequences of images during the past decade. Much less has been done on the analysis of sequences of stereo images. This book describes our work in this domain. As the relative positions of the cameras are known, we can obtain a set of three-dimensional representations of the environment sequentially reconstructed by a stereo system. Therefore, we estimate three-dimensional motion from three-dimensional structure, hence the title of the book.

An important feature of the work described in this book is that uncertainty is modeled as early as in the original 2D images and manipulated during all subsequent stages of processing. This distinguishes our approach from most of the approaches reported in the literature, which usually do not deal with this issue. Because of its power and simplicity, we have chosen a probabilistic framework to represent and process this uncertainty.

As described earlier, image sequence analysis has been motivated by a number of applications. The ultimate objective of the research in this field is to use vision as a sensory process to solve practical problems. The work described in this book has been carried out in the context of visual navigation of a mobile robot. The algorithms proposed have been tested using a large number of real images acquired by a trinocular stereo system. We have used some of these experiments in this book to illustrate the strong points and the shortcomings of the algorithms.

The authors would like to thank Nicholas Ayache and Francis Lustman for developing the trinocular stereo system, without which the work described in this book would not have been possible. The authors would also like to thank the other members, both past and present, of the INRIA Computer Vision and Robotics Group for generating an active and productive research environment, and for many fruitful discussions over many years.

Last but not least we would like to express our thanks for the support we have received during the course of this work through two Esprit projects, P940 and P2502. The financial support we have received from the EEC and the strong amount of collaboration with other European groups which has been initiated by these projects have played very important roles in stimulating and orienting our work. We would like to thank in particular Giorgio Musso and Giovanni Garibotto from Elsag, and Bernard Buxton from GEC whose roles as project leaders have been invaluable.

This book was typeset by the authors using the LATEX macropackage under the strict Springer instructions. We are indebted to Mrs. Urda Beiglböck and Dr. Helmut Lotsch of Springer-Verlag for their aid and encouragement.

Sophia Antipolis, France   Zhengyou Zhang
March 1992   Olivier Faugeras

Contents

Part II   Two-View Motion Analysis

1. Introduction

The world in which we live is three-dimensional. Everything (in macrocosm and in microcosm) in this world is in motion. So, our world can also be considered to be four-dimensional (with the fourth dimension — time). More concretely, our eyeballs are in continuous motion. An observer may move relative to a static environment, and some other objects in this environment may also move relative to the observer and to the environment. It is hard to visualize this 4D world. Figure 1.1 illustrates a sample of changing environment seen by a camera (called sometimes *spatiotemporal images*).

Although even the simplest creature has the ability to discern objects, ascertain their motion, and navigate in three-dimensional space using vision, it has proved extremely difficult to incorporate such vision capability into machines. Major efforts in the vision community are invested in understanding vision in man and machines [1.1]. Neurophysiologists and psychophysicists seek to understand biological vision systems at different but complementary levels. Computer vision scientists attempt to develop vision systems which can have certain features in common with the human visual system. The findings in those two domains have in fact fertilized each other. Neurophysiological and psychophysical experiments have influenced the design of computer algorithms for image analysis. At the same time, Computer Vision scientists have provided neurophysiologists and psychophysicists with a mathematical framework for modeling biological vision. However, since humans are subject to many robust illusions, *Hochberg* [1.2] asserts that "machines should not see as people do, but must know how people see". It will be a long time before we can design a complete model capable of accomplishing tasks as complicated as those done by human vision.

This monograph describes our computational investigation of motion analysis from a sequence of stereo frames. Motion analysis is a very active field of research in Computer and Robot Vision. Its applications include mobile robot navigation [1.3, 4], scene segmentation [1.5–7], world model construction [1.8, 9] and road following [1.10, 11]. We can easily find many other potential applications: e.g., automatic object tracking, which is of immense interest in the domain of automatic surveillance (traffic control, aerial attack and defense); data compression [1.12] for efficient transmission is another application of motion analysis, as we can see in Chap. 12 where we describe the integration of multiple views.

time

Fig. 1.1. Projection of a 4D world on a camera

1.1 Brief Overview of Motion Analysis

Movement is an inherently continuous process. The relative motions between objects in a scene and cameras, and between different moving objects in the scene produce the apparent motions of objects between frames. The objective of motion analysis is to infer the real motion of objects in the scene from the changes between adjacent frames.

Table 1.1 summarizes different approaches used for the computation of motion from a sequence of frames. The feature-based approach consists in using the information of changes in feature positions, while the optical-flow approach consists in using the information of changes in gray levels. Current approaches distinguish to each other in the number of frames used: short-sequence analysis uses only 2 or 3 frames, while long-sequence analysis uses much more frames. We can further classify the currently existing approaches into two categories ac-

Table 1.1. Summary of different approaches to motion determination

Change in frames	*Number* of frames	*Dimensions* of frames
features	**small** (2 or 3)	**2D**
gray levels	**big** (long sequence)	**3D**

cording to the dimension of data available. If we have one image at each time (monocular sequences), only 2D data are available. If we have two or three images at each instant (bi- or trinocular sequences, or simply, stereo sequences), we can use directly 3D data by stereo in motion analysis. From Table 1.1, we can roughly divide existing methods into 8 categories (2^3 combinations).

The feature-based approach is usually divided into three steps. The first step is the extraction of a set of salient two-dimensional features from a sequence of monocular (2D) frames. Features such as points, lines or curves correspond to corners, boundaries, region marks, occluding boundaries of surfaces, or shadows of objects in 3D space. Feature detection is an old but still very active field of research in Image Processing and Computer Vision [1.13–19]. In the case of a sequence of 3D frames, three-dimensional features are reconstructed from a set of already extracted two-dimensional features by a stereo range finder or other methods [1.20–23].

The second step is to find correspondences between features. Features on an object need to be tracked from one frame to another, prior to motion computation. This problem is recognized as extremely difficult. Some constraints such as rigidity are exploited in matching algorithms. Techniques, such as template matching, structural matching, matching by tree-searching, matching by constrained relaxation and hypothesis generation and testing, have been developed to solve this problem. Some related work includes [1.5, 6, 24–28]. We will return to this problem where appropriate.

Once the correspondence of features is established, the last step — Motion Estimation — can be realized. Over the past ten years, much work has been done in estimating three-dimensional motion parameters from 2D orthogonal and perspective views. Due to the nonlinearity of the relation between the projections and 3D objects, and due to the noise introduced in image formation and in image digitalization, the uniqueness of solutions and robustness of algorithms have attracted many researchers [1.29–37]. These results are theoretically very interesting but are limited to the estimation of the motion of a single object and to the reconstruction of the structure of the scene up to a scale factor unless considerable *a priori* information is available. Also, these solutions are reported in the literature to be very sensitive to noise, and thus have been so far of little practical use except, perhaps, for calibration. The complexity of motion estimation can be considerably reduced by using stereo frames or range data, since the depth information is available. However, the motion estimation is not trivial since the 3D data available are almost always corrupted by noise [1.8, 38–41]. Contrary to using purely 3D data obtained from stereo, another research direction called stereo-motion cooperation is pursued by several researchers [1.42]. That method consists in using 2D velocities of stereo matched lines, on a pair of cameras, to estimate the corresponding 3D line kinematics without prior 3D reconstruction.

The optical-flow based approach can be divided into two steps. No correspondence between features in adjacent frames is needed. The first step is the computation of an apparent velocity field or so-called optic flow from temporal change in gray levels in the image plane. The computation requires the evaluation of the first and often the second partial derivatives of the intensity images [1.43–47]. *Bouthemy* [1.48] proposed a method to detect moving edges and to estimate velocity field in a maximum likelihood framework. The second step is to infer with additional constraints the motion parameters and structure of objects in 3D space from optical flow [1.49–51]. Optical flow is also used in the analysis of short sequences of stereo frames. Optical flow is first computed for each view of the stereo system, stereo correspondence is then established, and finally three-dimensional velocity field is reconstructed [1.52,53]. As reported in the literature, optical-flow based techniques are very unstable and unreliable in the general real image case, since the evaluation of partial derivatives enhances the noise level, and the real velocity field is different from the apparent velocity field or optical flow [1.54,55]. However, several researchers follow a qualitative way using optical flow to obtain symbolic description of the scene [1.56–58]; the information is still rich enough in many applications such as obstacle avoidance.

The algorithm proposed by *Horn* and *Harris* [1.59] for recovering the motion of a vehicle from a sequence of range images of a static environment could be approximately considered as an optical-flow approach. Range images are taken by a range sensor rigidly attached to the vehicle. By assuming that most of the surface is smooth enough so that local tangent planes can be constructed, and that the motion between frames is small, the method first computes the slopes of the surface and the rate of change of the elevation at each pixel. The motion is then estimated utilizing a least-squares technique to minimize the differences between the measured and predicted (by the so-called *elevation rate constraint equation*) elevation change rates. If the motion is big, an initial estimate must be applied to approximately register range data before applying their algorithm.

Another approach to combat noise is the use of long sequences of frames. 2D or 3D feature based long sequence analysis has recently attracted several researchers [1.60–64]. Objects are assumed to move under certain kinematic models. We will discuss later this approach in more detail. With a long sequence of monocular images, the optical-flow approach can yield a more reliable solution than with a short sequence, as reported in [1.65]. To our knowledge, there is no report yet on the application of optical flow to long sequences of stereo frames.

For a full review of the development of motion analysis from sequences of frames up to 1987, the reader is referred to [1.66,67]. More recent developments have been reported in the proceedings of the biannual international conferences on pattern recognition, the proceedings of the biannual international conferences on computer vision, the proceedings of the 1989 IEEE workshop on visual motion, the proceedings of the annual international conference on computer vision and pattern recognition and a special issue on visual motion of the *IEEE Transaction on Pattern Analysis and Machine Intelligence* (Vol. 11, No. 5, 1989). We should note that motion analysis has much in common with another important field of Computer Vision: Model-based object recognition and localization [1.40, 68].

1.2 Statement of the "Motion from Stereo" Problem

As we have observed in the earlier section, there has been a tremendous amount of work on the analysis of monocular sequences of images. A lot of work has been published on the estimation of motion and structure using a minimum amount of information, on the development of linear algorithms and on the uniqueness and multiplicity of solutions [1.29–31, 33, 34, 69–74]. These results are theoretically very interesting but are limited to the estimation of the motion of a single object and to the reconstruction of the structure of the scene up to a scale factor unless considerable *a priori* information is available. Also, due to the complexity of image formation and to the nonlinear relation between 3D motion and changes in the images, the solutions have been reported to be very sensitive to noise, and thus have so far of little practical use except, perhaps, for calibration.

Much less has been done on the analysis of several, simultaneously acquired, sequences of images. With the development of passive stereovision systems [1.21–23] and range-finder systems [1.20], some work related to motion analysis from sequences of 3D frames has recently emerged. When stereo data (range data in general) are used, the motion computation problem can be significantly simplified. Clearly, the amount of information is much higher and one would hope that this would allow us to solve the problem in a more robust fashion. Since the structure of objects is known, the absolute value of the translation can be computed. In this monograph, we consider the problem called the *motion from stereo*. More concretely, we address the following problem:

> Given a stereo rig moving in an unknown environment where an arbitrary number of rigid mobile objects may be present, we want to segment the various mobile objects from the background and determine the motion of the stereo rig and that of the objects.

Although the structure of the environment is unknown, it is assumed to be composed of straight lines. We assume that the stereovision system works fast enough to provide 3D maps of the environment at a sufficient rate. The rate required depends, of course, upon the speed of the vehicle and that of the moving obstacles. The current stereo algorithm uses three cameras and has been described in [1.22, 23]. The 3D primitives (tokens) used in the stereo system are

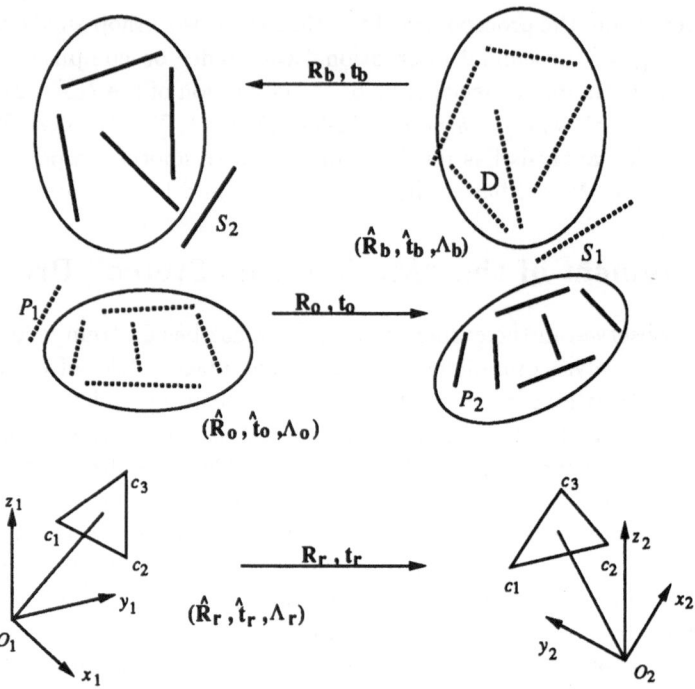

Fig. 1.2. The problem of motion from stereo
$(\hat{\mathbf{R}}_r, \hat{\mathbf{t}}_r, \Lambda_r)$: Predicted motion of the robot
$(\hat{\mathbf{R}}_b, \hat{\mathbf{t}}_b, \Lambda_b)$: Predicted motion of the background
which is the inverse of the robot motion
$(\hat{\mathbf{R}}_o, \hat{\mathbf{t}}_o, \Lambda_o)$: Predicted motion of the object

at the moment line segments corresponding to significant intensity discontinuities in the images obtained by the stereo rig.

At instants t, $t + \Delta t$, $t + 2\Delta t$, ..., we obtain sets of such 3D tokens which we call 3D snapshots. Δt is the time that separates the adjacent frames. It is assumed, for convenience and without loss of generality, that t is nonnegative, integer-valued, and that $\Delta t = 1$ (unit of time). For some of these 3D tokens (the ones we have been observing for a while), we may have some indication of their kinematics while for some of them (the ones which just appeared in the field of view), we have no information except their position in the coordinate system attached to the stereo rig at the given instant.

The situation is as indicated in Fig. 1.2. In this figure, the mobile vehicle is represented at times t_1 and t_2 by the two frames of reference (x_1, y_1, z_1) and (x_2, y_2, z_2). The three cameras are symbolized by the triangles $c_1c_2c_3$. Their position and orientation are fixed (and known, thanks to stereo calibration [1.75,76]) in the two previous frames of reference. At time t_1, we have an estimate

$(\hat{\mathbf{R}}_\mathbf{r}, \ \hat{\mathbf{t}}_\mathbf{r})$ of the predicted rigid motion of the mobile vehicle between t_1 and t_2, as well as some measure of our confidence in this estimate. This confidence is expressed as a covariance matrix $\boldsymbol{\Lambda}_\mathbf{r}$. The actual motion of the vehicle (usually called *egomotion*) between t_1 and t_2 is $(\mathbf{R}_\mathbf{r}, \ \mathbf{t}_\mathbf{r})$. In this figure, we have also grouped the 3D tokens into classes, each class corresponding to an hypothesized rigid object. Tokens are represented in dotted lines at time t_1 and in continuous lines at time t_2.

The figure has also been drawn assuming that the background (indexed as b) and one mobile object (indexed as o) are present in the scene. The estimation $(\hat{\mathbf{R}}_\mathbf{b}, \ \hat{\mathbf{t}}_\mathbf{b})$ of the predicted rigid motion of the background is equal to the inverse $(\hat{\mathbf{R}}_\mathbf{r}^\mathbf{T}, \ -\hat{\mathbf{R}}_\mathbf{r}^\mathbf{T}\hat{\mathbf{t}}_\mathbf{r})$ of the predicted motion of the mobile vehicle. The actual motion of the background is the inverse of the actual vehicle motion.

The segment S_1 is also part of the background but was not visible before time t_1, therefore it has not been grouped with the other segments of the background, although it matches with segment S_2, measured at time t_2. Similarly some 3D tokens which were present at time t_1 have disappeared from the field of view at time t_2, because of motion (for example, the segment D in the figure).

In the same way, we have an estimate $(\hat{\mathbf{R}}_\mathbf{o}, \ \hat{\mathbf{t}}_\mathbf{o})$ of the predicted object motion between t_1 and t_2 and a measure of the confidence in this measure, $\boldsymbol{\Lambda}_\mathbf{o}$. The actual object motion is $(\mathbf{R}_\mathbf{o}, \ \mathbf{t}_\mathbf{o})$ and segment P_1, also belonging to the object has just appeared and, even though it matches P_2 at time t_2, we do not have any estimation of its kinematics.

Figure 1.2 displays the situation in the so-called *steady-state mode* where some structure of the scene has been captured by the previous observations. In the *bootstrapping mode*, whenwe start analyzing the scene, no such information is available, no grouping of 3D tokens has yet taken place, and no estimation of their kinematics has been computed, with the exception of the motion of the vehicle which can in practice be partially estimated from odometry. From frame to frame, there may exist significant changes. In order to determine ego- and object motions, we have to bring into correspondence some 3D tokens in successive frames using available constraints. The rigidity constraint is, of course, a very powerful one which will be put into use.

Notice that even in the steady-state mode, there is always a little bit of bootstrapping in the sense that new 3D tokens appear in the field of view for which no *a priori* information is available.

1.3   Organization of This Book

This book is divided into four parts.

The first part consists of three chapters, dedicated to the fundamental elements required in the following three parts. Chapter 2 presents some mathematical notations in probability, uncertainty geometry and parameter estimation. Chapter 3 gives a brief review of how 3D line segments, the only primitives we have used so far, can be reconstructed from a stereo system. Chapter 4 describes how to represent 3D rigid motion and 3D line segments.

The second part describes our motion determination algorithm in the bootstrapping mode. We first compare different methods for determining motion from 3D line segment matches, using both synthetic and real data. The method based on the extended Kalman filter, which has been originally proposed in [1.8,77,78], is used in our motion analysis algorithm for its efficiency and robustness. We then develop an algorithm based on the "hypothesize-and-verify" paradigm to match two 3D views and to compute the motion between them. Since the rigidity constraints are extremely important in matching, one chapter is dedicated to it. We show theoretically that the rigidity constraints we use are complete for 3D line segments and allow us to recover a unique and rigid motion. The motion algorithm is then generalized to determine multiple object motions and to recognize and localize an object from the model base in the observed scene.

The third part describes several important applications to the mobile robot. We develop an algorithm to calibrate the stereo and odometric coordinate systems of the INRIA mobile robot, using only visual information from stereo. This calibration is essential for visual navigation. A simple experiment is provided to show its application to visual navigation. We demonstrate also how to integrate the information from the odometric system into our motion-determination algorithm, which allows us to speed up considerably the motion determination process. We develop a fusion module which, together with the motion algorithm, is used to integrate multiple stereo views, in order to build a global and precise world model of the environment. Interesting results including a global map of a room built by our robot are described.

The fourth part describes our motion determination algorithm in steady-state mode. We first present a framework which consists of two levels: at the low level, each individual token is tracked from frame to frame and its kinematic parameters can be computed by filtering in time; at the high level, the disordered tokens are grouped together into objects based on the similarity in kinematic parameters, and the low-level tracking is controlled by a supervisor. This framework intends to cope with the problems such as occlusions, appearances, disappearances and absences of tokens, which are very important when dealing with long sequences. The classical kinematics is used to model the motion of tokens. We then derive closed-form solutions for some special motions. Finally, we describe the matching and grouping techniques and provide the experimental results.

2. Uncertainty Manipulation and Parameter Estimation

I t is very important to explicitly manipulate the uncertainty of measurements. The observations made by a sensor are always corrupted by noise, sometimes they are even spurious or incorrect. In a stereovision system, for example, the uncertainty arises from five sources:

- Edge detection process (pixel noise),
- Edge fitting (for example, polygonal approximation),
- Stereo calibration process,
- Stereo matching process,
- 3D reconstruction process.

In order to use correctly and effectively the information provided by a sensor, one should model the sensor by its abilities and its limitations. Sensor modeling is considered one of the important problems to be solved in Robotics by *Brady* [2.1]. Currently, researchers follow both qualitative and quantitative approaches to model the capabilities of sensors.

Sensors can be modeled qualitatively. For example, *Henderson* and *Shilcrat* [2.2] have introduced the logical sensor concept. Figure 2.1 gives a pictorial description of a logical sensor. The *Logical Sensor Name* uniquely identifies the logical sensor. Sensor data flow up through the currently executing program (one of $Program_1$ to $Program_n$) whose output is characterized by the *Characteristic Output Vector*. Control commands are accepted by the *Control Command Interpreter*, which then issues the appropriate control commands to the logical sensors currently providing input to the selected program. Programs 1 through n provide alternative ways of producing the same characteristic output vector for logical sensor. The role of the *Selector* is to monitor the data produced by the currently selected program and the control commands. If a failure of the program or a failure of an input logical sensor at a lower level is detected, the selector must undertake the appropriate error recovery mechanism and choose an alternative method (if possible) to produce the characteristic output vector. In addition, the selector must determine if the control commands require the execution of a different program to compute the characteristic output vector (i.e., whether dynamic reconfiguration is necessary). The primary motivation for logical sensor

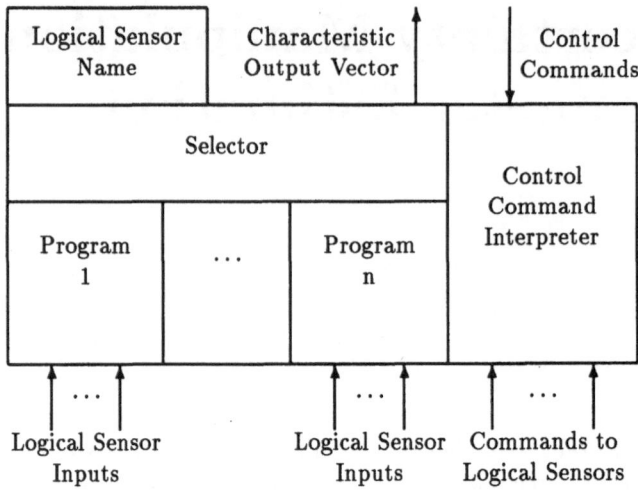

Fig. 2.1. Description of a logical Sensor

specifications was to develop a method to permit an implementation-independent description of the required sensors and algorithms in a multisensor system. Since the measurements we obtain in our system are inherently numeric, this technique has not been adopted.

Sensors can also be modeled quantitatively. There exist several methods for dealing explicitly with the uncertainty problem in Robotics and Computer Vision:

- Empirical method, by giving some *a priori* error bound on the measurement data [2.3,4],
- Fuzzy sets, by modeling uncertainty as a fuzzy subset [2.5],
- Probabilistic modeling, by modeling sensor as a statistical process [2.6–13].

Although all these methods have been reported to have success to some degree, the last one is preferred for our problem, and is also widely used by other researchers (as can be observed in the number of references). This is for several reasons:

- Probability theory is well studied and well developed in applied mathematics, and has been used in many areas. It is generally well understood by the researchers.
- Probabilistic models provide a natural way to specify the uncertainties in measurements.
- Probabilistic modeling can provide us a simple but consistent mechanism to manipulate and propagate uncertainties.

2.1 Probability Theory and Geometric Probability

This section recapitulates some fundamental notions of probability theory and applies them to the geometry. For more details and proofs, the reader is referred to [2.14, 15].

Let us consider a random vector $\mathbf{x} = [x_1, \ x_2, \ \cdots, \ x_n]^T$, where T denotes the transpose of a vector (or a matrix).

Definition 2.1. *The* (**joint**) **probability density function,** $p(x_1, \cdots, x_n)$ *(or $p(\mathbf{x})$ for simplicity), is defined as a function that $p(x_1, \cdots, x_n)dx_1 \cdots dx_n$ is the probability that the value of random vector* $\mathbf{x}$ *will lie in the differential volume $dx_1 \cdots dx_n$ with center at $[x_1, \ \cdots, \ x_n]^T$.* ◇

From the above definition, we have $p(\mathbf{x}) \geq 0$ and

$$\int_{-\infty}^{\infty} \cdots \int_{-\infty}^{\infty} p(x_1, \cdots, x_n)dx_1 \cdots dx_n = 1 , \qquad (2.1)$$

or simply

$$\int_{-\infty}^{\infty} p(\mathbf{x})d\mathbf{x} = 1 . \qquad (2.2)$$

Here and in the rest of this chapter, for reason of simplicity, we use the integral over a vector to denote the multiple integral over the volume defined by the elements of the vector. That is

$$\int \cdots \int \cdots dx_1 \cdots dx_n \overset{\triangle}{=} \int \cdots d\mathbf{x} . \qquad (2.3)$$

Two important probability density functions are:

• **Uniform density function**

$$p(\mathbf{x}) = \begin{cases} \frac{1}{V} & \bar{\mathbf{x}} - \frac{\mathbf{w}}{2} \leq \mathbf{x} \leq \bar{\mathbf{x}} + \frac{\mathbf{w}}{2} \\ 0 & \text{otherwise,} \end{cases} \qquad (2.4)$$

where V is the volume of the element in $\mathbf{x}$-space defined by $\bar{\mathbf{x}} - \mathbf{w}/2 \leq \mathbf{x} \leq \bar{\mathbf{x}} + \mathbf{w}/2$, i.e.,

$$V = w_1 \cdots w_n ,$$

where w_i is the ith element of vector $\mathbf{w}$.

• **Gaussian density function**

$$p(\mathbf{x}) = \frac{1}{(2\pi)^{n/2}|\Lambda_{\mathbf{x}}|^{1/2}} \exp[-\frac{1}{2}(\mathbf{x} - \bar{\mathbf{x}})^T \Lambda_{\mathbf{x}}^{-1}(\mathbf{x} - \bar{\mathbf{x}})] . \qquad (2.5)$$

This is also called the *normal density function*. A conventional notation is "$\mathbf{x} \sim N(\bar{\mathbf{x}}, \Lambda_{\mathbf{x}})$".

The following defines the expected value and the variance-covariance matrix of a random vector. These concepts can similarly be defined for a random scalar.

Definition 2.2. *The* **expected value** (*or* **Expectation**) *of the random vector* x *is defined as*

$$E[\mathbf{x}] = \int_{-\infty}^{\infty} \mathbf{x}p(\mathbf{x})d\mathbf{x} \triangleq \bar{\mathbf{x}} . \qquad (2.6)$$

The **covariance matrix** *of the vector* x *is defined as*

$$\Lambda_{\mathbf{x}} = E[(\mathbf{x} - \bar{\mathbf{x}})(\mathbf{x} - \bar{\mathbf{x}})^T] = \int_{-\infty}^{\infty} (\mathbf{x} - \bar{\mathbf{x}})(\mathbf{x} - \bar{\mathbf{x}})^T p(\mathbf{x})d\mathbf{x} . \qquad (2.7)$$

$\Diamond$

Note that $E[\mathbf{x}]$ is the *first moment* of the probability density function $p(\mathbf{x})$ which determines its center of gravity, and $\Lambda_{\mathbf{x}}$ is the matrix of *second moments* which indicates the dispersion of the values of x from its mean $E[\mathbf{x}]$.

Theorem 2.1. *If* x *is a random vector with probability density function* $p(\mathbf{x})$ *and* $\mathbf{y} = \mathbf{f}(\mathbf{x})$ *is a one-to-one transformation that is everywhere differentiable, the density function of* y *is given by*

$$p(\mathbf{y}) = p(\mathbf{x})|J|^{-1} \quad with \quad \mathbf{x} = \mathbf{f}^{-1}(\mathbf{y}) , \qquad (2.8)$$

where J *is the Jacobian matrix* $\partial \mathbf{f}(\mathbf{x})/\partial \mathbf{x}$. ∎

Theorem 2.2. *A linear combination of Gaussian random vectors is also a Gaussian random vector.* ∎

Stated analytically, if x is a Gaussian random vector with mean $\bar{\mathbf{x}}$ and covariance matrix $\Lambda_{\mathbf{x}}$, and $\mathbf{y} = A\mathbf{x} + \mathbf{b}$[1] where A is a constant matrix and b is constant vector, then y is a Gaussian vector with mean $\bar{\mathbf{y}}$ and covariance $\Lambda_{\mathbf{y}}$ given by

$$\begin{aligned} \bar{\mathbf{y}} &= A\bar{\mathbf{x}} + \mathbf{b} , \\ \Lambda_{\mathbf{y}} &= A\Lambda_{\mathbf{x}}A^T . \end{aligned} \qquad (2.9)$$

In the above, we have presented several important notions of the probability theory in continuous-time case. In practice, we usually encounter the problem of continuous-time system with discrete-time observations. We can directly extend the above concepts to the discrete case.

---

[1]Any linear combination can be expressed in this form. Take an example. Let $\mathbf{x}_1$ and $\mathbf{x}_2$ be two 2D vectors. If $\mathbf{y} = a\mathbf{x}_1 + b\mathbf{x}_2$, then we can define

$$\mathbf{x} = \begin{bmatrix} \mathbf{x}_1 \\ \mathbf{x}_2 \end{bmatrix} \quad and \quad A = \begin{bmatrix} a & 0 & b & 0 \\ 0 & a & 0 & b \end{bmatrix}$$

so that $\mathbf{y} = A\mathbf{x}$.

Most sensors in Robotics and Robot Vision are geometric, in the sense that they give essentially geometric description about the environment. Our trinocular stereo system [2.16, 17], for example, gives a description based on 3D line segments about the environment. This implies that geometry must be the basis in most domains of Robotics: environment modeling, sensor integration, motion planning, kinematics, etc.. As we described earlier, geometric information provided by sensors is inherently uncertain. We shall apply probability theory to represent an uncertain geometry (the so-called *Geometric Probability* [2.18]). This approach is adopted in the Robotics Community, such as [2.7, 19, 20], and is well formulated in [2.8, 21].

Any geometric objects (features, locations and relations)(which are described by a parameter vector **p**) can be considered as random variables with a probability density function (p.d.f., for abbreviation) $p(\mathbf{p})$. The p.d.f. $p(\mathbf{p})$ describes the probability or likelihood of a particular instance **p** in the parameter space. Then, the analysis of uncertain geometry can be reduced to the problem of manipulation of random variables. A p.d.f. on a feature available in one form (or in one coordinate system) can be transformed into another using Theorem 2.1. However, geometric probability differs from conventional probability in requiring a physical interpretation to be placed on random variables, resulting in physical (geometric) constraints on functions and relations. This makes the analysis of geometric probability a more difficult problem than just the manipulation of random variable [2.21].

Two sources of difficulty arise in manipulating geometric uncertainties:

- P.d.f.'s are usually multi-dimensional,
- The transformations of geometric objects are usually nonlinear.

The p.d.f. depends, of course, on many physical properties of sensors. We are unlikely to have an exact description of the probability distribution of observations, and it is, in fact, not desirable since the exact p.d.f. can be expected to be very complex. A reasonable assumption is that the p.d.f. is jointly Gaussian. This assumption can be justified if measurement errors are relatively small, or if they can be considered as the sum of a large number of independent errors (law of large numbers). Under this assumption, the p.d.f. is well specified by the mean $\bar{\mathbf{p}}$ and the covariance matrix $\Lambda_{\mathbf{p}}$. This modelization provides us with a simple and efficient procedure to manipulate geometric uncertainties. The transformation of a p.d.f. is reduced to that of its mean and its covariance matrix. Let the transformation equation be

$$\mathbf{p}' = \mathbf{h}(\mathbf{p}) . \tag{2.10}$$

When the transformation function **h** is linear, Theorem 2.2 is directly applicable. When **h** is nonlinear, Theorem 2.3 is used by approximation:

Theorem 2.3. *Given a Gaussian variable* **p** *with its mean* $\bar{\mathbf{p}}$, *its covariance matrix* $\Lambda_{\mathbf{p}}$, *and a transformation equation* $\mathbf{p}' = \mathbf{h}(\mathbf{p})$, *then the transformed*

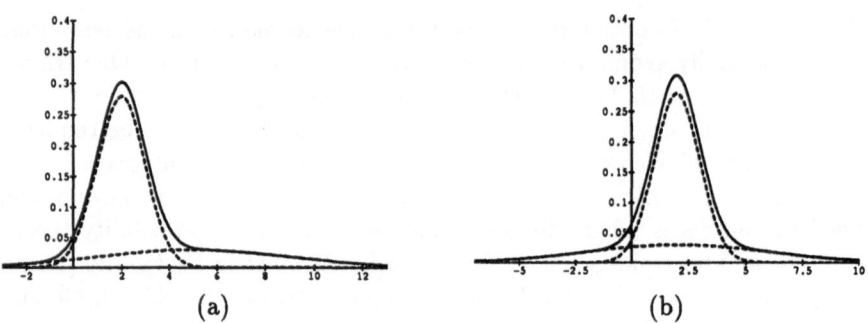

Fig. 2.2. Examples of the one-dimensional ϵ-contaminated normal distribution

variable $\mathbf{p}'$ has the following characteristics up to second order:

$$\begin{cases} \bar{\mathbf{p}}' = \mathbf{h}(\bar{\mathbf{p}}) \\ \Lambda_{\mathbf{p}'} = J\Lambda_{\mathbf{p}}J^T \end{cases} \tag{2.11}$$

where J is the Jacobian matrix $\partial\mathbf{h}/\partial\mathbf{p}$ at $\mathbf{p} = \bar{\mathbf{p}}$.   ∎

Proof. We expand $\mathbf{p}'$ into a Taylor series at $\mathbf{p} = \bar{\mathbf{p}}$:

$$\mathbf{p}' = \mathbf{h}(\bar{\mathbf{p}}) + \frac{\partial\mathbf{h}}{\partial\mathbf{p}}(\mathbf{p} - \bar{\mathbf{p}}) + O((\mathbf{p} - \bar{\mathbf{p}})^2) .$$

By ignoring the second order terms, we have

$$\begin{aligned} \bar{\mathbf{p}}' &= E[\mathbf{p}'] \approx \mathbf{h}(\bar{\mathbf{p}}) , \\ \Lambda_{\mathbf{p}'} &= E[(\mathbf{p}' - \bar{\mathbf{p}}')(\mathbf{p}' - \bar{\mathbf{p}}')^T] \\ &\approx E[\frac{\partial\mathbf{h}}{\partial\mathbf{p}}(\mathbf{p} - \bar{\mathbf{p}})(\mathbf{p} - \bar{\mathbf{p}})^T\frac{\partial\mathbf{h}}{\partial\mathbf{p}}^T] \\ &= \frac{\partial\mathbf{h}}{\partial\mathbf{p}}E[(\mathbf{p} - \bar{\mathbf{p}})(\mathbf{p} - \bar{\mathbf{p}})^T]\frac{\partial\mathbf{h}}{\partial\mathbf{p}}^T \\ &= J\Lambda_{\mathbf{p}}J^T . \end{aligned}$$

This ends our proof.   □

Note that when $\mathbf{h}$ is linear, Theorem 2.3 is the same as Theorem 2.2, as expected.

A more robust modelization is to use the so-called ϵ-*contaminated normal distribution* [2.22] (or *contaminated Gaussians*) to account for a small number of large deviations (gross errors). The ϵ-contaminated normal distribution has the form:

$$(1 - \epsilon)N(\boldsymbol{\mu}, \Lambda) + \epsilon N(\boldsymbol{\mu}_1, \Lambda_1) . \tag{2.12}$$

In Figure 2.2a, we display the distribution of $0.7N(2, 1) + 0.3N(5, 4)$. One observes that the distribution is not symmetric. The symmetry is retained for $\boldsymbol{\mu} =$

$\boldsymbol{\mu_1}$. Figure 2.2b shows such a case (the distribution of $0.7N(2,1) + 0.3N(2,4)$). Usually we set $0.01 < \epsilon < 0.3$ and $|\Lambda| << |\Lambda_1|$. This is equivalent to say that the sensor behaves as $N(\boldsymbol{\mu}, \Lambda)$ most of time but provides occasional spurious measurements described by $N(\boldsymbol{\mu_1}, \Lambda_1)$.

In [2.23], *Szeliski* also followed the probabilistic approach to mainly model dense fields and their associated uncertainty. Such fields are used in low-level vision to represent, for example, visible surfaces. Assume that visible surfaces are smooth, Markov Random Fields (MRFs) are used to model the *a priori* knowledge of the visible surfaces. The MRFs are generated by a Gibbs sampler. The sensor errors are modeled, as what we use, as Gaussians or contaminated Gaussians. The visible surfaces are then reconstructed using Bayes' rule. Although MRFs exhibit some attractive computational properties, they have some limitations [2.23]. A MRF is suitable for representing a smooth surface, but not for complicated structures such as piecewise planar surfaces. An accurate reconstruction of such surfaces necessitates a precise localization of surface discontinuities. Discontinuity detection is a very difficult problem (as it requires the knowledge of the surface which is also to be solved).

Another aspect of the uncertain geometry is *stochastic topology*. Stochastic topology describes geometric relations between different probabilistic geometric objects. We do not intend to address this subject in this monograph. The interested reader is referred to [2.8].

2.2 Parameter Estimation

Most of the problems invoked in this monograph will be formulated as the problem of parameter estimation: Given several noisy observations, how does one optimally recover the information required? Kalman filtering, as pointed out by *Lowe* [2.24], is likely to have applications throughout Computer Vision as a general method for integrating noisy measurements. It will be intensively used in this monograph.

The behavior of a dynamic system can be described by the evolution of a set of variables, called *state variables*. In practice, the individual state variables of a dynamic system cannot be determined exactly by direct measurements; instead, we usually find that the measurements that we make are functions of the state variables and that these measurements are corrupted by random noise. The system itself may also be subjected to random disturbances. It is then required to estimate the state variables from the noisy observations.

If we denote the state vector by $\mathbf{s}$ and denote the measurement vector by $\mathbf{x}'$, a dynamic system (in discrete-time form) can be described by

$$\mathbf{s}_{i+1} = \mathbf{h}_i(\mathbf{s}_i) + \mathbf{n}_i , \quad i = 0, 1, \cdots , \tag{2.13}$$

$$\mathbf{f}_i(\mathbf{x}'_i, \mathbf{s}_i) = 0 , \quad i = 0, 1, \cdots , \tag{2.14}$$

where $\mathbf{n}_i$ is the vector of random disturbance of the dynamic system and is usually modeled as white noise:

$$E[\mathbf{n}_i] = \mathbf{0} \quad \text{and} \quad E[\mathbf{n}_i\mathbf{n}_i^T] = Q_i \ .$$

In practice, the system noise covariance Q_i is usually determined on the basis of experience and intuition (i.e., it is guessed). In (2.14), the vector $\mathbf{x}'_i$ is called the measurement vector. In practice, the measurements that can be made contain random errors. We assume the measurement system is disturbed by additive white noise, i.e., the real observed measurement $\mathbf{x}_i$ is expressed as

$$\mathbf{x}_i = \mathbf{x}'_i + \boldsymbol{\eta}_i \ , \tag{2.15}$$

where

$$E[\boldsymbol{\eta}_i] = \mathbf{0} \ ,$$
$$E[\boldsymbol{\eta}_i\boldsymbol{\eta}_j^T] = \begin{cases} \Lambda_{\eta_i} & \text{for } i = j \ , \\ \mathbf{0} & \text{for } i \neq j \ . \end{cases}$$

The measurement noise covariance Λ_{η_i} is either provided by some signal processing algorithm or guessed in the same manner as the system noise. In general, these noise levels are determined independently. We assume then there is no correlation between the noise process of the system and that of the observation, that is

$$E[\boldsymbol{\eta}_i\mathbf{n}_j^T] = \mathbf{0} \quad \text{for every } i \text{ and } j.$$

2.2.1 Standard Kalman Filter

When $\mathbf{h}_i(\mathbf{s}_i)$ is a linear function

$$\mathbf{s}_{i+1} = H_i\mathbf{s}_i + \mathbf{n}_i$$

and we are able to write down explicitly a linear relationship

$$\mathbf{x}_i = F_i\mathbf{s}_i + \boldsymbol{\eta}_i$$

from

$$\mathbf{f}_i(\mathbf{x}'_i, \mathbf{s}_i) = \mathbf{0} \ ,$$

then the standard Kalman filter (Algorithm 2.1) is directly applicable.

Figure 2.3 is a block diagram for the Kalman filter. At time t_i, the system model inherently in the filter structure generates $\hat{\mathbf{s}}_{i|i-1}$, the best prediction of the state, using the previous state estimate $\hat{\mathbf{s}}_{i-1}$. The previous state covariance matrix P_{i-1} is extrapolated to the predicted state covariance matrix $P_{i|i-1}$. $P_{i|i-1}$ is then used to compute the Kalman gain matrix K_i and to update the covariance matrix P_i. The system model generates also $F_i\hat{\mathbf{s}}_{i|i-1}$ which is the best prediction of what the measurement at time t_i will be. The real measurement $\mathbf{x}_i$ is then read in, and the measurement residual (also called *innovation*)

$$\mathbf{r}_i = \mathbf{x}_i - F_i\hat{\mathbf{s}}_{i|i-1}$$

Algorithm 2.1: Kalman Filter

- Prediction of states:
$$\hat{s}_{i|i-1} = H_{i-1}\hat{s}_{i-1}$$
- Prediction of the covariance matrix of states:
$$P_{i|i-1} = H_{i-1}P_{i-1}H_{i-1}^T + Q_{i-1}$$
- Kalman gain matrix:
$$K_i = P_{i|i-1}F_i^T(F_iP_{i|i-1}F_i^T + \Lambda_{\eta_i})^{-1}$$
- Update of the state estimation:
$$\hat{s}_i = \hat{s}_{i|i-1} + K_i(x_i - F_i\hat{s}_{i|i-1})$$
- Update of the covariance matrix of states:
$$P_i = (\mathbf{I} - K_iF_i)P_{i|i-1}$$
- Initialization:
$$P_{0|0} = \Lambda_{s_0}$$
$$\hat{s}_{0|0} = E[s_0]$$

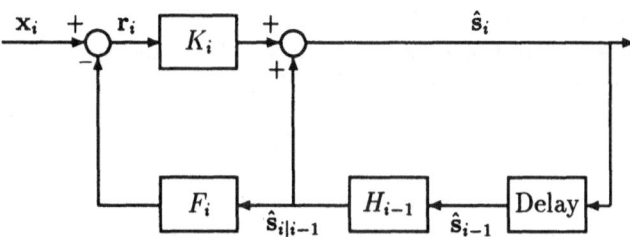

Fig. 2.3. Kalman filter block diagram

is computed. Finally, the residual r_i is weighted by the Kalman gain matrix K_i to generate a correction term and is added to $\hat{s}_{i|i-1}$ to obtain the updated state $\hat{s}_i$.

The Kalman filter gives a linear, unbiased, and minimum error variance recursive algorithm to optimally estimate the unknown state of a linear dynamic system from noisy data taken at discrete real-time intervals. Without entering into the theoretical justification of the Kalman filter, for which the reader is referred to many existing books such as [2.14, 15], we insist here on the point that the Kalman filter yields at t_i an optimal estimate of s_i, optimal in the sense that the spread of the estimate-error probability density is minimized. In other words, the estimate $\hat{s}_i$ given by the Kalman filter minimizes the following cost function

$$\mathcal{F}_i(\hat{s}_i) = E[(\hat{s}_i - s_i)^T M(\hat{s}_i - s_i)] \ ,$$

where M is an arbitrary, positive semidefinite matrix. The optimal estimate $\hat{s}_i$ of the state vector s_i is easily understood to be a least-squares estimate of s_i

with the properties that [2.25]:

1. the transformation that yields $\hat{\mathbf{s}}_i$ from $[\mathbf{x}_0^T \cdots \mathbf{x}_i^T]^T$ is linear,
2. $\hat{\mathbf{s}}_i$ is unbiased in the sense that $E[\hat{\mathbf{s}}_i] = E[\mathbf{s}_i]$,
3. it yields a minimum variance estimate with the inverse of covariance matrix of measurement as the optimal weight.

By inspecting the Kalman filter equations, the behavior of the filter agrees with our intuition. First, let us look at the Kalman gain K_i. After some matrix manipulation, we express the gain matrix in the form:

$$K_i = P_i F_i^T \Lambda_{\eta_i}^{-1} . \tag{2.16}$$

Thus, the gain matrix is "proportional" to the uncertainty in the estimate and "inversely proportional" to that in the measurement. If the measurement is very uncertain and the state estimate is relatively precise, then the residual $\mathbf{r}_i$ is resulted mainly by the noise and little change in the state estimate should be made. On the other hand, if the uncertainty in the measurement is small and that in the state estimate is big, then the residual $\mathbf{r}_i$ contains considerable information about errors in the state estimate and strong correction should be made to the state estimate. All these are exactly reflected in (2.16).

Now, let us examine the covariance matrix P_i of the state estimate. By inverting P_i and replacing K_i by its explicit form (2.16), we obtain:

$$P_i^{-1} = P_{i|i-1}^{-1} + F_i^T \Lambda_{\eta_i}^{-1} F_i . \tag{2.17}$$

From this equation, we observe that if a measurement is very uncertain (Λ_{η_i} is big), the covariance matrix P_i will decrease only a little if this measurement is used. That is, the measurement contributes little to reducing the estimation error. On the other hand, if a measurement is very precise (Λ_{η_i} is small), the covariance P_i will decrease considerably. This is logic. As described in the previous paragraph, such measurement contributes considerably to reducing the estimation error. Note that Equation (2.17) should not be used when measurements are noise free because $\Lambda_{\eta_i}^{-1}$ is not defined.

2.2.2 Extended Kalman Filter

If $h_i(\mathbf{s}_i)$ is not linear or a linear relationship between $\mathbf{x}_i$ and $\mathbf{s}_i$ cannot be written down, the so-called *Extended Kalman Filter* (EKF for abbreviation) can be applied[2].

The EKF approach is to apply the standard Kalman filter (for *linear* systems) to *nonlinear* systems with additive white noise by continually updating a *linearization* around the previous state estimate, starting with an initial guess. In other words, we only consider a linear Taylor approximation of the system

---

[2]Note that in the usual formulation of the EKF, the measurement (observation) function $f_i(\mathbf{x}'_i, \mathbf{s}_i)$ is of the form $\mathbf{x}'_i - \mathbf{g}_i(\mathbf{s}_i)$. Unfortunately, that formulation is not general enough to deal with the problems addressed in this monograph.

function at the previous state estimate and that of the observation function at the corresponding predicted position. This approach gives a simple and efficient algorithm to handle a nonlinear model. However, convergence to a reasonable estimate may *not* be obtained if the initial guess is poor or if the disturbances are so large that the linearization is inadequate to describe the system.

We expand $\mathbf{f}_i(\mathbf{x}'_i, \mathbf{s}_i)$ into a Taylor series about $(\mathbf{x}_i, \hat{\mathbf{s}}_{i|i-1})$:

$$\mathbf{f}_i(\mathbf{x}'_i, \mathbf{s}_i) = \mathbf{f}_i(\mathbf{x}_i, \hat{\mathbf{s}}_{i|i-1}) + \frac{\partial \mathbf{f}_i(\mathbf{x}_i, \hat{\mathbf{s}}_{i|i-1})}{\partial \mathbf{x}'_i}(\mathbf{x}'_i - \mathbf{x}_i) + \frac{\partial \mathbf{f}_i(\mathbf{x}_i, \hat{\mathbf{s}}_{i|i-1})}{\partial \mathbf{s}_i}(\mathbf{s}_i - \hat{\mathbf{s}}_{i|i-1})$$
$$+ O((\mathbf{x}'_i - \mathbf{x}_i)^2) + O((\mathbf{s}_i - \hat{\mathbf{s}}_{i|i-1})^2) . \tag{2.18}$$

By ignoring the second order terms, we get a linearized measurement equation:

$$\mathbf{y}_i = M_i \mathbf{s}_i + \boldsymbol{\xi}_i , \tag{2.19}$$

where $\mathbf{y}_i$ is the new measurement vector, $\boldsymbol{\xi}_i$ is the noise vector of the new measurement, and M_i is the linearized transformation matrix. They are given by

$$M_i = \frac{\partial \mathbf{f}_i(\mathbf{x}_i, \hat{\mathbf{s}}_{i|i-1})}{\partial \mathbf{s}_i} ,$$

$$\mathbf{y}_i = -\mathbf{f}_i(\mathbf{x}_i, \hat{\mathbf{s}}_{i|i-1}) + \frac{\partial \mathbf{f}_i(\mathbf{x}_i, \hat{\mathbf{s}}_{i|i-1})}{\partial \mathbf{s}_i}\hat{\mathbf{s}}_{i|i-1} ,$$

$$\boldsymbol{\xi}_i = \frac{\partial \mathbf{f}_i(\mathbf{x}_i, \hat{\mathbf{s}}_{i|i-1})}{\partial \mathbf{x}'_i}(\mathbf{x}'_i - \mathbf{x}_i) = -\frac{\partial \mathbf{f}_i(\mathbf{x}_i, \hat{\mathbf{s}}_{i|i-1})}{\partial \mathbf{x}'_i}\boldsymbol{\eta}_i ,$$

$$E[\boldsymbol{\xi}_i] = \mathbf{0} ,$$

$$E[\boldsymbol{\xi}_i\boldsymbol{\xi}_i^T] = \frac{\partial \mathbf{f}_i(\mathbf{x}_i, \hat{\mathbf{s}}_{i|i-1})}{\partial \mathbf{x}'_i}\Lambda_{\eta_i}\frac{\partial \mathbf{f}_i(\mathbf{x}_i, \hat{\mathbf{s}}_{i|i-1})}{\partial \mathbf{x}'_i}^T \triangleq \Lambda_{\xi_i} .$$

The extended Kalman filter equations are given in Algorithm 2.2, where the derivative $\frac{\partial \mathbf{h}_i}{\partial \mathbf{s}_i}$ is computed at $\mathbf{s}_i = \hat{\mathbf{s}}_{i-1}$.

2.2.3 Discussion

The above Kalman filter formalism is under the assumptions that the system-noise process and the measurement-noise process are uncorrelated and that they are all Gaussian white noise sequences. These assumptions are adequate in solving the problems addressed in this monograph. In the case that noise processes are correlated or they are not white (i.e., colored), the reader is referred to [2.25] for the derivation of the Kalman filter equations. The numerical unstability of Kalman filter implementation is well known. Several techniques are developed to overcome those problems, such as square-root filtering and U-D factorization. See [2.15] for a thorough discussion.

There exist many other methods to solve the parameter estimation problem: general minimization procedures, weighted least-squares method, and the Bayesian decision-theoretic approach. In the appendix to this chapter, we review

---

Algorithm 2.2: Extended Kalman Filter

- Prediction of states:
$$\hat{\mathbf{s}}_{i|i-1} = \mathbf{h}_i(\hat{\mathbf{s}}_{i-1})$$
- Prediction of the covariance matrix of states:
$$P_{i|i-1} = \frac{\partial \mathbf{h}_i}{\partial \mathbf{s}_i} P_{i-1} \frac{\partial \mathbf{h}_i}{\partial \mathbf{s}_i}^T + Q_{i-1}$$
- Kalman gain matrix:
$$K_i = P_{i|i-1} M_i^T (M_i P_{i|i-1} M_i^T + \Lambda_{\xi_i})^{-1}$$
- Update of the state estimation:
$$\hat{\mathbf{s}}_i = \hat{\mathbf{s}}_{i|i-1} + K_i(\mathbf{y}_i - M_i\hat{\mathbf{s}}_{i|i-1})$$
$$= \hat{\mathbf{s}}_{i|i-1} - K_i \mathbf{f}_i(\mathbf{x}_i, \hat{\mathbf{s}}_{i|i-1})$$
- Update of the covariance matrix of states:
$$P_i = (\mathbf{I} - K_i M_i) P_{i|i-1}$$
- Initialization:
$$P_{0|0} = \Lambda_{\mathbf{s}_0}$$
$$\hat{\mathbf{s}}_{0|0} = E[\mathbf{s}_0]$$

---

briefly several least-squares techniques. We choose the Kalman filter approach as our main tool to solve the parameter estimation problem. This is for the following reasons:

- the Kalman filter takes explicitly into account the measurement uncertainties,
- the Kalman filter takes measurements into account incrementally (recursivity),
- the Kalman filter is a simple and efficient procedure to solve the problem (computational tractability),
- the Kalman filter can take into account *a priori* information, if any.

The linearization of a nonlinear model leads to small errors in the estimates, which in general can be neglected, especially if the relative accuracy is better than 10% [2.26,27]. However, as pointed by *Maybank* [2.28], the extended Kalman filter seriously *underestimates* covariance. Furthermore, if the current estimate $\hat{\mathbf{s}}_{i|i-1}$ is very different from the true one, the first-order approximation, (2.18 and 2.19), is not good anymore, and the final estimate given by the filter may be significantly different from the true one. One approach to reduce the effect of nonlinearities is to apply iteratively the Kalman filter (called the *iterated extended Kalman filter*).

2.2.4 Iterated Extended Kalman Filter

The Iterated Extended Kalman Filter (IEKF) could be applied either globally or locally.

The global IEKF is applied to the whole observed data. Given a set of n observations $\{\mathbf{x}_i, \ i = 1 \cdots n\}$. The initial state estimate is $\hat{\mathbf{s}}_0$ with covariance matrix $\Lambda_{\hat{s}_0}$. After applying the EKF to the set $\{x_i\}$, we get an estimate $\hat{s}_n^1$ with covariance matrix P_n^1 (the superscript, 1 here, denotes the number of iteration). Before performing the next iteration, we must back propagate $\hat{s}_n^1$ to time t_0, denoted by $_0\hat{s}_n^1$. At iteration 2, $_0\hat{s}_n^1$ is used as the initial state estimate, but the original initial covariance matrix $\Lambda_{\hat{s}_0}$ is again used as the initial covariance matrix at this iteration. This is because if we use the new covariance matrix, it would mean we have two identical sets of measurements. Due to the requirement of the back propagation of the state estimate, the application of the global IEKF is very limited. Maybe it is interesting only when the state does not evolve over time [2.17]. In that case, no back propagation is required. In the problem of estimating 3D motion between two frames, the EKF is applied spatially, i.e., it is applied to a number of matches. The 3D motion (the state) does not change from one match to another, thus the global IEKF can be applied.

The local IEKF [2.14, 29] is applied to a single sample data by redefining the nominal trajectory and relinearizing the measurement equation. It is capable of providing better performance than the basic EKF, especially in the case of significant nonlinearity in the measurement function $f_i(x'_i, s)$. This is because when $\hat{s}_i$ is generated after measurement incorporation, this value can serve as a better state estimate than $\hat{s}_{i|i-1}$ for evaluating f_i and M_i in the measurement update relations. Then the state estimate after measurement incorporation could be recomputed, iteratively if desired. Thus, in IEKF, the measurement update relations are replaced by setting $\hat{s}_i^0 = \hat{s}_{i|i-1}$ (here, the superscript denotes again the number of iteration) and doing iteration on

$$K_i = P_{i|i-1}M_i^{k^T}(M_i^k P_{i|i-1}M_i^{k^T} + \Lambda_{\xi_i})^{-1} , \qquad (2.20)$$

$$\hat{s}_i^{k+1} = \hat{s}_i^k - K_i f_i^k(x_i, \hat{s}_i^k) \qquad (2.21)$$

for iteration number $k = 0, 1, \cdots, N - 1$ and then setting $\hat{s}_i = \hat{s}_i^N$. The iteration could be stopped when consecutive values $\hat{s}_i^k$ and $\hat{s}_i^{k+1}$ differ by less than a preselected threshold. The covariance matrix is then updated based on $\hat{s}_i^N$.

2.2.5 Robustness and Confidence Procedure

Although the least-squares techniques (including Kalman filtering) are simple to apply and are in fact standard for solving most parameter estimation problems, the robustness of the estimates given by such techniques are an issue. The basic assumptions of least-squares techniques are: they exclude gross errors, systematic errors and correlations in observations, and assume exact models. Such assumptions are, of course, not realistic. Gross errors are not avoidable; models are usually only approximate. The effects of those errors can prevent least-squares techniques from yielding acceptable results. Many robust estimators have been developed to eliminate such effects [2.22], several of which have been applied to pose estimation in [2.30] from 3D-2D line correspondences. We do not intend to adopt such robust estimators, because of the advantages of the EKF mentioned

in the previous section and also because we can obtain a quasi-robust estimate
by applying the following simple statistical test.

If the ϵ-contaminated normal distribution model is used to account for gross
errors, see (2.12), the least squares techniques (including Kalman filtering) can
yield a robust estimate by first applying confidence procedures. The confidence
procedure we use is based on the measure called the *Mahalanobis distance* which
is developed hereinbelow.

Assume that we have two independent estimates μ and x_i (in many cases,
μ is a prediction and x_i is an observation (measurement)). The corresponding
covariance matrices are denoted as Λ and Λ_{x_i}. Let x and x_i' be the corresponding
true values. Denote their difference as

$$y_i' \overset{\Delta}{=} x - x_i' \ . \tag{2.22}$$

Its estimate is then

$$y_i = \mu - x_i \ . \tag{2.23}$$

We want to test the hypothesis that they represent the same geometric object,
i.e.,

$$H_0 : \quad y_i' = 0$$

against the alternate hypothesis

$$H_1 : \quad y_i' \neq 0 \ .$$

Under hypothesis H_0 ("same geometric object") and the assumption that the
two estimates have statistically independent errors, the covariance matrix of the
difference y_i' is given by

$$\Lambda_{y_i} = \Lambda + \Lambda_{x_i} \ .$$

Define a measure δ_i^M as

$$\delta_i^M = y_i^T \Lambda_{y_i}^{-1} y_i = (x_i - \mu)^T (\Lambda_{x_i} + \Lambda)^{-1} (x_i - \mu) \ , \tag{2.24}$$

which is usually called the *(squared) Mahalanobis distance* between the estimates
μ and x_i. We can simply interpret δ_i^M as the square of the Euclidean distance
between μ and x_i weighted by the sum of their covariances. The statistical test
of H_0 versus H_1 is as follows:

$$\text{accept } H_0 \text{ if } \delta_i^M \leq \epsilon \ ,$$

where ϵ is the threshold such that $P\{\delta_i^M \leq \epsilon | H_0\} = \alpha$. Here α is the probability
that a true hypothesis H_0 will be accepted and it is taken, typically, as 95%.

In fact, the distance δ_i^M has a χ^2 distribution with the degrees of freedom n
equal to the dimension of y_i or the rank of Λ_{y_i}. This can be seen as follows. Let

$$x^2 = y_i'^T \Lambda_{y_i}^{-1} y_i'$$

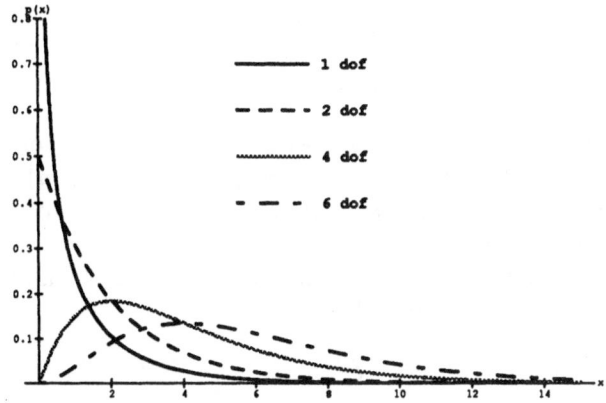

Fig. 2.4. The χ^2 distribution

Table 2.1. Table of the χ^2 distribution

$$\int_0^x p_{\chi^2}(t,n)dt = P(x,n)$$

The values in the table are x

probability	degrees of freedom (n)					
$P(x,n)$	1	2	3	4	5	6
50%	0.45	1.39	2.37	3.35	4.35	5.34
80%	1.64	3.21	4.64	5.99	7.29	8.56
90%	2.71	4.61	6.25	7.78	9.24	10.64
95%	3.84	5.99	7.81	9.49	11.07	12.59
99%	6.63	9.21	11.34	13.28	15.09	16.81

where $\mathbf{y}_i' = 0$ under hypothesis H_0. If we define

$$\chi = \Lambda_{\mathbf{y}_i}^{-1/2}\mathbf{y}_i' \, ,$$

then χ is Gaussian with mean zero and covariance

$$E[\chi\chi^T] = E[\Lambda_{\mathbf{y}_i}^{-1/2}\mathbf{y}_i'\mathbf{y}_i'^T\Lambda_{\mathbf{y}_i}^{-1/2}] = \Lambda_{\mathbf{y}_i}^{-1/2}\Lambda_{\mathbf{y}_i}\Lambda_{\mathbf{y}_i}^{-1/2} = \mathbf{I} \, ,$$

where $\mathbf{I}$ is the identity matrix. By definition, $x^2 = \chi^T\chi$ follows a χ^2 distribution with n degrees of freedom

$$p_{\chi^2}(x,n) = \begin{cases} \dfrac{1}{2^{\frac{n}{2}}\Gamma(\frac{n}{2})}x^{\frac{n}{2}-1}e^{-\frac{x}{2}} & x > 0 \\ 0 & x \leq 0 \, , \end{cases} \qquad (2.25)$$

where $\Gamma(.)$ is the Γ-function, and

$$\Gamma(n+1) = n! \, ,$$

$$\Gamma(n + 1/2) = \frac{(2n - 1)!!}{2^n} \sqrt{\pi} \, ,$$

$$\Gamma(1/2) = \sqrt{\pi} \, .$$

Figure 2.4 draws the χ^2 distribution with different degrees of freedom and Table 2.1 gives the relation between x and the probability $P(x, n) = \int_0^x p(t, n)dt$. For example, the probability that a χ^2 variable x appears between 0 and 9.49 is 95% if $\mathbf{y}_i$ has 4 degrees of freedom.

We can then choose an appropriate threshold ε on the Mahalanobis distance by looking at the χ^2 distribution table. If

$$\delta_i^M > \varepsilon \, , \tag{2.26}$$

then the measurement $\mathbf{x}_i$ is considered to be spurious and is rejected. In other words, the measurement $\mathbf{x}_i$ was sampled from $N(\boldsymbol{\mu}_1, \Lambda_1)$. A measurement $\mathbf{x}_i$ with $\delta_i^M \le \varepsilon$ is considered as coming from $N(\boldsymbol{\mu}, \Lambda)$, i.e., it is nonspurious. The Kalman filter is applied only to the nonspurious measurements, i.e., gross errors in measurements do not contribute to the parameter estimation. This confidence procedure amounts to replacing the ε-contaminated normal distribution by a single Gaussian. A quasi-robust estimate can then be obtained.

2.3 Summary

In this chapter, we have discussed that it is important to represent and manipulate explicitly uncertainty of measurements in order to use correctly and effectively the information provided by a sensor. To this end, we first recalled some fundamental concepts in Probability Theory and then applied them to the geometric world in which our mobile robot operates. A geometric object is represented by a parameter vector and a probability distribution on it. Since most problems addressed in this monograph can be formulated as parameter estimation problems, we have presented and motivated the Kalman filter and its extension to deal with nonlinear systems (the so-called extended Kalman filter) as powerful tools to deal with these problems. By first applying a confidence procedure based on Mahalanobis distance to the observations, gross errors can be discarded and the Kalman filtering approach can yield a quasi-robust estimate. As we will see later, thanks to the explicit modeling of uncertainty and to the tools described in this chapter, we shall be able to optimally integrate information from different viewpoints to compute motion or to build a global model of the environment.

2.4 Appendix: Least-Squares Techniques

As described at the beginning of this chapter, due to the limited accuracy of measurements and to, sometimes, poor understanding of a system, it is not likely

to be possible to obtain an exact solution. Given sufficient data to overdetermine
a solution, we are content to use some type of approximation method to solve the
problem. Most frequently, *least-squares* is the approximation criterion chosen.
More formally, a basic least-squares problem can be stated as follows:

Definition 2.3. *Given a linear observation equation*

$$\mathbf{y} = H\mathbf{x}$$

*where H is a $m \times n$ observation matrix relating the constant state n-vector $\mathbf{x}$ with
the error-free output m-vector $\mathbf{y}$, and given the noisy measurement m-vector $\mathbf{z}$.
The least-squares problem is to find a n-vector $\hat{\mathbf{x}}$ minimizing a quadratic error
function (objective function)*

$$J(\mathbf{x}) = \frac{1}{2}(\mathbf{z} - H\mathbf{x})^T(\mathbf{z} - H\mathbf{x}) \ . \tag{2.27}$$

*That is, find $\hat{\mathbf{x}}$ so that for all $\mathbf{x}$ we have $J(\hat{\mathbf{x}}) \leq J(\mathbf{x})$. For simplicity, we use
$H\mathbf{x} \cong \mathbf{z}$ to denote this problem.*   ◇

The minimum of $J(\mathbf{x})$ and its corresponding state vector $\hat{\mathbf{x}}$ are found by
setting the gradient to zero, i.e.,

$$\left.\frac{\partial J(\mathbf{x})}{\partial \mathbf{x}}\right|_{\mathbf{x}=\hat{\mathbf{x}}} = (\mathbf{z} - H\hat{\mathbf{x}})^T H = 0 \ .$$

Solving for $\hat{\mathbf{x}}$, the least-squares estimation is found to be

$$\hat{\mathbf{x}} = (H^T H)^{-1} H^T \mathbf{z} = L\mathbf{z} \ ,$$

where L is usually called the *left pseudoinverse* of H. This is the well-known
pseudoinverse method. The sufficient condition for a minimum is that the $n \times n$
Hessian matrix (the second derivative of J with respect to $\mathbf{x}$)

$$\left.\frac{\partial^2 J(\mathbf{x})}{\partial \mathbf{x}^2}\right|_{\mathbf{x}=\hat{\mathbf{x}}} = H^T H$$

is positive definite. This can be guaranteed if H has rank n.

Other methods exist to solve the least-squares problem. One of them is the
Singular Value Decomposition (SVD) method. In fact, the $m \times n$ matrix H can
be put through SVD in the form

$$H = USV^T \ ,$$

where U is an $m \times m$ orthogonal matrix containing the *left singular vectors* of H,
V is an $n \times n$ orthogonal matrix containing the *right singular vectors* of H, and S
is an $m \times n$ quasi-diagonal matrix containing the *singular values* of H. Without

loss of generality, the diagonal entries of S are arranged to be nonincreasing. The singular values of H are defined as the positive square roots of the eigenvalues of $H^T H$. The columns of U are the normalized eigenvectors of HH^T, and the columns of V are the normalized eigenvectors of $H^T H$.

Using the singular value decomposition, the least-squares problem $H\mathbf{x} \cong \mathbf{z}$ is then replaced by its equivalent form

$$SV^T\mathbf{x} \cong U^T\mathbf{z} .$$

Since the $m - n$ last rows of S are zeros, the above problem can be transformed into the following exact problem by removing the last $m - n$ elements of $\mathbf{b} \triangleq U^T\mathbf{z} = [\mathbf{b}_n^T, \ \mathbf{b}_{m-n}^T]^T$:

$$S_{nn}V^T\mathbf{x} = \mathbf{b}_n ,$$

where S_{nn} is the submatrix of S containing the n first rows. The solution is then given by $\hat{\mathbf{x}} = VS_{nn}^{-1}\mathbf{b}_n$.

In the above description, all measurements are treated in the same way. However, the measurement error may vary from one point to another. It is thus desirable that the good measurements contribute more heavily than the poor ones to estimating state vector. This can be achieved by minimizing a quadratic cost function of a normalized measurement residual. If the measurement error covariance is known as

$$E[(\mathbf{z} - \mathbf{y})(\mathbf{z} - \mathbf{y})^T] \triangleq R ,$$

an appropriate cost function can be defined as the following weighted quadratic function

$$J'(\mathbf{x}) = \tfrac{1}{2}(\mathbf{z} - H\mathbf{x})^T R^{-1}(\mathbf{z} - H\mathbf{x}) .$$

$J'(\mathbf{x})$ is nothing but the Euclidean distance weighted by its error covariance. By setting $\partial J'(\mathbf{x})/\partial\mathbf{x}$ to zero, the *weighted least-squares estimation* is found to be

$$\hat{\mathbf{x}} = (H^T R^{-1} H)^{-1} H^T R^{-1}\mathbf{z} .$$

It is worth noting that the Kalman filter can be shown to be a recursive implementation of the weighted least-squares estimation technique.

In the nonlinear case, that is, a linear relation cannot be written down between the observations and the vector to be estimated, we can still define a suitable cost function. However, the vector minimizing the cost function usually cannot be derived by an analytic analysis. Numerical methods must be exploited which are usually iterative procedures starting from a given initial estimate of the state vector. The Newton-Raphson algorithm is one of such methods. Let $J(\mathbf{x})$ be the cost function. Given an initial estimate $\mathbf{x}_0$ of $\mathbf{x}$. Using the Taylor-series expansion around $\mathbf{x}_0$ and retaining terms up to second order, we have

$$J(\mathbf{x}) = J(\mathbf{x}_0) + \left.\frac{\partial J}{\partial\mathbf{x}}\right|_{\mathbf{x}=\mathbf{x}_0} (\mathbf{x} - \mathbf{x}_0) + \frac{1}{2}(\mathbf{x} - \mathbf{x}_0)^T \left.\frac{\partial^2 J}{\partial\mathbf{x}^2}\right|_{\mathbf{x}=\mathbf{x}_0} (\mathbf{x} - \mathbf{x}_0) .$$

If the above equation fits J exactly in the vicinity of the minimum, where $\mathbf{x} = \hat{\mathbf{x}}$, then the following should hold

$$\left.\frac{\partial J}{\partial \mathbf{x}}\right|_{\mathbf{x}=\hat{\mathbf{x}}} = 0 = \left.\frac{\partial J}{\partial \mathbf{x}}\right|_{\mathbf{x}=\mathbf{x}_0} + (\hat{\mathbf{x}} - \mathbf{x}_0)\left.\frac{\partial^2 J}{\partial \mathbf{x}^2}\right|_{\mathbf{x}=\mathbf{x}_0} \quad,$$

which leads to

$$\hat{\mathbf{x}} = \mathbf{x}_0 - \left(\frac{\partial^2 J}{\partial \mathbf{x}^2}\right)^{-1}_{\mathbf{x}=\mathbf{x}_0} \left(\frac{\partial J}{\partial \mathbf{x}}\right)^T_{\mathbf{x}=\mathbf{x}_0} \quad.$$

When J is not precisely quadratic and when $\mathbf{x}_0$ is only a poor initial guess, the above equation will not yield $\hat{\mathbf{x}}$ on the first try. However, doing an iteration based on this form

$$\mathbf{x}_k = \mathbf{x}_{k-1} - \left(\frac{\partial^2 J}{\partial \mathbf{x}^2}\right)^{-1}_{\mathbf{x}=\mathbf{x}_{k-1}} \left(\frac{\partial J}{\partial \mathbf{x}}\right)^T_{\mathbf{x}=\mathbf{x}_{k-1}} \quad,$$

where k is an iteration index, a solution could be obtained when $\mathbf{x}_k$ and $\mathbf{x}_{k-1}$ differ by less than a preselected threshold. It is worth noting that the solution found by an iterative method could not be guaranteed to be the global minimum.

3. Reconstruction of 3D Line Segments

As we described earlier, our motion analysis is based on 3D data, particularly 3D line segments. 3D data can be acquired with active range finders (using, for example, laser and ultrasound to measure depth directly) [3.1], or with passive stereo systems (algorithms from binocular or trinocular images) [3.2–9], or even with motion stereo systems (using a moving camera under constrained motion) [3.10]. Although the work described in this monograph should be applicable to 3D data obtained from various different sources, we are especially interested in 3D data from stereo.

One static camera does not allow us to recover the 3-dimensional structure of the environment. This is because the lost information in projection is not invertible in the usual sense: a projection maps an infinite number of 3D points onto a 2D point in the image. We, human beings, use the difference between the images in our left and right eyes to perceive the depth. This motivates the use of two cameras in different positions: binocular stereovision. However, for a given point in one image, there exist probably many homologous points in the other image if only using the geometric constraint — epipolar constraint (see below). One must solicit other constraints such as continuity, order and unicity [3.8] to resolve the matching ambiguities. By introducing a third camera, one reinforces the geometric constraint, which simplifies considerably the matching process. The stereo system we use is trinocular, developed by *Ayache* and *Lustman* [3.7,8].

3.1 Why 3D Line Segments

The type of 3D primitive depends, of course, on the choice of the extracted 2D tokens in each image. Those tokens must satisfy some important properties. The tokens must be [3.11]:

1. **Compact**, to allow for a concise description of the images, and to reduce the complexity of the stereo-matching algorithm,
2. **Intrinsic**, i.e., they must correspond to the projection of some physical object, in order to make the stereo-matching process meaningful,
3. **Robust** to the acquisition noise,

4. **Discriminant,** i.e., they possess some properties which distinguish them from one another, to facilitate the matching process,
5. **Precisely located,** to allow for accurate 3D reconstruction,
6. **Dense** in the image, to allow for the computation of 3D points in every part of the scene.

We chose as tokens the line segments coming from a polygonal approxima- tion [3.12] of the edges which are detected by gradient computation, using the technique developed by *Canny* [3.13] and further improved by *Deriche* [3.14]. In our opinion, these tokens fulfill very well the first 5 properties, and relatively well the 6th property, depending on the amount of texture in the observed scene. Anyhow, as it is not possible to derive meaningful 3D measurements within uni- form regions, we believe that these tokens represent an almost optimal solution.

Line segments are very important in Robotics, especially in man-made en- vironments. Our stereo system reconstructs a set of 3D line segments in its environment. These segments may correspond to the contours of objects, or may correspond to shadows and region marks.

While line segments suffice in many cases to describe, either exactly or ap- proximatively, objects in the environment of a robot, there are many objects which are better represented by high order representations. For example, it is difficult to represent objects such as coffeecups, some chairs and tables uniquely by line segments. They can be precisely represented by including curved primi- tives such as circular arcs. Our current research aims at extending the techniques described in this monograph to include curved primitives.

In the following sections, we describe succinctly how to build 3D line segments by our trinocular stereo system.

3.2   Stereo Calibration

The three cameras of our stereovision system are first calibrated [3.15, 16].

3.2.1   Camera Calibration

Each camera is modeled by its optical center C and its image plane Π (pinhole model) (see Fig. 3.1). A perspective transformation $\mathbf{M}$ is sufficient to model the camera. This transformation $\mathbf{M}$ is linear in homogeneous coordinates and can be represented by a 3×4 matrix that maps a scene point $M = [x, y, z]^T$ to its image point $\mathbf{m} = [u, v]^T$:

$$\begin{bmatrix} su \\ sv \\ s \end{bmatrix} = \mathbf{M} \begin{bmatrix} x \\ y \\ z \\ 1 \end{bmatrix}, \qquad (3.1)$$

where s is an arbitrary nonzero scalar. Each matrix $\mathbf{M}$ depends upon 11 un- knowns, since it is defined only up to a scale factor. The matrix $\mathbf{M}$ can model the

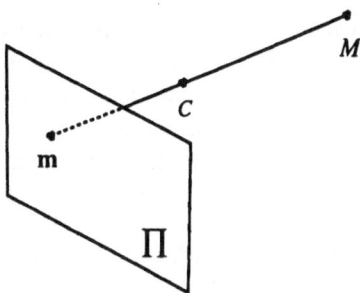

Fig. 3.1. Pinhole Model for a Camera

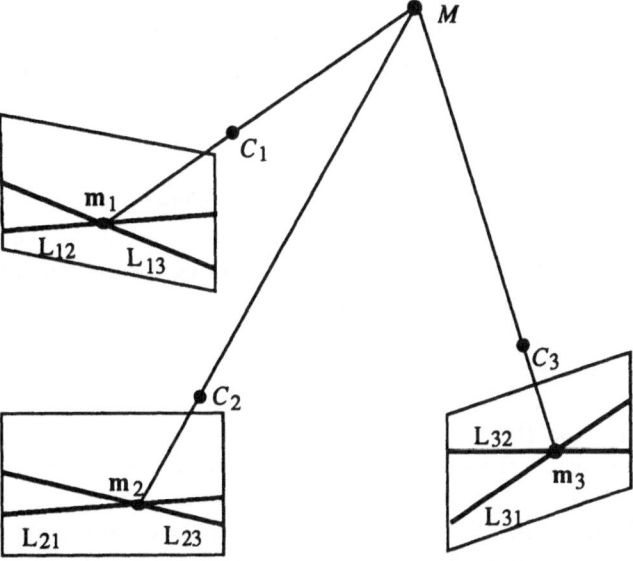

Fig. 3.2. The geometry of the trinocular stereo system

following parameters: the scale factors in the two axes of the image coordinate system, the angle between the two axes, the coordinates of the image center, and the six parameters describing the transformation between the world and camera coordinate systems. The focal length of the camera (the distance between the optical center C and the image plane Π) is indistinguishable from the scale factors in the two image axes.

The geometry of our trinocular stereo system is illustrated in Fig. 3.2 (see [3.7, 8]). Our stereo calibration consists in determining the perspective transformation matrices $\mathbf{M}_1$, $\mathbf{M}_2$ and $\mathbf{M}_3$ of the three cameras with respect to a common world coordinate system, without knowing *a priori* information of the cameras. That is, a camera is treated as a block box. The perspective transformations $\mathbf{M}_1$, $\mathbf{M}_2$ and $\mathbf{M}_3$ of the cameras are automatically estimated using a calibration apparatus which consists of two orthogonal checkerboard patterns. The calibration apparatus is placed in front of the cameras so that all three see it completely. Images of the lines in the patterns are extracted and their equations are accurately computed by a least-squares estimator. The intersection points of the checkerboard are then computed. Since the precise positions of their correspondences in 3D space are known, the perspective transformation can then be estimated for each camera by a least-squares estimator. The original work of *Faugeras* and *Toscani* on the calibration has been improved by *Vaillant* [3.17].

A wealth of work has been carried out on the camera calibration during the past decade. The above work consists in computing directly perspective transformation matrix. Other techniques exist in the literature. *Tsai* and *Lenz* [3.18,19], for example, divide the calibration parameters into two groups; each group can be solved easily and rapidly. Their method uses exact camera modeling including lens distortion and at the same time avoids large scale nonlinear optimization.

3.2.2 Epipolar Constraint

Since the perspective transformations $\mathbf{M}_1$, $\mathbf{M}_2$ and $\mathbf{M}_3$ are estimated at the same time from the calibration apparatus, the pairwise relations of the cameras are known. The pairwise relations of the cameras can be used through the so-called *epipolar constraint*. Consider two cameras shown in Fig. 3.3. Given a point $\mathbf{m}_1$ in the first image, its corresponding point in the second image is constrained to lie within a line called the *epipolar line* of $\mathbf{m}_1$, denoted by L_{21}. L_{21} is the intersection of the plane $\wp$ defined by $\mathbf{m}_1$, C_1, C_2 (called also the *epipolar plane*), with the second image plane Π_2. This is because image point $\mathbf{m}_1$ may correspond to an arbitrary point on the semi-line C_1M (M may be at infinity) and that the projection of C_1M in Π_2 is the line L_{21}. The epipolar constraint is symmetric for camera 1. The corresponding point in the first image of each point $\mathbf{m}_2$ lying on L_{21} must lie on its epipolar line L_{12}, which is the intersection of the same plane $\wp$ with the first image plane Π_1.

Furthermore, one observes that all epipolar lines in image i of points in image j pass through a common point E_{ij}, called *epipole*. E_{ij} is the intersection of the line C_iC_j with the image plane Π_i. Consider Fig. 3.3. For each point $\mathbf{m}_1^k$ in Π_1, its epipolar line L_{21}^k in Π_2 is the intersection of the plane $\wp^k$ defined by $\mathbf{m}_1^k$, C_1, C_2, with image plane Π_2. Since all planes $\wp^k$s contain the line C_1C_2, they must intersect in Π_2 at a common point, denoted by E_{21}. E_{21} is called the epipole of image 1 in image 2. E_{ij} can be computed from the perspective transformations in the calibration phase. The epipolar line in image i of a point $[u_j, v_j]^T$ in image j is then characterized by its direction vector, which can be computed by the multiplication of a 2×3 matrix $\mathbf{M}_{ij}$ and the point $[u_j, v_j, 1]^T$. The matrix $\mathbf{M}_{ij}$ depends only on $\mathbf{M}_i$ and $\mathbf{M}_j$, and thus can also be computed in the calibration phase.

Now we show how to compute the epipole E_{ij}. First, we must compute the coordinates of the optical center C_j, which can be obtained by setting s in (3.1)

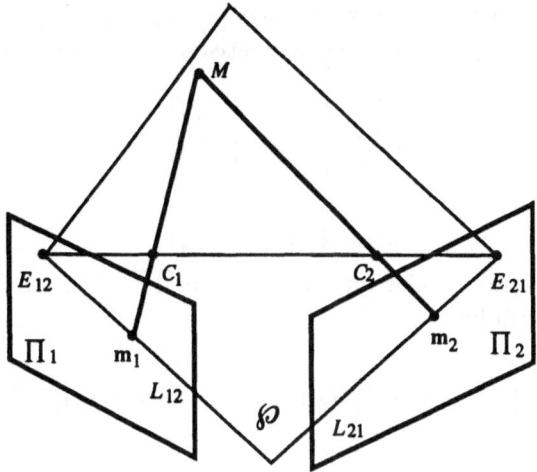

Fig. 3.3. Epipolar constraint in stereovision

to zero, which yields the following equation:

$$\begin{bmatrix} 0 \\ 0 \\ 0 \end{bmatrix} = \mathbf{M}_j \begin{bmatrix} C_j \\ 1 \end{bmatrix} .$$

If we denote the matrix formed by the first three columns of $\mathbf{M}_j$ by A_j and the last column of $\mathbf{M}_j$ by $\mathbf{b}_j$, i.e., $\mathbf{M}_j = [A_j \ \mathbf{b}_j]$, then we have

$$C_j = -A_j^{-1}\mathbf{b}_j .$$

As the epipole E_{ij} is the projection of C_j on image i, its homogeneous coordinates are given by

$$\begin{bmatrix} u_{ij} \\ v_{ij} \\ s_{ij} \end{bmatrix} = \mathbf{M}_i \begin{bmatrix} C_j \\ 1 \end{bmatrix} = -A_i A_j^{-1}\mathbf{b}_j + \mathbf{b}_i .$$

If $s_{ij} \neq 0$, then

$$E_{ij} = \begin{bmatrix} u_{ij}/s_{ij} \\ v_{ij}/s_{ij} \end{bmatrix} .$$

If $s_{ij} = 0$, then E_{ij} is in infinity. This means that C_j is in the *focal plane*[1] of the camera i. In this case, the epipolar lines in image i are parallel to each other.

The matrix $\mathbf{M}_{ij}$ is given, if $s_{ij} \neq 0$, by the following formula

$$\mathbf{M}_{ij} = \begin{bmatrix} s_{ij} & 0 & -u_{ij} \\ 0 & s_{ij} & -v_{ij} \end{bmatrix} A_i A_j^{-1} ,$$

---

[1]The focal plane of a camera is the plane parallel to the image plane Π and passing through its optical center C.

which can be obtained as follows. A point $[u_j, \ v_j]^T$ in image j corresponds to an infinite number of points M in space, satisfying the following equation

$$\begin{bmatrix} su_j \\ sv_j \\ s \end{bmatrix} = \mathbf{M}_j \begin{bmatrix} M \\ 1 \end{bmatrix} ,$$

i.e.,

$$M = A_j^{-1}(s\mathbf{m}_j - \mathbf{b}_j) ,$$

where s is an arbitrary scalar and $\mathbf{m}_j = [u_j, v_j, 1]^T$. The projection of M on image i is then given by

$$\begin{bmatrix} s'u_i \\ s'v_i \\ s' \end{bmatrix} = \mathbf{M}_i \begin{bmatrix} M \\ 1 \end{bmatrix} = A_i A_j^{-1}(s\mathbf{m}_j - \mathbf{b}_j) + \mathbf{b}_i = sA_i A_j^{-1}\mathbf{m}_j + \begin{bmatrix} u_{ij} \\ v_{ij} \\ s_{ij} \end{bmatrix} .$$

Explicitly, we have

$$\begin{aligned} u_i &= (s\mathbf{a}_1^T\mathbf{m}_j + u_{ij})/(s\mathbf{a}_3^T\mathbf{m}_j + s_{ij}) \\ v_i &= (s\mathbf{a}_2^T\mathbf{m}_j + v_{ij})/(s\mathbf{a}_3^T\mathbf{m}_j + s_{ij}) \end{aligned}$$

where $\mathbf{a}_1$, $\mathbf{a}_2$ and $\mathbf{a}_3$ are the first, second and third row vectors of the matrix $A_i A_j^{-1}$, respectively. These two equations define in fact the epipolar line in image i of the point $[u_j, \ v_j]^T$, and we can verify that the epipolar lines passes through the epipole by setting $s = 0$. Deriving the above equation with respect to s and multiplying the common denominator, we get the direction vector of the epipolar line:

$$\begin{bmatrix} \Delta u_i \\ \Delta v_i \end{bmatrix} = \begin{bmatrix} s_{ij}\mathbf{a}_1^T - u_{ij}\mathbf{a}_3^T \\ s_{ij}\mathbf{a}_2^T - v_{ij}\mathbf{a}_3^T \end{bmatrix} \mathbf{m}_j \triangleq \mathbf{M}_{ij}\mathbf{m}_j .$$

We observe that the matrix $\mathbf{M}_{ij}$ is independent of $\mathbf{m}_j$. If $s_{ij} = 0$, that is, if the epipolar is at infinity, all epipolar lines are parallel to each other, i.e., they have the same direction vector. In fact, the direction vector is $[u_{ij}, \ v_{ij}]^T$.

3.3   Algorithm of the Trinocular Stereovision

3D line segments about the environment are obtained by the following steps: edge detection [3.14], nonmaxima suppression, threshold by hysteresis, edge linking [3.20], polygonal approximation [3.12], stereo matching and 3D reconstruction [3.7,8]. The implementation on the parallel machine CAPITAN is described in [3.21].

The epipolar constraint plays an essential role in our trinocular stereovision algorithm. Figure 3.2 shows the geometry of the stereovision system. Algorithm 3.1 shows how the trinocular stereo system works. This algorithm performs completely in parallel for each segment in the first image. For more details, the reader is referred to [3.7,8]. A new trinocular stereo algorithm is reported in [3.22] which performs a rectification of images before stereo matching.

Algorithm 3.1: Trinocular Stereovision

- **For** each segment S_1 in the first image whose midpoint is m_1
  1. Compute the epipolar segment L_{21} of m_1 in image 2
     and also its epipolar segment L_{31} in image 3
  2. Consider the segments S_2 in image 2 which intersect L_{21} at m_2
     $\oplus$ **For** each of these segments
       - Compare the geometric attributes such as
             orientation, length and intensity contrast
       - **If** these attributes are similar to those of S_1
             then S_2 is considered as a potential match of S_1
       $\uparrow$ **endif**
     $\Uparrow$ **endfor**
  3. Consider those potential matches
     $\oplus$ **For** each potential match (S_1, S_2)
       - Compute the epipolar segment L_{32} of m_2 in image 3
       - Compute the intersection m_3 of L_{32} with L_{31}
       - Consider segments S_3 in the neighbor of m_3
         $\odot$ **For** each of those segments
             $\hookrightarrow$ Compare their geometric attributes
             $\hookrightarrow$ **If** S_3 is compatible with S_1 and S_2
                 then (S_1, S_2, S_3) is considered as
                     a potential stereo match
           $\uparrow$ **endif**
         $\Uparrow$ **endfor**
     $\Uparrow$ **endfor**
  $\Uparrow$ **endfor**

3.4 Reconstruction of 3D Segments

At this point, we have a set of stereo matches (S_1, S_2, S_3). We should recover their corresponding segments in 3D space. Figure 3.4 explains the problem. Each segment S_i determines a plane with the corresponding optical center C_i, in which the 3D segment should lie. When there is no noise, the intersection of the three planes must be the segment desired. In the real case, the three planes do not intersect in a line, as shown in Fig. 3.4. The extended Kalman filter is then used to estimate the best segment S. *Best* means that the sum of the squared distances from S to the three planes is minimal.

More precisely, we use the **abpq** representation described in Sect. 4.2 to describe 3D lines to be reconstructed. The reconstruction must be performed for the three maps (4.19). We describe the computation for the first map (4.18), given that it is almost similar for the other two. Thus, the line S is described by $x = az + p$ and $y = bz + q$.

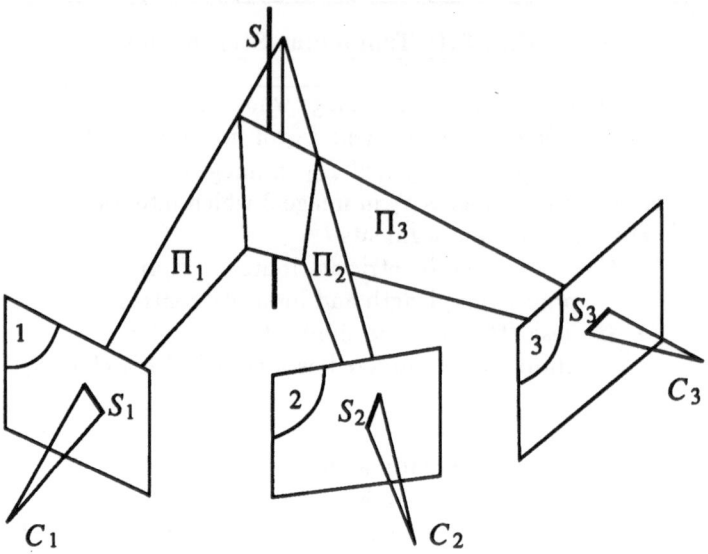

Fig. 3.4. Reconstruction of 3D segments

From the calibration result, it is easy to compute the representation of the plane Π_i formed by S_i and C_i. Let the plane be represented by a 4-vector $\mathbf{v}_i$ such that any 3D point $\mathbf{m}$ on the plane satisfies

$$[\mathbf{m}, \ 1]^T \mathbf{v}_i = 0 \ .$$

This in turn implies that any point on S must verify the relation:

$$[az + p, \ bz + q, \ z, \ 1]^T \mathbf{v}_i = 0 \ .$$

Six equations can then be derived for the three planes:

$$[a, b, 1, 0]^T \mathbf{v}_i = 0 \quad \text{and} \quad [p, q, 0, 1]^T \mathbf{v}_i = 0 \ , \quad \text{for } i = 1, 2, 3. \qquad (3.2)$$

According to different minimization criteria, several methods exist to compute the 4 unknowns. Here, we present a simple least-squares technique. Let $\mathbf{v}_i = [\alpha_i, \ \beta_i, \ \gamma_i, \ \delta_i]^T$. From (3.2), we can write down two independent linear equations in $\mathbf{x}_{ab} = [a, b]^T$ and $\mathbf{x}_{pq} = [p, q]^T$, i.e.,

$$A\mathbf{x}_{ab} = \mathbf{b}_{ab} \quad \text{and} \quad A\mathbf{x}_{pq} = \mathbf{b}_{pq} \ , \qquad (3.3)$$

where A, $\mathbf{b}_{ab}$ and $\mathbf{b}_{pq}$ are given by

$$A = \begin{bmatrix} \alpha_1 & \beta_1 \\ \alpha_2 & \beta_2 \\ \alpha_3 & \beta_3 \end{bmatrix} , \quad \mathbf{b}_{ab} = \begin{bmatrix} -\gamma_1 \\ -\gamma_2 \\ -\gamma_3 \end{bmatrix} \quad \text{and} \quad \mathbf{b}_{pq} = \begin{bmatrix} -\delta_1 \\ -\delta_2 \\ -\delta_3 \end{bmatrix} .$$

The least-squares solutions (see Sect. 2.4) are

$$\mathbf{x}_{ab} = (A^T A)^{-1} A^T \mathbf{b}_{ab} \quad \text{and} \quad \mathbf{x}_{pq} = (A^T A)^{-1} A^T \mathbf{b}_{pq} \; . \tag{3.4}$$

The covariance matrices of $\mathbf{v}_i$ $(i = 1, 2, 3)$, denoted by $\Lambda_{\mathbf{v}_i}$, can be estimated in the edge detection and polygonal approximation procedure [3.23]. Let $\mathbf{x}_{abpq} = [a, b, p, q]^T$ be the parameter vector of the reconstructed line S, we have $\mathbf{x}_{abpq} = [\mathbf{x}_{ab}^T, \mathbf{x}_{pq}^T]^T$ from (3.4). Assuming the three $\mathbf{v}_i$'s are independent, the covariance matrix of $\mathbf{x}_{abpq}$ is given, up to the first order approximation (Theorem 2.3), by:

$$\Lambda_{abpq} = \sum_{i=1}^{3} \frac{\partial \mathbf{x}_{abpq}}{\partial \mathbf{v}_i} \Lambda_{\mathbf{v}_i} \frac{\partial \mathbf{x}_{abpq}}{\partial \mathbf{v}_i}^T \; . \tag{3.5}$$

See the appendix to this monograph for the definition of the vector differentiation. Furthermore, the reader will find the following relations useful:

$$\frac{\partial A^T}{\partial v_{ij}} = \frac{\partial A}{\partial v_{ij}}^T \; , \tag{3.6}$$

$$\frac{\partial A^T A}{\partial v_{ij}} = \frac{\partial A}{\partial v_{ij}}^T A + A^T \frac{\partial A}{\partial v_{ij}} \; , \tag{3.7}$$

$$\frac{\partial (A^T A)^{-1}}{\partial v_{ij}} = -(A^T A)^{-1} \frac{\partial A^T A}{\partial v_{ij}} (A^T A)^{-1} \; , \tag{3.8}$$

where v_{ij} is the jth element of $\mathbf{v}_i$. As A, $\mathbf{b}_{ab}$ and $\mathbf{b}_{pq}$ are affine functions of the v_{ij}'s $(i, j = 1, 2, 3)$, those partial derivatives are very simple to compute.

In fact, what we have estimated at this stage is the infinite line S supporting the segment we want. To recover the 3D segment, several approaches can be considered. We describe one of them. We consider only one endpoint of the segment. The same computation is applied for the other. As the endpoint is restrained on the line, it has only one parameter, see z, and is characterized by $[az + p, bz + q, z]^T$. From (3.1), its image in camera i is given by

$$[u_i(at_{31}^i + bt_{32}^i + t_{33}^i) - (at_{11}^i + bt_{12}^i + t_{13}^i)]z =$$
$$-u_i(pt_{31}^i + qt_{32}^i + t_{34}^i) + (pt_{11}^i + qt_{12}^i + t_{14}^i) \; , \tag{3.9}$$
$$[v_i(at_{31}^i + bt_{32}^i + t_{33}^i) - (at_{21}^i + bt_{22}^i + t_{23}^i)]z =$$
$$-v_i(pt_{31}^i + qt_{32}^i + t_{34}^i) + (pt_{21}^i + qt_{22}^i + t_{24}^i) \; . \tag{3.10}$$

where t_{jk}^i is the (j, k) element of the perspective transformation $\mathbf{M}_i$. Therefore, we have 6 equations for one unknown z, which can be expressed as

$$\mathbf{a}z = \mathbf{b} \; , \tag{3.11}$$

where $\mathbf{a}$ and $\mathbf{b}$ are 6-dimensional vectors and can be easily obtained from (3.9) and (3.10). The least-squares solution is given by

$$z = (\mathbf{a}^T \mathbf{a})^{-1} \mathbf{a}^T \mathbf{b} \; . \tag{3.12}$$

Therefore, the coordinates of the endpoint are $\mathbf{p} = [az + p, \; bz + q, \; z]^T$.
Let $\mathbf{w} = [a, b, p, q, u_1, v_1, u_2, v_2, u_3, v_3]^T$. Its covariance matrix is

$$\Lambda_{\mathbf{w}} = \mathrm{diag}(\Lambda_{abpq}, \; \Lambda_1, \; \Lambda_2, \; \Lambda_3) \, ,$$

where Λ_i is the covariance matrix of the corresponding endpoint in camera i.
Here we assume there is no correlation between $\mathbf{x}_{abpq}$ and 2D endpoints. Strictly
speaking, there does exist a little correlation because $\mathbf{x}_{abpq}$ have been computed
from 2D endpoints. Similar to the computation of the covariance matrix Λ_{abpq},
we now compute, under the first order approximation, the covariance matrix of
the reconstructed point $\mathbf{p}$ as

$$\Lambda_{\mathbf{p}} = \frac{\partial \mathbf{p}}{\partial \mathbf{w}} \Lambda_{\mathbf{w}} \frac{\partial \mathbf{p}}{\partial \mathbf{w}}^T \, . \tag{3.13}$$

For the details of the 3D reconstruction process, the reader is referred to [3.7,
8, 23].

3.5 Summary

In this chapter, we have first discussed why line segments are chosen as our
basic primitives for motion analysis. Using line segments allows us to obtain a
relatively compact and structured representation of the environment. Including
other primitives, especially curves, remains the center of our current research.
3D line segments are reconstructed using a trinocular stereo system. We have
presented the camera model and described very succinctly how to calibrate a
trinocular stereovision system. A geometric constraint called epipolar constraint
has been made explicit, which allows the search space in stereo matching to be
reduced from 2D to 1D. Finally, the stereo matching algorithm and the process
of 3D reconstruction have been presented.

4. Representations of Geometric Objects

W e study in this chapter the representations of the geometric objects which we encounter in the application of 3D vision and robotics. The geometric objects of interest here include rigid motion and line segments.

The change in a dynamic scene is usually very complex. Rigid motion is of particular interest, not only because of its simplicity and well understanding, but also because of the consistency with the human visual perception system. The change between two time instants can be interpreted in many ways, but our visual system only accepts a few interpretation, often only one. The ability to reject unlikely interpretations is consistent with the rigidity assumption [4.1]. As we choose 3D line segments as our basic primitives, it is important to investigate how to represent them conveniently and efficiently. We shall review the existing representations for 3D line segments, discuss their deficiencies, and propose a new one which we think is well adapted to a 3D vision system.

4.1 Rigid Motion

By an object undergoing a rigid motion , we mean that the object's geometric relations (for example, distance between two points) remain constant throughout the motion. It is well known that any 3D rigid motion can be *uniquely* decomposed into a rotation around an axis passing through the origin of the coordinate system, and a translation. We assume that the translation is applied after the rotation. Let P and P' be the position vectors of the same 3D point before and after motion, the following relation holds:

$$P' = \mathbf{R}P + \mathbf{t} , \qquad (4.1)$$

where $\mathbf{R}$ is a 3×3 matrix called the **rotation matrix** and $\mathbf{t}$ is a 3-dimensional vector called the **translation vector**.

4.1.1 Definition

If we use $(\mathbf{R}, \mathbf{t})$ to represent a rigid motion, 12 parameters are required. However, it has long since been known that a rigid motion has only *six* Degrees Of

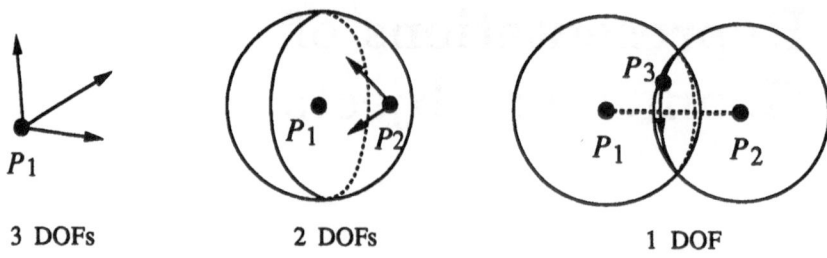

3 DOFs   2 DOFs   1 DOF

Fig. 4.1. A rigid motion has 6 degrees of freedom

Freedom (DOFs). This can be easily understood from the following consideration (see Fig. 4.1). Given two sets of points of a rigid object before and after
motion in 3-space. To establish the correspondence between the first point (arbitrarily chosen) from one set and one point from the other set, three parameters
must be supplied. Once that correspondence is fixed, two more parameters are
required to establish the second correspondence, because the rigidity constrains
the correspondence of the second point to move only on the surface of the sphere
centered at the first point (which has two DOFs). If two correspondences have
been established, the correspondence of the third point has only one degree of
freedom, as it can only rotate around the axis joining the first two points. These
situations are illustrated in Fig. 4.1. After three correspondences have been established, no degree of freedom is left. Hence six parameters are sufficient to
describe a 3D rigid motion.

The above consideration implies that the nine elements of the rotation matrix
are not independent to each other. In order that $(\mathbf{R}, \mathbf{t})$ might describe a rigid
motion, they must satisfy the following requirements.

Definition 4.1. $(\mathbf{R}, \mathbf{t})$ *represents a* **rigid motion** *if and only if the following
requirements*

- $\mathbf{R}$ *is a* 3×3 *orthogonal matrix and its determinant equals +1,*
- $\mathbf{t}$ *is a real 3-dimensional vector,*

are satisfied.   ◇

Theorem 4.1 gives some important properties of the rotation matrix $\mathbf{R}$ (the
proof is omitted).

Theorem 4.1. *The rotation matrix* $\mathbf{R}$ *has the following properties:*

1. $\mathbf{R}\mathbf{R}^T = \mathbf{R}^T\mathbf{R} = \mathbf{I}_3$, *and* $\mathbf{R}^{-1} = \mathbf{R}^T$,
2. $\|\mathbf{R}\mathbf{u}\| = \|\mathbf{u}\|$,
3. $\mathbf{R}\mathbf{u} \cdot \mathbf{R}\mathbf{v} = \mathbf{u} \cdot \mathbf{v}$,
4. $\mathbf{R}\mathbf{u} \wedge \mathbf{R}\mathbf{v} = \mathbf{R}(\mathbf{u} \wedge \mathbf{v})$,

where $\mathbf{u}$ *and* $\mathbf{v}$ *are two arbitrary 3-dimensional vectors,* ^{T} *denotes the transpose*

of a matrix, $^{-1}$ denotes the inverse of a matrix, $\| \ \|$ denotes the norm of a vector,
· denotes the inner (dot) product of two vectors, and $\wedge$ denotes the cross product
of two vectors.   ∎

4.1.2 Representations

The rotation matrix gives a simple way of representing a 3D rotation, but leads to
a high dimensional space of constraints. Several other representations of rotation
are available [4.2]. Here, we present two of them: using the rotation axis and
using a quaternion.

Representing a rotation by its rotation axis

A rotation can be defined as a 3-dimensional vector $\mathbf{r} = [a, \ b, \ c]^T$ whose direction
is that of the rotation axis and whose norm is equal to the rotation angle.

For convenience, we note $\tilde{v}$ as the antisymmetric matrix defined by $\mathbf{v}$. More
precisely, if $\mathbf{v} = [x, \ y, \ z]^T$, then

$$\tilde{v} = \begin{bmatrix} 0 & -z & y \\ z & 0 & -x \\ -y & x & 0 \end{bmatrix} . \tag{4.2}$$

In fact, for any three-dimensional vectors $\mathbf{u}$ and $\mathbf{v}$, we have $\mathbf{u} \wedge \mathbf{v} = \tilde{u} \mathbf{v}$.

The following notation (exponential of a matrix) will be used later

$$e^M \triangleq \mathbf{I} + \frac{1}{1!}M + \frac{1}{2!}M^2 + \cdots + \frac{1}{n!}M^n + \cdots , \tag{4.3}$$

where M is an $m \times m$ matrix, $\mathbf{I}$ is an $m \times m$ identity matrix, and M^n denotes
the multiplication of n matrices M, i.e.,

$$M^n \triangleq \overbrace{M \cdots M}^{n} .$$

The following useful theorem is known as *Rodrigues' formula.*

Theorem 4.2. *Given a three-dimensional vector* $\mathbf{r}$. *The following relation
holds:*

$$e^{\tilde{r}} = \mathbf{I}_3 + \frac{\sin \theta}{\theta}\tilde{r} + \frac{1 - \cos \theta}{\theta^2}\tilde{r}^2 , \tag{4.4}$$

where $\tilde{r}$ *is defined as in (4.2) and* θ *is the norm of* $\mathbf{r}$ *(i.e.,* $\theta = \|\mathbf{r}\|$*).*   ∎

Proof. From (4.3), $e^{\tilde{r}}$ can be developed as:

$$e^{\tilde{r}} = \mathbf{I}_3 + \frac{1}{1!}\tilde{r} + \frac{1}{2!}\tilde{r}^2 + \cdots + \frac{1}{n!}\tilde{r}^n + \cdots .$$

It can be easily verified that

$$\tilde{\mathbf{r}}^{2n-1} = (-1)^{n-1}\theta^{2(n-1)}\tilde{\mathbf{r}} \quad \text{for } n \geq 1 \ ,$$
$$\tilde{\mathbf{r}}^{2n} = (-1)^{n-1}\theta^{2(n-1)}\tilde{\mathbf{r}}^2 \quad \text{for } n \geq 1 \ .$$

Therefore we have

$$e^{\tilde{\mathbf{r}}} = \mathbf{I}_3 + \left(\frac{1}{1!} - \frac{\theta^2}{3!} + \cdots + \frac{(-1)^{n-1}\theta^{2(n-1)}}{(2n-1)!} + \cdots\right)\tilde{\mathbf{r}}$$
$$+ \left(\frac{1}{2!} - \frac{\theta^2}{4!} + \cdots + \frac{(-1)^{n-1}\theta^{2(n-1)}}{(2n)!} + \cdots\right)\tilde{\mathbf{r}}^2 \ .$$

Recall that

$$\sin\theta = \frac{\theta}{1!} - \frac{\theta^3}{3!} + \frac{\theta^5}{5!} - \cdots + (-1)^{n-1}\frac{\theta^{2n-1}}{(2n-1)!} + \cdots \ ,$$

and

$$\cos\theta = 1 - \frac{\theta^2}{2!} + \frac{\theta^4}{4!} - \cdots + (-1)^n\frac{\theta^{2n}}{(2n)!} + \cdots \ ,$$

we thus get (4.4).   □

The relation between $\mathbf{R}$ and $\mathbf{r}$ is given exactly by the Rodrigues' formula:

$$\mathbf{R} = e^{\tilde{\mathbf{r}}} = \mathbf{I}_3 + f(\theta)\tilde{\mathbf{r}} + g(\theta)\tilde{\mathbf{r}}^2 \ , \tag{4.5}$$

where $\theta = \sqrt{a^2 + b^2 + c^2}$ is the rotation angle, $f(\theta) = \sin\theta/\theta$ and $g(\theta) = (1 - \cos\theta)/\theta^2$.

A 3D rigid motion is then represented by a 6-dimensional vector, noted as $\mathbf{s}$:

$$\mathbf{s} = \begin{bmatrix} \mathbf{r} \\ \mathbf{t} \end{bmatrix} \ . \tag{4.6}$$

Representing a rotation by a quaternion

Quaternions have been found useful in Robotics and Vision [4.3, 4]. They are elements of a vector space endowed with multiplication. A quaternion $\mathbf{q}$ can be considered as being either a 4-dimensional vector $[\lambda_0, \lambda_1, \lambda_2, \lambda_3]^T$ or as a pair $(\alpha, \boldsymbol{\gamma})$ where α is a real number equal to λ_0, and $\boldsymbol{\gamma}$ is the vector $[\lambda_1, \lambda_2, \lambda_3]^T$. We define the multiplication $\times$ of two quaternions $\mathbf{q}$ and $\mathbf{q}'$ as follows:

$$\mathbf{q} \times \mathbf{q}' = (\alpha\alpha' - \boldsymbol{\gamma} \cdot \boldsymbol{\gamma}', \alpha\boldsymbol{\gamma}' + \alpha'\boldsymbol{\gamma} + \boldsymbol{\gamma} \wedge \boldsymbol{\gamma}') \ . \tag{4.7}$$

Multiplication is associative but not commutative. The conjugate and the magnitude of a quaternion $\mathbf{q}$ are defined as follows:

$$\bar{\mathbf{q}} = (\alpha, -\boldsymbol{\gamma}) \ , \tag{4.8}$$
$$|\mathbf{q}|^2 = \mathbf{q} \times \bar{\mathbf{q}} = (\alpha^2 + \|\boldsymbol{\gamma}\|^2, 0) = (\|\mathbf{q}\|^2, 0) \ . \tag{4.9}$$

We can easily verify that $|q \times q'| = |q| \, |q'|$. A real number x is identified with the quaternion $(x, \mathbf{0})$ and a 3-dimensional vector $\mathbf{v}$ is identified with the quaternion $(0, \mathbf{v})$.

A rotation can then be represented by two quaternions $\mathbf{q} = (\alpha, \boldsymbol{\gamma})$ and $-\mathbf{q}$, with $|\mathbf{q}| = 1$. The relation between this representation and the rotation axis one is:

$$\alpha = \cos(\theta/2) \quad \text{and} \quad \boldsymbol{\gamma} = \sin(\theta/2)\mathbf{u} \,, \tag{4.10}$$

where $\theta = \|\mathbf{r}\|$ and $\mathbf{u} = \mathbf{r}/\|\mathbf{r}\|$. Note that there are two quaternions for one rotation. It is not surprising since a rotation of angle θ around an axis $\mathbf{u}$ is the same as a rotation of angle $2\pi - \theta$ around the axis $-\mathbf{u}$. Usually, the rotation angle between two successive views does not go beyond π, so we can impose that the first element α of a quaternion $\mathbf{q}$ must be positive. Thus the mapping between rotation and quaternion is unique under this new constraint.

The relation between $\mathbf{R}$ and a unit quaternion $\mathbf{q} = [\lambda_0, \lambda_1, \lambda_2, \lambda_3]^T$ is given as follows:

$$\mathbf{R} = \begin{bmatrix} \lambda_0^2 + \lambda_1^2 - \lambda_2^2 - \lambda_3^2 & 2(\lambda_1\lambda_2 - \lambda_0\lambda_3) & 2(\lambda_1\lambda_3 + \lambda_0\lambda_2) \\ 2(\lambda_1\lambda_2 + \lambda_0\lambda_3) & \lambda_0^2 - \lambda_1^2 + \lambda_2^2 - \lambda_3^2 & 2(\lambda_2\lambda_3 - \lambda_0\lambda_1) \\ 2(\lambda_1\lambda_3 - \lambda_0\lambda_2) & 2(\lambda_2\lambda_3 + \lambda_0\lambda_1) & \lambda_0^2 - \lambda_1^2 - \lambda_2^2 + \lambda_3^2 \end{bmatrix}. \tag{4.11}$$

The product $\mathbf{Rv}$ can be identified as the product of quaternions:

$$\mathbf{Rv} = \mathbf{q} \times \mathbf{v} \times \overline{\mathbf{q}} \,. \tag{4.12}$$

Composition of rotations corresponds to multiplication of quaternions, that is,

$$\mathbf{R}_2\mathbf{R}_1\mathbf{v} = \mathbf{q}_2 \times (\mathbf{q}_1 \times \mathbf{v} \times \overline{\mathbf{q}}_1) \times \overline{\mathbf{q}}_2 = (\mathbf{q}_2 \times \mathbf{q}_1) \times \mathbf{v} \times (\overline{\mathbf{q}_2 \times \mathbf{q}_1}) \,. \tag{4.13}$$

A 3D rigid motion is then represented by a 7-dimensional vector, noted as $\mathbf{s}_q$:

$$\mathbf{s}_q = \begin{bmatrix} \mathbf{q} \\ \mathbf{t} \end{bmatrix} \,, \tag{4.14}$$

under the constraint $\|\mathbf{q}\| = 1$.

4.2 3D Line Segments

As described in Chap. 2, it is consensus among many researchers in Computer Vision and Robotics that uncertainty should be *explicitly* represented and manipulated [4.5,6]. For the reason of computational efficiency, the error is usually modeled as Gaussian. Then any geometric object can be considered as a random vector and is completely characterized by its mean and its covariance matrix. In [4.7], stereo triangulation error is modeled as three-dimensional Gaussian distribution for stereo navigation. For the same reason, measurement errors are also modeled as Gaussian in our representation. The Gaussian assumption is also used in [4.8] to modelize the uncertainty. However, we should remember that this assumption can only be justified for small errors, but not for gross or systematic errors.

In this section, after a brief review of previous representations for a line segment and their deficiencies, we present a new representation which we think is well adapted for the task at hand.

4.2.1 Previous Representations and Deficiencies

A line segment in 3D space is usually represented by its endpoints M_1 and M_2, which requires 6 parameters, and its covariance matrices Λ_1 and Λ_2. Λ_1 and Λ_2 are estimated by stereo triangulation from point correspondences (see the last chapter). Equivalently, a line segment can be represented by its direction vector $\mathbf{v}$ and its midpoint M, and their covariance matrices $\Lambda_{\mathbf{v}}$, Λ_M, and cross-correlation matrix $\Lambda_{\mathbf{v}M}$. The latter, sometimes referred as $(\mathbf{v}, M)$ representation, was used in a previous version of our motion algorithm [4.9], and will be used in Chap. 5 to compare different motion-determination methods. The same representation is also used in [4.8] and [4.10]. The relation between all those vectors and matrices is simple:

$$\mathbf{v} = M_2 - M_1 , \quad M = (M_1 + M_2)/2 ,$$
$$\Lambda_{\mathbf{v}} = \Lambda_1 + \Lambda_2 , \quad \Lambda_M = (\Lambda_1 + \Lambda_2)/4 , \quad \Lambda_{\mathbf{v}M} = (\Lambda_2 - \Lambda_1)/2 . \tag{4.15}$$

But we cannot use directly those parameters in most cases. The endpoints or midpoint of a segment are not reliable. This is mainly for two reasons:

- The first is purely algorithmic: because of noise in the images and because we sometimes approximate significantly curved contours with line segments, the polygonal approximation may vary from frame to frame inducing a variation in the segments endpoints.
- The second is physical: because of partial occlusion in the scene, a segment can be considerably shortened or lengthened, and the occluded part may change over time.

Thus, instead of the line segment, the infinite line supporting the segment is usually used, as in [4.11]. In a previous version of our motion-determination algorithm [4.9, 12, 13], a line segment was treated in a mixed way. The infinite supporting line was used in estimating motion, and the line segment was used in matching.

There are a lot of representations proposed in the literature for a line. The Plücker coordinate representation is a very elegant one which has 6 parameters and is, of course, not a minimal representation. The Plücker coordinates are given by two vectors $(\mathbf{u}, \mathbf{d})$, where

$$\mathbf{u} = \mathbf{v}/\|\mathbf{v}\| ,$$
$$\mathbf{d} = \mathbf{u} \wedge \mathbf{m} . \tag{4.16}$$

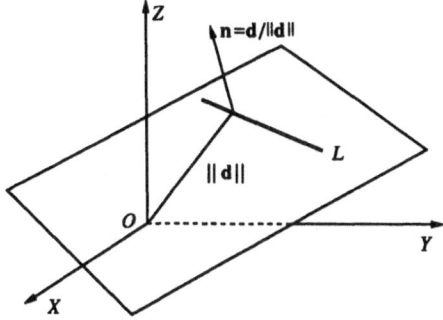

Fig. 4.2. A Plücker coordinate representation

Here, **m** may be any point on the line. In reality, **u** is the unit direction vector, the norm of **d** is the distance from the origin to the line and **d** is parallel to the normal of the plane passing through the line and the origin (see Fig. 4.2). This representation is used in Chap. 5 to derive an analytical solution of a 3D motion from line correspondences. A variant of the Plücker coordinates is given by

$$\mathbf{p}_1 = M_2 - M_1 \quad \text{and} \quad \mathbf{p}_1 = M_1 \wedge M_2 , \tag{4.17}$$

where M_1 and M_2 are two points on the line.

A minimal representation of a straight line is four-dimensional. In [4.14, 15], a representation called **abpq** was adopted. Under this representation, a line is represented as the intersection of two planes:

$$x = az + p , \qquad y = bz + q . \tag{4.18}$$

This is a quasi-minimal representation. The four parameters are $[a,\ b,\ p,\ q]^T$. The line goes through the point $[p,\ q,\ 0]^T$ and is parallel to the vector $[a,\ b,\ 1]^T$. But Equation (4.18) cannot represent lines parallel to the xy-plane. To fully define a 3D line, two other representations are required. The full **abpq** representation is then given by:

$$\begin{cases} \text{Map 1:} & x = az + p , \quad y = bz + q , \\ \text{Map 2:} & y = ax + p , \quad z = bx + q , \\ \text{Map 3:} & x = ay + p , \quad z = by + q . \end{cases} \tag{4.19}$$

This representation works well when all manipulations are done in the same coordinate system, for example, examination of the perpendicularity, parallelism and coplanarity. When we must deal with dynamic scenes, a difficulty occurs: a segment may change its representation in different views. This means that 9 different functions are required to transform a line segment from one view to another. Another difficulty with this representation is that the covariance matrix of $[a,\ b,\ p,\ q]^T$ cannot correctly characterize the uncertainty of the segment which it supports. For example, let $[p,\ q,\ 0]^T$ be the intersection of the line with the

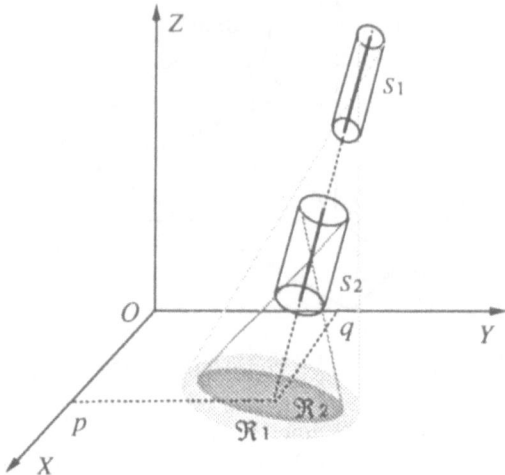

Fig. 4.3. The uncertainty in (p, q) does not reflect that of a segment

xy-plane. A segment with small uncertainty but far away from that point may give an uncertainty in p and q greater than a segment with bigger uncertainty but nearby that point. Figure 4.3 illustrates such problem. Suppose two collinear line segments S_1 and S_2 (they have the same (p, q) values). The uncertainty of S_2 is bigger than that of S_1, but S_2 is closer to the xy-plane than S_1. To simplify the illustration, the uncertainties of the segments are represented by cylinders. As can be observed, the uncertainty region $\Re_2$ of S_2 is smaller than the uncertainty region $\Re_1$ of S_1.

Roberts [4.16] proposed a minimal representation for a line. He did not consider oriented lines. A (nonoriented) line L is represented by the four-dimensional vector:

$$[u_x,\ u_y,\ x',\ y']^T$$

where u_x and u_y are the x and y components of the unit direction vector **u**. x' and y' are coordinates of the intersection of line L with the L-plane, in a 2D Cartesian frame defined on the L-plane. The L-plane is defined as the plane which contains the origin, and whose normal is **u**. One difficulty with this representation is that the position parameters $(x',\ y')$ are expressed in a purely local coordinate frame, which is unique for each possible orientation of a line. Positions must first be converted to a common frame before we can proceed with most computations. Another difficulty, as with the **abpq** representation, is that the covariance matrix of $[u_x,\ u_y,\ x',\ y']^T$ cannot correctly characterize the uncertainty of the segment which it supports.

4.2.2 A New Representation

Because of the deficiencies of the previous representations for a line or a line segment, we use a five parameter representation for a line segment: two for the orientation, three for the position of the segment. This is a trade-off between an infinite line and a line segment. If we add the length, a line segment can then be fully specified. Special attention is given to the representation of uncertainty.

Representing the orientation by its Euler angles ϕ and θ

Let us consider the spherical coordinates (see Fig. 4.4). Let $\mathbf{u} = [u_x, \ u_y, \ u_z]^T$ be a unit orientation vector, we have:

$$\begin{cases} u_x &= \cos\phi \sin\theta \\ u_y &= \sin\phi \sin\theta \\ u_z &= \cos\theta \end{cases} \tag{4.20}$$

with $0 \leq \phi < 2\pi, \ 0 \leq \theta \leq \pi$.

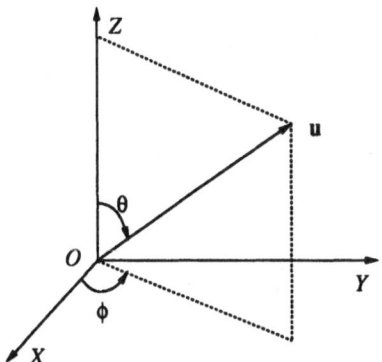

Fig. 4.4. Spherical coordinates

From $\mathbf{u}$, we can compute ϕ, θ:

$$\phi = \begin{cases} \arccos \frac{u_x}{\sqrt{1-u_z^2}} & \text{if } u_y \geq 0 \\ 2\pi - \arccos \frac{u_x}{\sqrt{1-u_z^2}} & \text{otherwise,} \end{cases} \tag{4.21}$$

$$\theta = \arccos u_z \ .$$

If we denote $[\phi, \ \theta]^T$ by ψ, then the mapping between ψ and $\mathbf{u}$ is 1-to-1 correspondence, except when $\theta = 0$. When $\theta = 0$, ϕ is not defined. A special treatment in covariance matrix of ψ is applied for this case, as described below. Another problem with this representation is the discontinuity in ϕ when a segment varies nearly parallel to the plane $y = 0$. This discontinuity must be dealt with in matching and data fusion.

In the following, we assume that the direction vector $\mathbf{v} = [x, \, y, \, z]^T$ and its covariance matrix $\Lambda_{\mathbf{v}}$ of a given segment are known. We want to compute ψ and its covariance matrix Λ_{ψ} from $\mathbf{v}$ and $\Lambda_{\mathbf{v}}$. ϕ and θ are simply given by:

$$\phi = \begin{cases} \arccos \dfrac{x}{\sqrt{x^2+y^2}} & \text{if } y \geq 0 \\ 2\pi - \arccos \dfrac{x}{\sqrt{x^2+y^2}} & \text{otherwise,} \end{cases}$$

$$\theta = \arccos \dfrac{z}{\sqrt{x^2+y^2+z^2}} \, . \tag{4.22}$$

Since the relation between ψ and $\mathbf{v}$ is not linear, we use the first order approximation to compute the covariance matrix Λ_{ψ} from $\Lambda_{\mathbf{v}}$ (see Theorem 2.3). That is

$$\Lambda_{\psi} = \frac{\partial \psi}{\partial \mathbf{v}} \Lambda_{\mathbf{v}} \frac{\partial \psi}{\partial \mathbf{v}}^T \, , \tag{4.23}$$

where the Jacobian matrix

$$\frac{\partial \psi}{\partial \mathbf{v}} = \begin{bmatrix} \dfrac{\partial \phi}{\partial x} & \dfrac{\partial \phi}{\partial y} & \dfrac{\partial \phi}{\partial z} \\[2mm] \dfrac{\partial \theta}{\partial x} & \dfrac{\partial \theta}{\partial y} & \dfrac{\partial \theta}{\partial z} \end{bmatrix} \, . \tag{4.24}$$

Note that

$$\frac{\partial \arccos x}{\partial x} = -\frac{1}{\sqrt{1-x^2}} \, .$$

After some simple computation, we have

$$\frac{\partial \phi}{\partial x} = -\frac{y}{x^2+y^2} \, , \qquad\qquad \text{for all } y$$

$$\frac{\partial \phi}{\partial y} = \frac{x}{x^2+y^2} \, , \qquad\qquad \text{for all } y$$

$$\frac{\partial \phi}{\partial z} = 0 \, , \qquad\qquad\qquad \text{for all } y$$

$$\frac{\partial \theta}{\partial x} = \frac{xz}{(x^2+y^2+z^2)\sqrt{x^2+y^2}} \, ,$$

$$\frac{\partial \theta}{\partial y} = \frac{yz}{(x^2+y^2+z^2)\sqrt{x^2+y^2}} \, ,$$

$$\frac{\partial \theta}{\partial z} = -\frac{\sqrt{x^2+y^2}}{x^2+y^2+z^2} \, .$$

Notice that when $x^2+y^2 = 0$, i.e., $x = 0$ and $y = 0$,

$$\frac{\partial \theta}{\partial x} = \frac{\partial \theta}{\partial y} = \frac{1}{z} \quad \text{and} \quad \frac{\partial \theta}{\partial z} = 0 \, ,$$

but $\frac{\partial \phi}{\partial x}$ and $\frac{\partial \phi}{\partial y}$ are not differentiable. It is reasonable since ϕ is not defined when the vector is parallel to the z-axis, and a slight change in x or y may provoke a drastic change in ϕ. To deal with this problem, we replace $\frac{\partial \phi}{\partial x}$ and $\frac{\partial \phi}{\partial y}$ by a very big number when $x^2 + y^2 = 0$, so that the components of ϕ in Λ_ψ are very big. That is to say that the measurement of ϕ has no information content.

Modeling the midpoint of a 3D line segment

We choose the midpoint as the three parameters to localize the segment, but a special treatment on the covariance is introduced to characterize its uncertainty.

As in (4.15), the midpoint M and its covariance matrix Λ_M can be computed from the endpoints of the segment by:

$$M = (M_1 + M_2)/2 \text{ and } \Lambda_M = (\Lambda_1 + \Lambda_2)/4 .$$

However, the way the uncertainty of the endpoints of a three-dimensional segment is computed takes only into account the uncertainty of the pixel coordinates due to the edge detection process and the uncertainty of the calibration of the stereo rig. It does not take into account the uncertainty due to the variations in the different images of the stereo triplet of the polygonal approximations of corresponding contours, as described in Sect. 4.2.1. In an attempt to cope with all this, we model the midpoint m of a segment $M_1 M_2$ as:

$$\mathbf{m} = (M_1 + M_2)/2 + n\mathbf{u} , \tag{4.25}$$

where $\mathbf{u}$ is the unit direction vector of the segment and n is a random scalar. Equation (4.25) says in fact that the midpoint has some extra uncertainty attached to it. It may vary randomly along the line supporting it in successive views. Remark that this modelization is in accordance with the definition of a line. If a point $\mathbf{p_0}$ on a line and its orientation $\mathbf{u}$ are given, the line L may be defined as a set of points in 3D space parametrized by a real variable t:

$$L = \{\mathbf{p} \mid \mathbf{p} = \mathbf{p_0} + \mathbf{u}t, \ -\infty < t < \infty\} . \tag{4.26}$$

The random variable n in (4.25) is modeled as Gaussian with mean zero and standard deviation σ_n, a positive scalar. If a segment is reliable, σ_n may be chosen to be a small number; if not, it may be chosen to be a big one. In our implementation, σ_n is related to the length l of the segment, i.e., $\sigma_n = \kappa l$, where κ is some constant. That is to say that the longer a segment is, the bigger the deviation σ_n is. This is reasonable since a long segment is much likely to be broken into smaller segments in other views. In our experiments, we found that $\kappa = 0.2$ gives us very good results.

In order to compute the covariance of $\mathbf{m}$, we should first compute the unit direction vector $\mathbf{u}$ and its covariance $\Lambda_{\mathbf{u}}$. They can be computed either from the nonnormalized direction vector $\mathbf{v}$ and its covariance matrix $\Lambda_{\mathbf{v}}$, or from the representation ψ and Λ_ψ. Indeed, from (4.15), we have

$$\mathbf{u} = \frac{\mathbf{v}}{\|\mathbf{v}\|} , \qquad \Lambda_{\mathbf{u}} = \frac{\partial \mathbf{u}}{\partial \mathbf{v}} \Lambda_{\mathbf{v}} \frac{\partial \mathbf{u}}{\partial \mathbf{v}}^T , \tag{4.27}$$

where $\frac{\partial \mathbf{u}}{\partial \mathbf{v}}$ is a 3×3 matrix

$$\frac{\partial \mathbf{u}}{\partial \mathbf{v}} = \frac{\mathbf{I}_3}{\|\mathbf{v}\|} - \frac{\mathbf{v}\mathbf{v}^T}{\|\mathbf{v}\|^3} .$$

In another way, we can compute $\mathbf{u}$ from ψ based on (4.20), and $\Lambda_\mathbf{u}$ is given by

$$\Lambda_\mathbf{u} = \frac{\partial \mathbf{u}}{\partial \psi}\Lambda_\psi\frac{\partial \mathbf{u}}{\partial \psi}^T , \tag{4.28}$$

where $\frac{\partial \mathbf{u}}{\partial \psi}$ is a 3×2 matrix:

$$\frac{\partial \mathbf{u}}{\partial \psi} = \begin{bmatrix} -\sin\phi\sin\theta & \cos\phi\cos\theta \\ \cos\phi\sin\theta & \sin\phi\cos\theta \\ 0 & -\sin\theta \end{bmatrix} . \tag{4.29}$$

Note that the covariance matrix $\Lambda_\mathbf{u}$ is singular (the determinant is zero). This is reasonable since the three components of $\mathbf{u}$ are not independent ($\|\mathbf{u}\| = 1$).

At this point, the covariance of $\mathbf{m}$ can be computed. We start with the covariance of $n\mathbf{u}$. Since n and $\mathbf{u}$ are independent to each other, we have

$$E[n\mathbf{u}] = E[n]E[\mathbf{u}] = 0 , \tag{4.30}$$

$$\Lambda_{n\mathbf{u}} = E[(n\mathbf{u})(n\mathbf{u})^T] = E[n^2\mathbf{u}\mathbf{u}^T] = E[n^2]E[\mathbf{u}\mathbf{u}^T] = \sigma_n^2(\Lambda_\mathbf{u} + \bar{\mathbf{u}}\bar{\mathbf{u}}^T) , \tag{4.31}$$

where $\bar{\mathbf{u}} = E[\mathbf{u}]$. Now we have

$$E[\mathbf{m}] = E[M] , \tag{4.32}$$

and

$$\begin{aligned} \Lambda_\mathbf{m} &= E[(\mathbf{m} - E[\mathbf{m}])(\mathbf{m} - E[\mathbf{m}])^T] \\ &= E[(\mathbf{M} - E[\mathbf{M}])(\mathbf{M} - E[\mathbf{M}])^T] + E[(n\mathbf{u})(n\mathbf{u})^T] \\ &\quad + E[n(\mathbf{M} - E[\mathbf{M}])\mathbf{u}^T] . \end{aligned} \tag{4.33}$$

Since n is independent of $\mathbf{M}$ and $\mathbf{u}$ and has zero-mean, the last term is equal to 0 and

$$\Lambda_\mathbf{m} = \Lambda_M + \Lambda_{n\mathbf{u}} .$$

If we add another parameter l to denote the length of the segment, we can then represent exactly a line segment. The variance on the length does not need to be modeled, since this information is not required in this monograph. This ends our modelization of a line segment.

More rigorously, there exists a correlation between ψ and $\mathbf{m}$, but this correlation is negligible. We have computed the correlation for many segments, and we found that the coefficient of correlation between ψ and $\mathbf{m}$ is less than 0.01.

The same idea has been adopted in [4.17] to represent 2D segments obtained by projecting 3D segments on the ground plane. A similar idea is found in [4.10], where the representation (4.15) is used. Considering the unreliability of the midpoint or the endpoints of a segment, a lengthened covariance matrix along the direction of the segment is added to the original covariance matrix before using it and is removed later.

4.3   Summary

In this chapter, we have investigated the representations of rigid motions and 3D line segments, which will be used in the following chapters.

We have given the definition of a rigid motion and its properties. Two representations for rotation have been discussed. The first one has three parameters, describing directly the rotation axis and the rotation angle. The second one uses a quaternion, and thus has 4 parameters, together with a constraint.

We have reviewed several representations of lines or line segments, and their deficiencies have been discussed. We have proposed a new representation for a line segment. It has 5 parameters: two for the orientation and three for the midpoint. Length can be added to fully define a line segment, which, however, does not play an important role in the problems addressed in this monograph. Special consideration is devoted to the uncertainty representation and manipulation. We believe that our representation is very convenient for solving the problems addressed in this monograph, as highlighted later.

4.4   Appendix: Visualizing Uncertainty

It is usually useful to graphically display the uncertainty of a geometric object. As described in Chap. 2, a geometric object is represented by a parameter vector $\mathbf{p}$ with mean $\bar{\mathbf{p}}$ and associated covariance matrix $\Lambda_{\mathbf{p}}$. The problem is how to display the domain in which the geometric object $\mathbf{p}$ can be found with a predetermined probability. *Ayache* has explained in [4.15] how to visualize the uncertainty of a 2D point or an orthogonal projection of a 3D point by an ellipse. Here, we give more details.

The following equation describes a hyperellipsoid with center at $\bar{\mathbf{p}}$

$$(\mathbf{p} - \bar{\mathbf{p}})^T \Lambda_{\mathbf{p}}^{-1} (\mathbf{p} - \bar{\mathbf{p}}) = k^2 \ , \tag{4.34}$$

where k is a constant. As described in Sect. 2.2.5, $(\mathbf{p} - \bar{\mathbf{p}})^T \Lambda_{\mathbf{p}}^{-1} (\mathbf{p} - \bar{\mathbf{p}})$ follows a χ^2 distribution with n degrees of freedom, where n is the dimension of $\mathbf{p}$ or the rank of $\Lambda_{\mathbf{p}}$. By looking up the χ^2 distribution Table 2.1, we know what is the probability that $\mathbf{p}$ lies inside the hyperellipsoid.

Of particular interest are the ellipses for $n = 2$ and the ellipsoids for $n = 3$. Consider first how to draw ellipses. We first compute the eigenvalues of the covariance matrix $\Lambda_{\mathbf{p}}$, noted by σ_1^2, σ_2^2, and its eigenvectors, noted by $\mathbf{v}_1, \mathbf{v}_2$. Let $M = [\mathbf{v}_1, \mathbf{v}_2]$, then we have

$$M^T \Lambda_{\mathbf{p}} M = \begin{bmatrix} \sigma_1^2 & 0 \\ 0 & \sigma_2^2 \end{bmatrix} \ .$$

Let $\mathbf{p}' = [p_1', p_2']^T$, the following equation

$$\mathbf{p}'^T M^T \Lambda_{\mathbf{p}} M \mathbf{p}' = k^2 \tag{4.35}$$

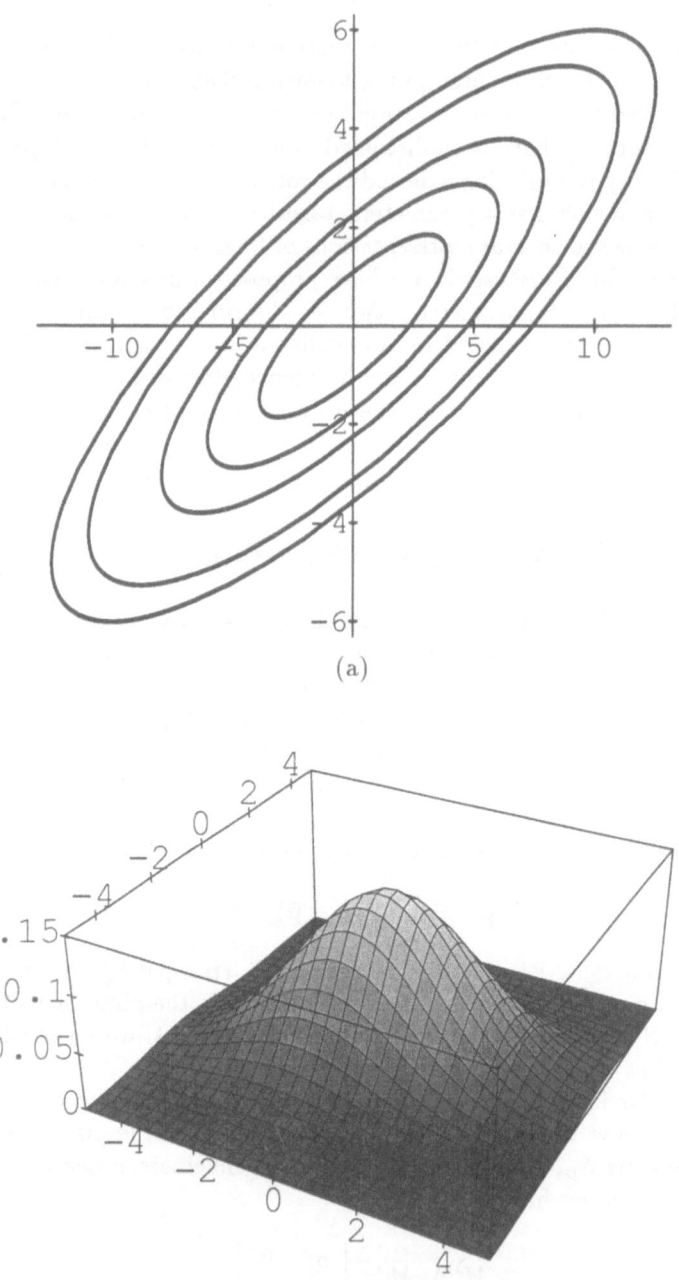

(a)

(b)

Fig. 4.5. Visualization of uncertainty of a 2D point

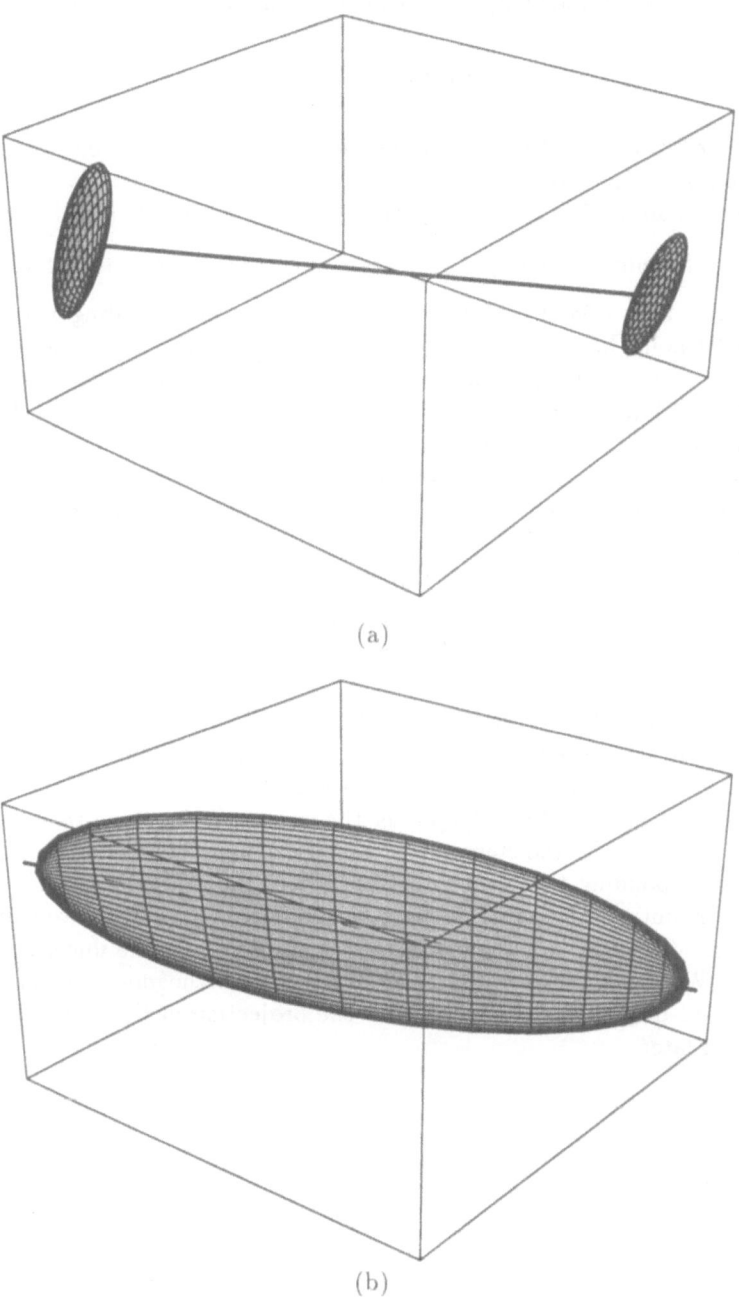

(a)

(b)

Fig. 4.6. Visualization of uncertainty of a 3D line segment

describes the transformed ellipse, which can be easily drawn by setting

$$p'_1 = \frac{k}{\sigma_1} \sin\theta , \qquad p'_2 = \frac{k}{\sigma_2} \cos\theta ,$$

and varying θ from 0 to 2π. The original ellipse can be recovered from $M\mathbf{p}' = \mathbf{p} - \bar{\mathbf{p}}$, i.e., $\mathbf{p} = M\mathbf{p}' + \bar{\mathbf{p}}$.

Figure 4.5 shows the uncertainty of a 2D point with $\bar{\mathbf{p}} = [0,0]^T$ and $\Lambda_{\mathbf{p}} = \begin{bmatrix} 4 & 1 \\ 1 & 2 \end{bmatrix}$. The ellipses in Fig. 4.5a correspond, from interior to exterior, to k equal to 0.76, 1.18, 1.55, 2.15, 2.45, respectively. The corresponding probabilities that the 2D point lies inside those ellipses are 25%, 50%, 70%, 90% and 95%. Figure 4.5b gives a 3D visualization of the uncertainty of a 2D point — the 2D Gaussian distribution. In the remainder of this monograph, we choose $k = 2.15$ for a probability of 90% in visualizing 2D uncertainties.

If $\mathbf{p}$ is 3-dimensional, we compute the eigenvalues of $\Lambda_{\mathbf{p}}$, noted by σ_1^2, σ_2^2, σ_3^2, and its eigenvectors, noted by $\mathbf{v}_1$, $\mathbf{v}_2$, $\mathbf{v}_3$. The matrix $M = [\mathbf{v}_1, \mathbf{v}_2, \mathbf{v}_3]$ transforms the original ellipsoid to principal axes. Let $\mathbf{p}' = [p'_1, p'_2, p'_3]^T$, the transformed ellipsoid can be drawn by setting

$$p'_1 = \frac{k}{\sigma_1} \sin\theta \cos\phi , \qquad p'_2 = \frac{k}{\sigma_2} \sin\theta \sin\phi , \qquad p'_3 = \frac{k}{\sigma_3} \cos\theta ,$$

and varying θ from 0 to π and ϕ from 0 to 2π. The original ellipsoid can be recovered from $\mathbf{p} = M\mathbf{p}' + \bar{\mathbf{p}}$. Figure 4.6 displays the uncertainty of a 3D line segment. Figure 4.6a shows the ellipsoids with $k = 2.5$ corresponding to a probability of 90% in the endpoints of the segment. The uncertainties in the endpoints are given by the stereo system. Figure 4.6b shows the ellipsoid with $k = 2.5$ corresponding to a probability of 90% in the midpoint of the segment whose uncertainty is computed as described in Sect. 4.2.2. In the remainder of this monograph, we do not show directly the ellipsoids for the uncertainty of a 3D point. We display always the uncertainty of its orthogonal projection to a plane, i.e., we display the ellipse which is the projection of the ellipsoid, as can be observed later.

5. A Comparative Study of 3D Motion Estimation

The motion analysis problem is usually divided into two subproblems: correspondence of primitives and estimation of motion, usually in that order. Typically used primitives are points, line segments or planar patches. For the correspondence problem, the reader is referred to the following chapters. Given correspondences of primitives, many methods exist to determine 3D motion.

If we have some appropriate constraints on the environment, motion can be determined without explicit correspondence. For example, the method of *Szeliski* [5.1] assumes that the sensed points lie on a piecewise smooth surface. Motion parameters are estimated by finding a rigid transformation which makes it most likely (in a Bayesian sense) that the points came from the same surface. Matching is taken here in a broader sense, and motion is estimated from the point-surface correspondences.

Although it is easier to estimate motion parameters from 3D data than from 2D (monocular) images, it is not trivial since the 3D data we have are almost always corrupted by noise. A number of methods are proposed to combat the noise. To our knowledge, no work has been carried out to compare those methods. We believe that this work is important for researchers working on motion analysis to choose an appropriate approach based on efficiency, accuracy and robustness. In this chapter, we present a comparative study of different methods for determining motion from correspondences of 3D line segments. For the problem of determining motion from 3D point correspondences, the reader is referred to [5.2–5]. *Faugeras* and *Hébert* [5.4] presented a method to determine motion from planar patches. *Haralick* et al. [5.6] presented solutions to pose estimation problems from *point* data including 3D – 3D and 2D (perspective projection) – 3D pose estimation problems.

The chapter is organized as follows. First, two representations for line segments are presented, and the problem we address are formulated. Then, we show how to estimate motion using the extended Kalman filter, minimization techniques, a singular value decomposition technique, and the method proposed by *Kim* and *Aggarwal* [5.7] but with some modification. Finally we provide the results of our comparative study of different methods.

While this part of the monograph emphasizes methods for determining 3D general motion from two 3D views, one can use a long sequence of frames to estimate motion parameters. The general approach is, for the moment, based on simple kinematic models: constant acceleration in translation, constant angular velocity or constant precession in rotation. This approach will be discussed in the last part of this monograph.

5.1 Problem Statement

In this section, we give two representations of 3D line segments and the transformation of line segments under rigid motion. Finally, we formalize the problem we should solve.

5.1.1 Line Segment Representations

For the historical reason, we shall use the $(\mathbf{v},\ M)$ representation and the Plücker coordinate one, described in Sect. 4.2.1. A 3D line segment L can be represented by two vectors $(\mathbf{v}, M)$, where $\mathbf{v}$ is the nonnormalized direction vector of L and M, the midpoint of L. The covariance matrix of $(\mathbf{v}, M)$ is given as follows:

$$\Lambda_L = \left[\begin{array}{cc} \Lambda_\mathbf{v} & \Lambda_{\mathbf{v}M} \\ \Lambda_{\mathbf{v}M} & \Lambda_M \end{array} \right] , \tag{5.1}$$

which is a 6×6 matrix. The reader is referred to (4.15) for more details.

We can also represent a 3D line segment by its infinite line of support. There are a number of ways to represent a 3D line, one of which is to represent the line by the Plücker coordinates $(\mathbf{u}, \mathbf{d})$. Equation (4.16) describes the relation between $(\mathbf{u}, \mathbf{d})$ and $(\mathbf{v}, M)$. In Sect. 5.4, we use this representation to derive an analytical solution of motion.

We should point out that segments addressed in this monograph are *oriented*, which is obtained in stereo through the information about the intensity contrast.

5.1.2 3D Line Segment Transformation

If a 3D line segment undergoes a rigid displacement $(\mathbf{R}, \mathbf{t})$ (see Sect. 4.1) and if we use the $(\mathbf{v},\ M)$ representation, let $(\mathbf{v}, M)$ be the parameters before motion and $(\mathbf{v}', M')$, that after motion, the following relations hold:

$$\begin{array}{rcl} \mathbf{v}' &=& \mathbf{R}\mathbf{v} , \\ M' &=& \mathbf{R}M + \mathbf{t} . \end{array} \tag{5.2}$$

These are the exact equations; in practice, they may not be satisfied because a segment could be segmented differently in successive views. As we know that its direction is relatively more stable, we insist on having the transformed segment parallel to the segment in the second view, which yields [5.8]:

$$\mathbf{v}' \wedge \mathbf{R}\mathbf{v} = 0 \ , \tag{5.3}$$

$$\mathbf{v}' \wedge (M' - \mathbf{R}M - \mathbf{t}) = 0 \ . \tag{5.4}$$

We shall use these two equations to compute the motion parameters. One remark is that Equations (5.3) and (5.4) provide each in fact only two scalar equations since the three coordinates of a cross product are not linearly independent. This is to be compared with the 6 equations provided by (5.2). Note that in general, two 3D noncollinear line segments are sufficient to determine motion, but using (5.3,5.4), we cannot recover fully all motion parameters if two segments are parallel.

If we note:

$$\mathbf{f}(\mathbf{x}', \mathbf{p}) = \begin{bmatrix} \tilde{\mathbf{v}}'\mathbf{R}\mathbf{v} \\ \tilde{\mathbf{v}}'(M' - \mathbf{R}M - \mathbf{t}) \end{bmatrix} \ , \tag{5.5}$$

where the tilde "~" is defined as in (4.2), Equations (5.3) and (5.4) become:

$$\mathbf{f}(\mathbf{x}', \mathbf{p}) = 0 \ , \tag{5.6}$$

where $\mathbf{x}' = [\mathbf{v}^T, M^T, \mathbf{v}'^T, M'^T]^T$ is a 12-dimensional vector which we call *measurement vector* and $\mathbf{p}$ is one of the motion parametrizations described in Sect. 4.1, i.e., either $\mathbf{s}$ (the rotation axis representation, see (4.6)) or $\mathbf{s}_q$ (the quaternion representation, see (4.14)). In filtering terminology, $\mathbf{p}$ is called *state vector*. In practice, the exact value of $\mathbf{x}'$ is seldomly known. We usually have noisy measurements and we denote the real observed measurement vectors by $\mathbf{x}$ (see Sect. 2.2). Hence our problem can be formulated as follows:

Given n measurement vectors: $\mathbf{x}_1, \mathbf{x}_2, ..., \mathbf{x}_n$,
 i.e., given a set of equations:
  $\mathbf{f}(\mathbf{x}', \mathbf{p}_i) = 0$,   for $i = 1, ..., n$,
Recover the motion parameters $\mathbf{p}$.

5.2 Extended Kalman Filter Approaches

We describe in this section how to apply the extended Kalman filter described in Sect. 2.2 to our problem.

5.2.1 Linearization of the Equations

As Equation (5.6) is nonlinear, we first linearize it, and then apply the Kalman filter to the linearized system. The linearized system is as follows:

$$\mathbf{p}_i = \mathbf{p}_{i-1} \ , \tag{5.7}$$

$$\mathbf{y}_i = M_i\mathbf{p}_i + \boldsymbol{\xi}_i \ , \tag{5.8}$$

where

$$\mathbf{y}_i = -\mathbf{f}(\mathbf{x}_i, \hat{\mathbf{p}}_{i-1}) + \frac{\partial \mathbf{f}(\mathbf{x}_i, \hat{\mathbf{p}}_{i-1})}{\partial \mathbf{p}} \hat{\mathbf{p}}_{i-1} ,$$

$$M_i = \frac{\partial \mathbf{f}(\mathbf{x}_i, \hat{\mathbf{p}}_{i-1})}{\partial \mathbf{p}} ,$$

$$\boldsymbol{\xi}_i = \frac{\partial \mathbf{f}(\mathbf{x}_i, \hat{\mathbf{p}}_{i-1})}{\partial \mathbf{x}'}(\mathbf{x}'_i - \mathbf{x}_i) .$$

$\hat{\mathbf{p}}_{i-1}$ is the current estimate of $\mathbf{p}$ before processing the linearized system, and $\mathbf{x}_i$ is the current measurement. The computation of the derivative of $\mathbf{f}(\mathbf{x}', \mathbf{p})$ with respect to $\mathbf{p}$ will be given in the following subsection (either (5.14) or (5.15)). What we need to compute here is the derivative of $\mathbf{f}(\mathbf{x}', \mathbf{p})$ with respect to $\mathbf{x}'$. We use the first representation in Sect. 5.1.1 for 3D line segments, i.e., the $(\mathbf{v}, M)$ one. The derivative of $\mathbf{f}(\mathbf{x}', \mathbf{p})$ with respect to $\mathbf{x}'$ is the same for both motion parametrizations:

$$\frac{\partial \mathbf{f}(\mathbf{x}', \mathbf{p})}{\partial \mathbf{x}'} = \begin{bmatrix} \tilde{\mathbf{v}}'\mathbf{R} & 0 & -(\widetilde{\mathbf{R}\mathbf{v}}) & 0 \\ 0 & -\tilde{\mathbf{v}}'\mathbf{R} & \mathbf{R}\widetilde{M} + \tilde{\mathbf{t}} - \widetilde{M}' & \tilde{\mathbf{v}}' \end{bmatrix} . \tag{5.9}$$

The expectation and the covariance of the new measurement noise $\boldsymbol{\xi}_i$ are easily derived from that of $\mathbf{x}_i$ as:

$$E[\boldsymbol{\xi}_i] = 0 ,$$

$$\Lambda_{\boldsymbol{\xi}_i} \overset{\text{def}}{=} E[\boldsymbol{\xi}_i\boldsymbol{\xi}_i^T] = \frac{\partial \mathbf{f}(\mathbf{x}_i, \hat{\mathbf{p}}_{i-1})}{\partial \mathbf{x}'}\Lambda_i\frac{\partial \mathbf{f}(\mathbf{x}_i, \hat{\mathbf{p}}_{i-1})}{\partial \mathbf{x}'}^T ,$$

where Λ_i is the covariance matrix of $\mathbf{x}_i$. If we suppose that the matched segments L_i and L'_i are independent, then:

$$\Lambda_i = \begin{bmatrix} \Lambda_{L_i} & 0 \\ 0 & \Lambda_{L'_i} \end{bmatrix} .$$

One should note that the covariance matrix $\Lambda_{\boldsymbol{\xi}_i}$ thus computed is singular, because not all components of $\mathbf{f}(\mathbf{x}', \mathbf{p})$ are linearly independent. This is because a cross product of two 3D vectors gives only two independent components, see Sect. 5.1. In order to cope with this problem, we eliminate one component with least value among the first three components of $\mathbf{f}(\mathbf{x}', \mathbf{p})$ and one with least value among the last three components. The corresponding rows of the Jacobian matrix $\frac{\partial \mathbf{f}(\hat{\mathbf{p}}, \mathbf{x}')}{\partial \mathbf{x}'}$ are also eliminated and finally we get a 4×4 covariance matrix of full rank.

An important remark is that when we use the quaternion to represent the rotation, we have added the constraint $\|\mathbf{q}\| = 1$, i.e., $\mathbf{q}^T\mathbf{q} - 1 = 0$ in (5.6) as an additional measurement. The derivative of $\mathbf{q}^T\mathbf{q} - 1$ with respect to $\mathbf{s}_q$ is

$$\frac{\partial}{\partial \mathbf{s}_q}(\mathbf{q}^T\mathbf{q} - 1) = [2\mathbf{q}^T \ 0] , \tag{5.10}$$

and the component in the covariance matrix corresponding to this equation is given by

$$4\mathbf{q}^T \Lambda_\mathbf{q} \mathbf{q}$$

where $\Lambda_\mathbf{q}$ is the covariance matrix of $\mathbf{q}$.

5.2.2 Derivation of a Rotation Matrix

We have two parametrizations for a rotation: rotation axis and quaternion, as described in Sect. 4.1. In this section, we compute the first derivatives of the rotation matrix with respect to each of its parametrizations, which are used in the EKF method just described and in the method described in the next section. More precisely, we are interested in computing the derivative of $\mathbf{Rv}$ with respect to $\mathbf{r}$ and that with respect to $\mathbf{q}$, where $\mathbf{v} = [v_0, \ v_1, \ v_2]^T$ is an arbitrary 3-dimensional vector.

The derivative of $\mathbf{Rv}$ with respect to $\mathbf{r}$ is computed from (4.5) and is given by

$$\frac{\partial \mathbf{Rv}}{\partial \mathbf{r}} = \frac{\cos(\theta) - \mathrm{f}(\theta)}{\theta^2}(\mathbf{r} \wedge \mathbf{v})\mathbf{r}^T + \frac{\sin(\theta) - 2\theta \mathrm{g}(\theta)}{\theta^3}(\mathbf{r} \wedge (\mathbf{r} \wedge \mathbf{v}))\mathbf{r}^T$$
$$- \mathrm{f}(\theta)\tilde{\mathbf{v}} + \mathrm{g}(\theta)(-(\widetilde{\mathbf{r} \wedge \mathbf{v}}) + (\mathbf{r}.\mathbf{v})\mathbf{I}_3 - \mathbf{v}\mathbf{r}^T) \ . \tag{5.11}$$

where $\mathrm{f}(\theta)$ and $\mathrm{g}(\theta)$ have the same definition as before. The vector differentiation presented in the appendix to this monograph has been used in the above derivation.

We define $E(\mathbf{R}, \mathbf{v})$ to be the 3 × 3 matrix $\frac{\partial \mathbf{Rv}}{\partial \mathbf{r}}$.

The derivative of $\mathbf{Rv}$ with respect to $\mathbf{q}$ is simpler. Indeed, we have:

$$\frac{\partial \mathbf{Rv}}{\partial \mathbf{q}} = \begin{bmatrix} d_0 & d_1 & d_2 & d_3 \\ -d_3 & -d_2 & d_1 & d_0 \\ d_2 & -d_3 & -d_0 & d_1 \end{bmatrix} , \tag{5.12}$$

where

$$\begin{aligned} d_0 &= 2(\lambda_0 v_0 - \lambda_3 v_1 + \lambda_2 v_2) \ , \\ d_1 &= 2(\lambda_1 v_0 + \lambda_2 v_1 + \lambda_3 v_2) \ , \\ d_2 &= 2(-\lambda_2 v_0 + \lambda_1 v_1 + \lambda_0 v_2) \ , \\ d_3 &= 2(-\lambda_3 v_0 - \lambda_0 v_1 + \lambda_1 v_2) \ . \end{aligned} \tag{5.13}$$

We define $Q(\mathbf{R}, \mathbf{v})$ to be the 3 × 4 matrix $\frac{\partial \mathbf{Rv}}{\partial \mathbf{q}}$.

The derivative of $\mathbf{f}(\mathbf{x}', \mathbf{p})$ with respect to $\mathbf{s}$ can be easily computed as follows:

$$\frac{\partial \mathbf{f}(\mathbf{x}', \mathbf{p})}{\partial \mathbf{s}} = \begin{bmatrix} \tilde{\mathbf{v}}'E(\mathbf{R}, \mathbf{v}) & 0 \\ -\tilde{\mathbf{v}}'E(\mathbf{R}, M) & -\tilde{\mathbf{v}}' \end{bmatrix} , \tag{5.14}$$

and the derivative of $\mathbf{f}(\mathbf{x}', \mathbf{p})$ with respect to $\mathbf{s}_q$ is:

$$\frac{\partial \mathbf{f}(\mathbf{x}', \mathbf{p})}{\partial \mathbf{s}_q} = \begin{bmatrix} \tilde{\mathbf{v}}'Q(\mathbf{R}, \mathbf{v}) & 0 \\ -\tilde{\mathbf{v}}'Q(\mathbf{R}, M) & -\tilde{\mathbf{v}}' \end{bmatrix} . \tag{5.15}$$

5.3 Minimization Techniques

The method presented in this section is probably the most reasonable thing to do if we assume that the matches are given. We can restate the problem addressed in Sect. 5.1 as a minimization problem, i.e.,

> **Given** n measurement vectors: $\mathbf{x}_1, \mathbf{x}_2, \ldots, \mathbf{x}_n$,
> **Recover** the motion parameters $\mathbf{p}$ so that
> $$\sum_{i=1}^{n}[\mathbf{f}(\mathbf{x}_i, \mathbf{p})^T \mathbf{f}(\mathbf{x}_i, \mathbf{p})]$$
> is minimized.

Here $\sum_{i=1}^{n}[\mathbf{f}(\mathbf{x}_i, \mathbf{p})^T \mathbf{f}(\mathbf{x}_i, \mathbf{p})]$ is called the *objective function* in the minimization problem, which is denoted by $\mathcal{F}(\mathbf{p})$, i.e.,

$$\mathcal{F}(\mathbf{p}) = \sum_{i=1}^{n}[\|\mathbf{v}_i' \wedge \mathbf{R}\mathbf{v}_i\|^2 + \|\mathbf{v}_i' \wedge (M_i' - \mathbf{R}M_i - \mathbf{t})\|^2] \ . \tag{5.16}$$

If we use the rotation axis to represent the rotation, then the above problem is just an unconstrained minimization one. If we use a quaternion to represent the rotation, then it is a minimization problem subject to a nonlinear constraint $|\mathbf{q}| = 1$. There exist many minimization routines in mathematical libraries (like NAG [5.9]) to solve such problems. The minimization process can be speeded up if we also supply the first derivations. We adopt the routine E04GDF in the NAG library to minimize $\mathcal{F}(\mathbf{p})$ with the rotation axis representation, and the routine E04VCF to minimize $\mathcal{F}(\mathbf{p})$ with the quaternion representation under the constraint $|\mathbf{q}| = 1$. In either case, we need to compute the derivative of $\mathcal{F}(\mathbf{p})$ with respect to the motion parameters $\mathbf{p}$, which can be easily obtained using (5.14) or (5.15).

E04GDF is a modified Gauss-Newton algorithm for finding an unconstrained minimum of a sum of squares of m nonlinear functions in n variables $(m \geq n)$. E04VCF is a routine designed to minimize an arbitrary smooth function subject to constraints, which may include simple bounds on the variables, linear constraints and smooth nonlinear constraints. E04VCF uses a sequential quadratic programming algorithm, in which the search direction is the solution of a quadratic programming problem. A starting point (an initial estimate) is required in both routines.

Generally, it is preferable for the objective function to be of the order of one in the region of interest when using these numerical minimization routines. We have scaled the objective function by multiplying it by a positive constant. The solution of a given problem is unaltered if the objective function is multiplied by a positive constant, or if a constant value is added to the objective function.

We can adopt the weighted least-squares method to take into account the uncertainty of measurements. That is, we first use the general minimization algorithms to obtain a better estimate of motion by minimizing the simple criterion

$\mathcal{F}(\mathbf{p})$ (5.16). Then we compute the covariance matrix of $\mathbf{f}(\mathbf{x}_i, \mathbf{p})$ using the first order approximation, denoted by Λ_{ξ_i}. We again use the minimization algorithms to obtain a new estimate by minimizing:

$$\sum_{i=1}^{n} [\mathbf{f}(\mathbf{x}_i, \mathbf{p})^T \Lambda_{\xi_i}^{-1} \mathbf{f}(\mathbf{x}_i, \mathbf{p})] . \tag{5.17}$$

The new estimate is generally better than the previous one, but more time is needed to compute the weighting. We have not implemented the weighted least-squares technique using the general minimization algorithms. However, we note that the EKF presented in the previous section can be considered as a recursive implementation of the weighted least-squares technique.

5.4   Analytical Solution

In this section, we present an analytical method to recover the motion parameters. The second representation of line segments (the Plücker coordinate one (4.16)) and the quaternion representation of rotation are used. *Faugeras and Hébert* [5.4] have proposed an analytical method from point and plane correspondences. The method described here is directly inspired from theirs.

The relation between a line segment $(\mathbf{u}, \mathbf{d})$ and the transformed line segment $(\mathbf{u}', \mathbf{d}')$ is:

$$\mathbf{u}' = \mathbf{R}\mathbf{u} , \tag{5.18}$$
$$\mathbf{d}' = \mathbf{R}\mathbf{d} + \mathbf{u}' \wedge \mathbf{t} . \tag{5.19}$$

The first one is evident, while the second one can be easily verified using the definition of $\mathbf{d}$:

$$\mathbf{d}' \triangleq \mathbf{u}' \wedge M' = \mathbf{u}' \wedge (\mathbf{R}M + \mathbf{t}) = \mathbf{R}\mathbf{u} \wedge \mathbf{R}M + \mathbf{u}' \wedge \mathbf{t}$$
$$= \mathbf{R}(\mathbf{u} \wedge M) + \mathbf{u}' \wedge \mathbf{t} = \mathbf{R}\mathbf{d} + \mathbf{u}' \wedge \mathbf{t} . \tag{5.20}$$

Due to the fact that the orientation of a segment is more conservative than other parameters (for example, $\mathbf{d}$ or M), we divide the motion determination problem into two subproblems:

1. Determine first the rotation using (5.18), that is, recover the rotation parameters by minimizing the following criterion:

$$\text{Min} \sum_{i=1}^{n} \|\mathbf{u}'_i - \mathbf{R}\mathbf{u}_i\|^2 . \tag{5.21}$$

2. Determine then the translation using (5.19), that is, recover the translation parameters by minimizing the following criterion:

$$\text{Min} \sum_{i=1}^{n} \|\mathbf{d}'_i - \mathbf{R}^* \mathbf{d}_i - \mathbf{u}'_i \wedge \mathbf{t}\|^2 , \tag{5.22}$$

where $\mathbf{R}^*$ is the rotation matrix recovered in the first step.

5.4.1 Determining the Rotation

By using (4.12), we can restate the minimization problem of (5.21) in quaternion notation as:

$$\text{Min} \sum_{i=1}^{n} |\mathbf{u}'_i - \mathbf{q} \times \mathbf{u}_i \times \overline{\mathbf{q}}|^2 \tag{5.23}$$

subject to the constraint $|\mathbf{q}| = 1$. Here, $\times$ is the quaternion multiplication. By multiplying each term in (5.23) with $|\mathbf{q}|^2$, we get:

$$\text{Min} \sum_{i=1}^{n} |\mathbf{u}'_i \times \mathbf{q} - \mathbf{q} \times \mathbf{u}_i|^2 . \tag{5.24}$$

From the definition of the product of two quaternions, we can express $\mathbf{u}'_i \times \mathbf{q} - \mathbf{q} \times \mathbf{u}_i$ as a linear function of $\mathbf{q}$. Indeed, there exists a matrix A_i such that:

$$\mathbf{u}'_i \times \mathbf{q} - \mathbf{q} \times \mathbf{u}_i = A_i \mathbf{q}, \tag{5.25}$$

where

$$A_i = \begin{bmatrix} 0 & (\mathbf{u}_i - \mathbf{u}'_i)^T \\ -(\mathbf{u}_i - \mathbf{u}'_i) & (\tilde{\mathbf{u}}_i + \tilde{\mathbf{u}}'_i) \end{bmatrix} .$$

Then, Equation (5.23) can be further restated as:

$$\text{Min} \sum_{i=1}^{n} \mathbf{q}^T A_i^T A_i \mathbf{q} = \text{Min} \; \mathbf{q}^T A \mathbf{q} , \tag{5.26}$$

where $A = \sum_{i=1}^{n} A_i^T A_i$ and $|\mathbf{q}| = 1$. The matrix A can be computed incrementally.

Since A is a symmetric matrix, the solution to this problem is the 4-dimensional vector $\mathbf{q}_{\text{min}}$ corresponding to the smallest eigenvalue of A.

5.4.2 Determining the Translation

We can determine the translation using the standard minimization technique. We differentiate (5.22) with respect to $\mathbf{t}$ and set the result equal to zero, to obtain

$$\sum_{i=1}^{n} 2(\mathbf{d}'_i - \mathbf{R}^*\mathbf{d}_i - \tilde{\mathbf{u}}'_i\mathbf{t})^T \tilde{\mathbf{u}}'_i = 0 . \tag{5.27}$$

Transposing the above equation and after a little algebra we get

$$\left(\sum_{i=1}^{n} \tilde{\mathbf{u}}'_i(\tilde{\mathbf{u}}'_i)^T \right) \mathbf{t} = \sum_{i=1}^{n} (\tilde{\mathbf{u}}'_i)^T (\mathbf{d}'_i - \mathbf{R}^*\mathbf{d}_i) . \tag{5.28}$$

If $\left(\sum_{i=1}^{n} \tilde{\mathbf{u}}'_i(\tilde{\mathbf{u}}'_i)^T \right)$ is a full rank matrix, we have explicitly the translation vector:

$$\mathbf{t} = \left(\sum_{i=1}^{n} \tilde{\mathbf{u}}'_i(\tilde{\mathbf{u}}'_i)^T \right)^{-1} \left(\sum_{i=1}^{n} (\tilde{\mathbf{u}}'_i)^T (\mathbf{d}'_i - \mathbf{R}^*\mathbf{d}_i) \right) . \tag{5.29}$$

If not, $\mathbf{t}$ is unrecoverable. It can be shown that $\left(\sum_{i=1}^{n} \tilde{\mathbf{u}}'_i(\tilde{\mathbf{u}}'_i)^T \right)$ is always of full rank if two of $\mathbf{u}'_i$ ($i = 1 \ldots n$) are different (nonparallel).

5.5   Kim and Aggarwal's Method

In [5.7], *Kim* and *Aggarwal* have proposed a simple algorithm to determine the motion from line correspondences. This method is similar to one of the methods proposed by *Blostein* and *Huang* [5.2] to determine motion from 3D point correspondences. As in the method described in Sect. 5.4, their method is based on the same observation that we can determine rotation and translation separately.

5.5.1   Determining the Rotation

Kim and *Aggarwal* represent a rotation by the orientation vector of its rotation axis n and the rotation angle θ about this axis. This representation is equivalent to the rotation axis representation (see Sect. 4.1.2). Indeed, we have:

$$\theta = \|\mathbf{r}\| \quad \text{and} \quad \mathbf{n} = \frac{\mathbf{r}}{\|\mathbf{r}\|} .$$

Given two pairings of segments such that $\mathbf{u}_1$, $\mathbf{u}_2$ and $\mathbf{u}_1'$, $\mathbf{u}_2'$ are the unit direction vectors of two known lines before and after motion, we need not solve any system of equation using their method. Since the rotation axis should be perpendicular to both $(\mathbf{u}_1' - \mathbf{u}_1)$ and $(\mathbf{u}_2' - \mathbf{u}_2)$ (see Fig. 5.1a), the direction of the rotation axis can be computed directly:

$$\mathbf{n} = \frac{(\mathbf{u}_1' - \mathbf{u}_1) \wedge (\mathbf{u}_2' - \mathbf{u}_2)}{\|(\mathbf{u}_1' - \mathbf{u}_1) \wedge (\mathbf{u}_2' - \mathbf{u}_2)\|} . \tag{5.30}$$

If $(\mathbf{u}_1' - \mathbf{u}_1)$ and $(\mathbf{u}_2' - \mathbf{u}_2)$ are parallel to each other, the following equation should be used:

$$\mathbf{n} = \frac{(\mathbf{u}_1 \wedge \mathbf{u}_2) \wedge (\mathbf{u}_1' - \mathbf{u}_1)}{\|(\mathbf{u}_1 \wedge \mathbf{u}_2) \wedge (\mathbf{u}_1' - \mathbf{u}_1)\|} . \tag{5.31}$$

If $(\mathbf{u}_1' - \mathbf{u}_1) = 0$, then $\mathbf{n} = \mathbf{u}_1$. If $(\mathbf{u}_2' - \mathbf{u}_2) = 0$, then $\mathbf{n} = \mathbf{u}_2$.

Once $\mathbf{n}$ is found, the rotation angle is simply the angle between the plane formed by $\mathbf{n}$ and $\mathbf{u}_1$ and that formed by $\mathbf{n}$ and $\mathbf{u}_1'$, that is:

$$\cos \theta = \frac{(\mathbf{n} \wedge \mathbf{u}_1) \cdot (\mathbf{n} \wedge \mathbf{u}_1')}{\|\mathbf{n} \wedge \mathbf{u}_1\| \cdot \|\mathbf{n} \wedge \mathbf{u}_1'\|} . \tag{5.32}$$

If $\mathbf{n} = \mathbf{u}_1$, $\mathbf{u}_2$ and $\mathbf{u}_2'$ should be used instead of $\mathbf{u}_1$ and $\mathbf{u}_1'$.

But we should point out that the above computation of the rotation axis may give an orientation opposite to the real orientation and that the error can not be recovered if using (5.32) to compute θ. In fact, if two segments constitute a configuration like that in Fig. 5.1a, Equation (5.30) gives $\mathbf{n}$ pointing *down*, not *up*, that is, the computed $\mathbf{n}$ is opposite to the real orientation of rotation axis. This is due to the definition of the cross product of two vectors:

If $\mathbf{n} = \mathbf{n}_1 \wedge \mathbf{n}_2$, then $\mathbf{n}_1, \mathbf{n}_2$ and $\mathbf{n}$ form a right handed system such that the angle from $\mathbf{n}_1$ to $\mathbf{n}_2$ is in the range $[0, \pi]$.

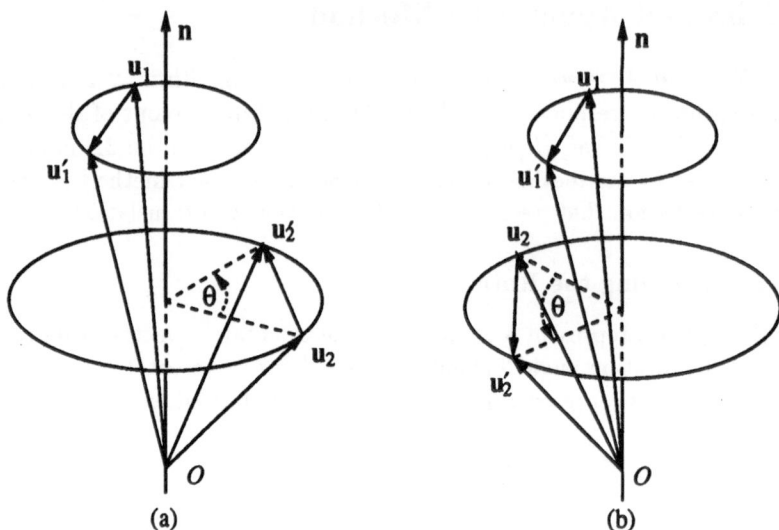

Fig. 5.1. Two configurations to illustrate the computation of the rotation axis

Since θ in (5.32) is implicitly restrained in the range $[0, \pi]$, an opposite $\mathbf{n}$ gives the same θ. That is, we cannot detect the error. Fig. 5.1b shows a configuration of two segments from which Equation (5.30) computes correctly the orientation of the rotation axis.

The above ambiguity cannot be resolved if the line segments are not oriented. Since the segments we obtain from stereo are oriented, we can compute correctly the orientation of the rotation axis in both cases by the modified algorithm described here. We use (5.30) to compute a vector, denoted by $\mathbf{n}'$ (as discussed earlier, $\mathbf{n}'$ may be the same as $\mathbf{n}$, or may be opposite to $\mathbf{n}$). Under the constraint that the rotation angle is less than π, the angle between the normal of the plane formed by $\mathbf{n}'$ and $\mathbf{u}_1$ and that of the plane formed by $\mathbf{n}'$ and $\mathbf{u}_1'$ is less than π. Let

$$\mathbf{n}_1 = \frac{\mathbf{n}' \wedge \mathbf{u}_1}{\|\mathbf{n}' \wedge \mathbf{u}_1\|} \quad \text{and} \quad \mathbf{n}_2 = \frac{\mathbf{n}' \wedge \mathbf{u}_1'}{\|\mathbf{n}' \wedge \mathbf{u}_1'\|} , \tag{5.33}$$

then their cross product

$$\mathbf{v} = \mathbf{n}_1 \wedge \mathbf{n}_2 , \tag{5.34}$$

gives the correct orientation of the rotation axis, i.e., the orientation of rotation axis $\mathbf{n}$ is equal to $\mathbf{v}/\|\mathbf{v}\|$. If the rotation angle θ is known to be less than $\pi/2$, then $\theta = \sin^{-1}(\|\mathbf{v}\|)$; if not, $\theta = \cos^{-1}(\mathbf{n}_1 \cdot \mathbf{n}_2)$.

5.5.2 Determining the Translation

Once the rotation matrix is found, we can find the translation vector by bringing two pairs of segments into overlap. Given $\mathbf{m}_1$ and $\mathbf{m}_2$, any points on the two line segments before motion and $\mathbf{m}_1'$ and $\mathbf{m}_2'$, any points on the two line segments after motion, there exist two arbitrary numbers β and σ so that

$$\mathbf{R}\mathbf{m}_1 + \mathbf{t} = \mathbf{m}_1' + \beta\mathbf{u}_1', \qquad (5.35)$$
$$\mathbf{R}\mathbf{m}_2 + \mathbf{t} = \mathbf{m}_2' + \sigma\mathbf{u}_2'. \qquad (5.36)$$

Thus we have six linear equations in five unknowns (three parameters of translation, plus β and σ). We have used the routines F01AXF and F04ANF in the NAG library to compute the approximate least-squares solution to the translation vector.

Using this algorithm, we compute the motion parameters for every possible combination of the matched segments, and then the average is considered as the optimal motion parameters. The average for the rotation is done on the $\mathbf{r}$ vectors (the rotation axis representation).

5.6 Experimental Results

The objective of our comparative study is to investigate the applicability of different methods to stereo data. Both synthetic and real stereo data are used to compare the methods described in the previous sections. In the following, we use some abbreviations, for simplicity, to refer to different methods. See Table 5.1.

Table 5.1. Abbreviations of different methods

Abbrev.	Rep. of Rotation	Approach
EKF-AXIS	Rotation axis	Extended Kalman Filter
EKF-QUAT	Quaternion	Extended Kalman Filter
MIN-AXIS	Rotation axis	Gauss-Newton Minimization
MIN-QUAT	Quaternion	Sequential Quadratic Programming
EIGEN	Quaternion	Singular Value Decomposition
KIM-AGGA	Rotation axis	*Kim* and *Aggarwal*'s Method

5.6.1 Results with Synthetic Data

The synthetic data we used contain 26 segments. One of the endpoints of each segment is fixed at the center of a sphere with radius 100 millimeters. The other endpoints are on the surface of the sphere. They are chosen so that the sphere surface is quasi-uniformly sampled. In other words, the orientation of segments are uniformly distributed in the space. Thus we obtain a set of noise-free 3D line segments in one position. Then we apply a motion, which is equal to $[0.4, 0.2, 0.5, 200.0, -150.0, 300.0]^T$ under the rotation axis representation, to this set and we obtain another set. The rotational components are in radian while

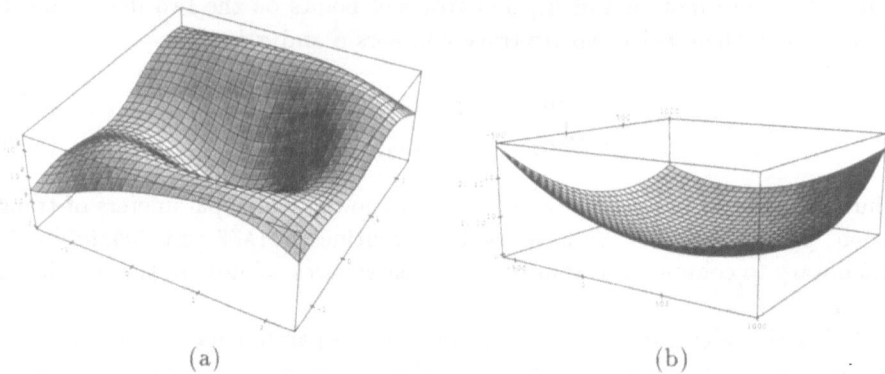

Fig. 5.2. Objective function for two matches of noisy line segments

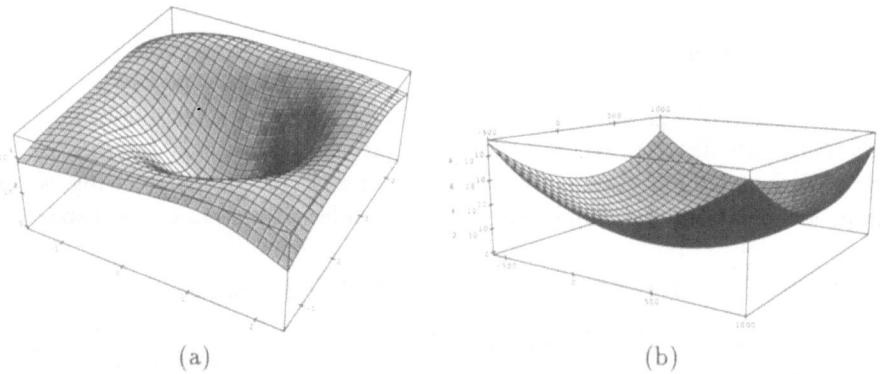

Fig. 5.3. Objective function for ten matches of noisy line segments

the translational components are in millimeters. Finally, independent Gaussian noise with mean zero and standard deviations σ_x, σ_y and σ_z is added to the 3D coordinates of each endpoint in the x, y and z directions to obtain the noisy measurements. In all experiments, the real motion is the same (equal to $[0.4, 0.2, 0.5, 200.0, -150.0, 300.0]^T$), but the noise level may differ.

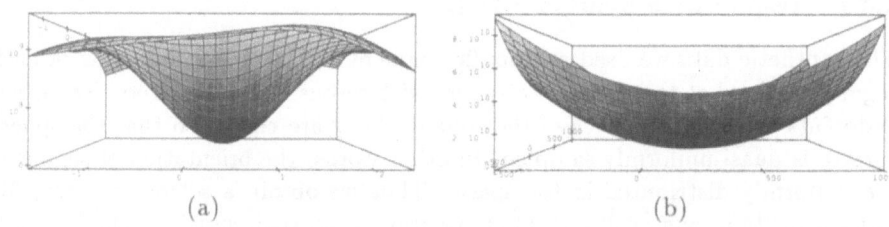

Fig. 5.4. Objective function for ten matches of noisy line segments (view in front)

Before going through the experimental results, it is useful to examine the shape of the objective function of the minimization problem (5.16). The noise level of the measurements is: $\sigma_x = \sigma_y = 2$, $\sigma_z = 6$. Fig. 5.2 displays the objective function for two matches of noisy line segments ($n = 2$ in (5.16)). In Fig. 5.2a, the objective function is plotted with respect to the x and z components of the rotation vector while fixing the translation at $[200.0, -150.0, 300.0]^T$ and the y component of the rotation vector at 0.2. The x and z components, r_x and r_z, vary around their real values, i.e., $-1.6 \le r_x \le 2.4$ and $-1.5 \le r_z \le 2.5$. In Fig. 5.2b, the objective function is plotted with respect to the x and z components of the translation vector while fixing the rotation at $[0.4, 0.2, 0.5]^T$ and the y component of the translation vector at -150.0. The x and z components, t_x and t_z, vary around their real values, i.e., $-600.0 \le t_x \le 1000.0$ and $-500.0 \le t_z \le 1000.0$. Figure 5.3 shows the same thing as Fig. 5.2, except the objective function is computed using ten matches ($n = 10$). Figure 5.4 draws the same objective function as in Fig. 5.3 from the viewpoint directly located in front of the box, in order to observe better the shape. From those drawings, we see that the objective function with respect to the rotational parameters is smooth, and that it has only one local minimum in a rather wide range near the true value. This implies that an iterative minimization procedure can yield the good solution if we start with an initial guess not *very* far from the true value, as we can observe later. If the initial guess differs from the real value about 0.6 radians in each direction (about 60 degrees of difference in rotation angle), the iterative minimization procedure can be expected to converge to the good solution. The objective function with respect to the translational parameters is also smooth, and it has only one minimum. This implies that we can always find the good solution even if the initial guess in translation is very bad. The difference between the objective functions plotted in Figs. 5.2 and 5.3 is that the latter is more symmetric in different directions than the former. This implies that the effects of Gaussian errors in measurements can be reduced by using more measurements and we can expect that the more data are used, the better an estimation can be obtained.

In the following, the motion error is given in two parts: rotation error and translation error. If we use the rotation axis representation, let $\mathbf{r}$ be the real rotation parameter ($[0.4, 0.2, 0.5]^T$ in our case) and $\hat{\mathbf{r}}$ be the estimated rotation parameter, the rotation error is defined as:

$$e_r = \|\mathbf{r} - \hat{\mathbf{r}}\| / \|\mathbf{r}\| \times 100\% . \qquad (5.37)$$

If we use the quaternion representation, we first transform it into the rotation axis representation, and then compute the error in the same way. Similarly, the translation error is defined as:

$$e_t = \|\mathbf{t} - \hat{\mathbf{t}}\| / \|\mathbf{t}\| \times 100\% . \qquad (5.38)$$

where $\mathbf{t}$ is the real translation parameter ($[200.0, -150.0, 300.0]^T$ in our case) and $\hat{\mathbf{t}}$ is the estimated one.

In the EKF approach, we use only the first order approximation to the non-linear equation. If the initial estimate is far from the true one, the linear approx-

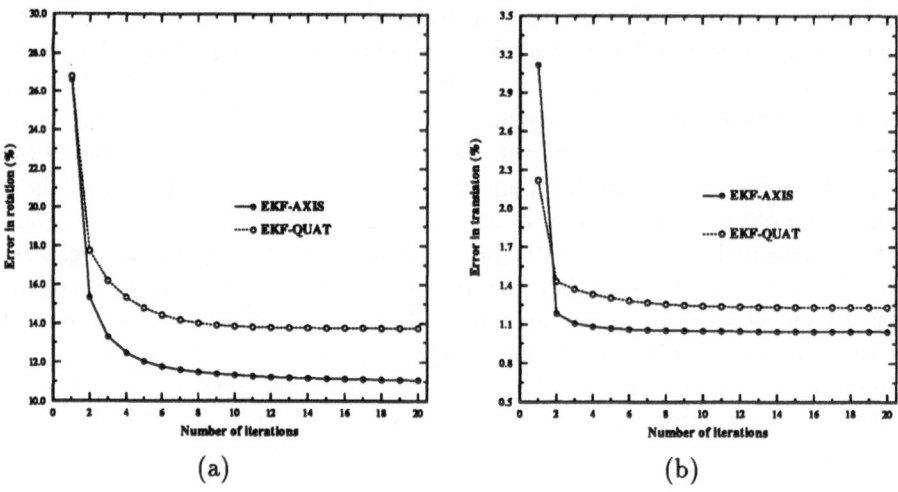

Fig. 5.5. Comparison between EKF-AXIS and EKF-QUAT: Errors in rotation (a) and in translation (b) versus number of iterations of EKF applied

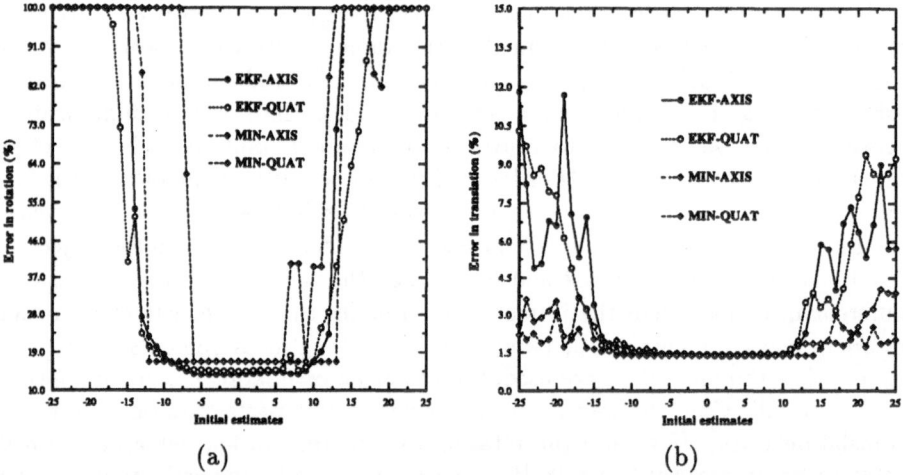

Fig. 5.6. Comparison between EKF-AXIS, EKF-QUAT, MIN-AXIS and MIN-QUAT: Errors in rotation (a) and in translation (b) versus different initial estimates

imation to the nonlinear equation is not very good. To reduce the effect of the nonlinearity, we iterate the EKF several times on the same measurements (see Chap. 2). At each iteration, we recompute the measurement y_i and the transformation matrix M_i in the linearized system using the estimate $\hat{p}_{i-1}$ computed at the previous iteration. Figure 5.5 shows the comparison between EKF-AXIS and EKF-QUAT with respect to the number of iterations of EKF we applied. The error shown is the average of twenty tries. The noise level of the measurements is: $\sigma_x = \sigma_y = 2$, $\sigma_z = 6$. The initial estimate $\hat{p}_0 = [-0.1, -0.1, -0.1, -200.0,$

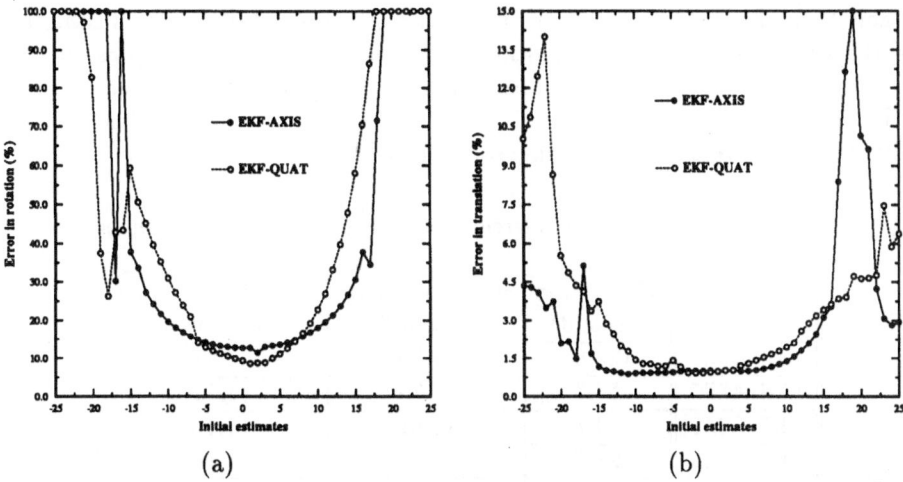

Fig. 5.7. Comparison (one sample) between `EKF-AXIS` and `EKF-QUAT`: Errors in rotation (a) and in translation (b) versus different initial estimates

$-200.0, -200.0]^T$. Two correspondences are used. We found that even when the initial estimate differs significantly from the true one (125% in rotation — the angle between the rotation axes is 108 degrees, the difference between the rotation angles is 32 degrees — and 165% in translation), the EKF converges to the true one after only a few iterations (5 or 6 iterations). One may remark that there is a difference between the final errors of `EKF-AXIS` and of `EKF-QUAT`. This is because among twenty tries done, `EKF-QUAT` diverges in a few tries. In reality, if `EKF-QUAT` converges, it converges to almost the same value as `EKF-AXIS`.

Since the system is nonlinear, recursive methods may give different solutions (even a wrong one) with different initial estimates. Figure 5.6 shows the motion errors of the four recursive methods with respect to different initial estimates. The noise level of the measurements is: $\sigma_x = \sigma_y = 2$, $\sigma_z = 6$. In `EKF-AXIS` and `EKF-QUAT`, five iterations of EKF are applied. In `MIN-AXIS` and `MIN-QUAT`, usually more than 30 iterations are needed to get a stable solution. The abscissa coordinates (from –25 to 25) correspond to different initial estimates. More precisely, the abscissa i corresponds to the initial estimate:

$$[0.4, 0.2, 0.5, 200.0, -150.0, 300.0]^T + i[0.05, 0.05, 0.05, 15.00, 15.00, 15.00]^T$$

For example, the abscissa –5 corresponds to the initial estimate [0.15, –0.05, 0.25, 125.0, –225.0, 225.0]^T. The errors are the average of twenty tries. Figure 5.7 shows the results of one of the twenty tries for `EKF-AXIS` and `EKF-QUAT`. From all these results, we can say that these methods (except `MIN-QUAT`) converge in a rather wide range (from –12 to 12 for EKF methods, i.e., from [–0.2, –0.4, –0.1, 20, –330, 120]^T to [1.0, 0.8, 1.1, 380, 30, 480]^T). This confirms our observation about the objective function of the minimization. We find also that the methods using the quaternion representation (especially `MIN-QUAT`) is less stable than the methods using the rotation axis representation. The instability of the EKF-QUAT can be observed at $i = 7$, but it is not as serious as that of MIN-QUAT. We observe also this instability of the quaternion representation in other experiments. Another remark is that if MIN-AXIS and MIN-QUAT do not diverge, they give exactly the same solution.

Table 5.2. Comparison of different methods: user time and motion error

Methods	User Time (second)	Rotation Error (percent %)	Translation Error (percent %)
MIN-QUAT	122.8	17.73	1.17
MIN-AXIS	57.7	17.73	1.17
EKF-QUAT	29.7	14.91	1.15
EKF-AXIS	27.7	14.26	1.16
EIGEN	0.07	20.72	1.12
KIM-AGGA	0.05	29.77	1.32

Table 5.2 shows the comparison on run time (on SUN 3/60 workstation), rotation error and translation error versus different methods. The results are the average of ten tries. Two line segment correspondences are used. The EKF is iterated five times. $\sigma_x = \sigma_y = 2$ and $\sigma_z = 6$. From Table 5.2, we observe that using a general minimization routine is very time expensive (this is because more than 30 iterations are required to get a stable solution) and that the EIGEN and KIM-AGGA methods are very fast. Note that KIM-AGGA becomes more time consuming than EIGEN when more line segment correspondences are available, because KIM-AGGA computes the rotation and translation for all pairs of lines and then averages the estimates. EKF gives smaller motion errors than other methods with a reasonable run time. This is expected since EKF takes into account the different uncertainty distributions of measurements, while the others treat equally each measurement and each component of a measurement. Another remark is that using the quaternion representation is more time consuming than using the rotation axis representation. This is for two reasons:

1. The quaternion representation has one parameter more than the rotation axis representation.
2. There is a constraint for the quaternion representation and in EKF-QUAT we add this constraint as an additional measurement.

One can expect that the weighted least-squares method using general minimization techniques gives the best estimation, but it would take even more time than MIN-AXIS or MIN-QUAT.

Tables 5.3 and 5.4 show the comparison of different methods versus different number of correspondences. The results are the average of ten tries. The initial estimate is $[0, 0, 0, 0, 0, 0]^T$ for the methods MIN-QUAT, MIN-AXIS, EKF-QUAT and

Table 5.3. Comparison: Rotation errors versus number of correspondences

Methods	number of correspondences								
	2	3	4	5	6	7	8	9	10
MIN-QUAT	16.61	10.77	56.19[1]	8.26	7.55	6.52	6.15	5.52	4.84
MIN-AXIS	16.61	10.77	8.69	8.26	7.55	6.52	6.15	5.52	4.84
EKF-QUAT	15.91	9.26	7.63	6.65	6.10	5.53	5.08	4.69	4.35
EKF-AXIS	15.13	9.70	7.72	6.48	5.55	4.62	4.34	4.11	3.80
EIGEN	18.66	11.97	9.40	8.72	7.46	6.27	5.90	5.22	4.55
KIM-AGGA [2]	22.12	19.40	23.14	25.32	25.94	27.35	27.24	28.29	27.84
	(10)	(30)	(57)	(94)	(142)	(200)	(266)	(341)	(427)

Table 5.4. Comparison: Translation errors versus number of correspondences

Methods	number of correspondences								
	2	3	4	5	6	7	8	9	10
MIN-QUAT	1.49	1.08	1.24[1]	0.71	0.58	0.56	0.56	0.60	0.61
MIN-AXIS	1.49	1.08	0.81	0.71	0.58	0.56	0.56	0.60	0.61
EKF-QUAT	1.34	1.03	0.87	0.69	0.55	0.46	0.48	0.41	0.44
EKF-AXIS	1.45	1.09	0.90	0.70	0.56	0.46	0.41	0.42	0.42
EIGEN	1.18	0.92	0.71	0.61	0.53	0.58	0.60	0.65	0.71
KIM-AGGA [2]	1.66	1.54	1.61	1.65	1.66	1.78	1.71	1.77	1.76
	(10)	(30)	(57)	(94)	(142)	(200)	(266)	(341)	(427)

[1]In one of the ten tries, MIN-QUAT converges to an incorrect minimum
[2]In this line, the number in the parentheses shows the number of combinations used to compute the motion with KIM-AGGA

EKF-AXIS and we only iterated 3 times the extended Kalman filter. $\sigma_x = \sigma_y = 2$ and $\sigma_z = 6$. We observe that with the increase of the number of correspondences, the motion error decreases except for the KIM-AGGA method. This phenomenon is not observed in KIM-AGGA even 427 combinations are used when 10 correspondences are available. In our opinion, this is because the KIM-AGGA method uses an arithmetical average of estimates from all pairs of segments, which is not robust.

Other experiments done are to compare the performance of each method with respect to different noise levels. Figure 5.8 shows the motion errors while the standard deviation in z direction (σ_z) varies from 1 to 20 and σ_x and σ_y are fixed at 1. The initial estimate for the recursive methods are all zero. Five iterations of EKF are applied. The error is the average of ten tries. Two correspondences are used. We can observe that error in rotation varies almost linearly with the deviation σ_z, but the slopes of the curves corresponding to EKF-AXIS and EKF-QUAT are much smaller than the others. This shows the advantage to take into account the uncertainty, especially when the uncertainty distribution is not

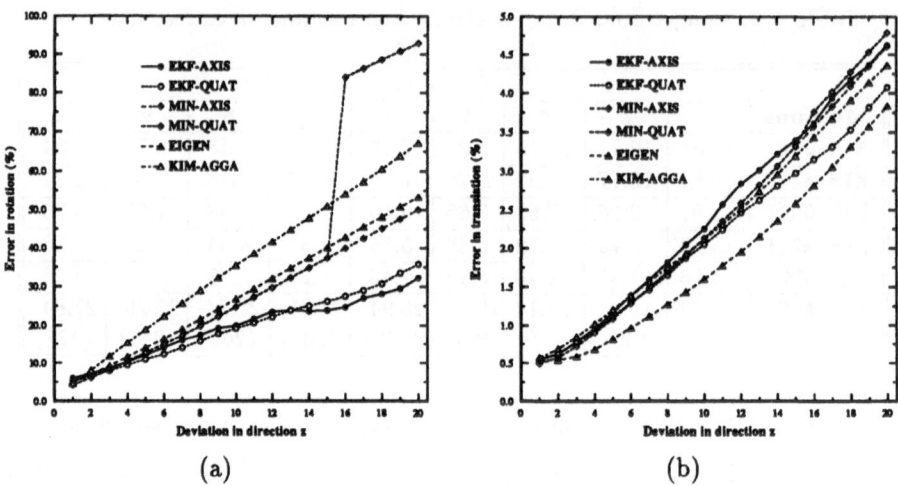

Fig. 5.8. Comparison: Errors in rotation (a) and in translation (b) versus σ_z while σ_x and σ_y are fixed at 1

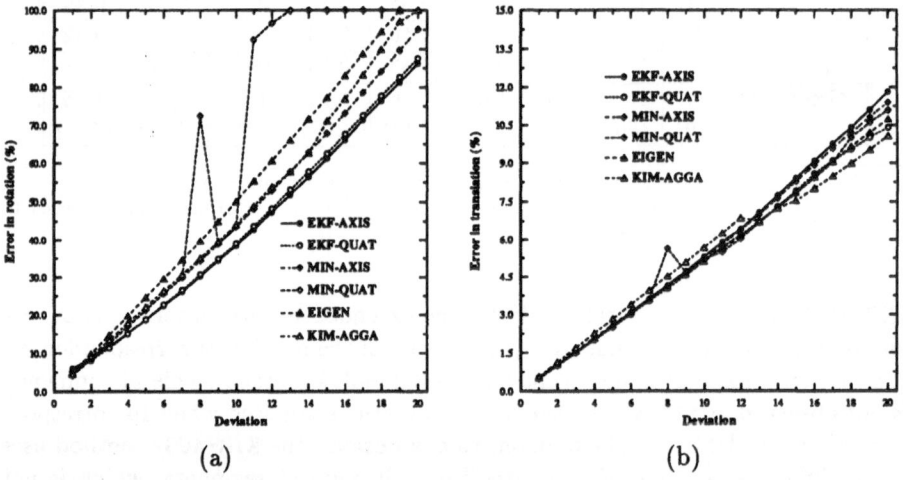

Fig. 5.9. Comparison: Errors in rotation (a) and in translation (b) versus deviation σ of measurements ($\sigma_x = \sigma_y = \sigma_z = \sigma$, they vary equally)

uniform. MIN-QUAT gives the same error as MIN-AXIS, but it diverges when σ_z becomes big. From Fig. 5.8b, we see that all methods, except EIGEN, give almost the same error in translation. (The reason why EIGEN gives a smaller error in translation is not yet clear.)

Figure 5.9 shows the motion errors while the standard deviation of measurements varies equally in x, y and z directions. We observe that the error in rotation given by EKF-AXIS and EKF-QUAT is a little smaller than the error given by other methods, even when the uncertainty distribution is uniform in

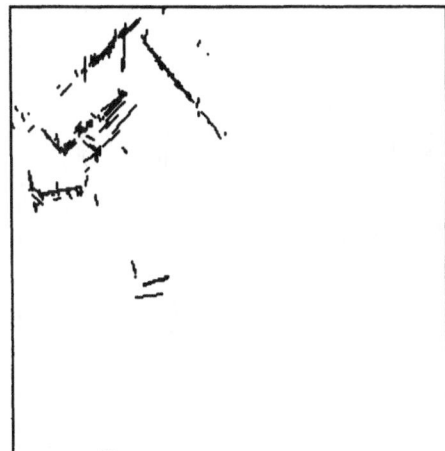

Fig. 5.10. Front and top views of the first 3D scene

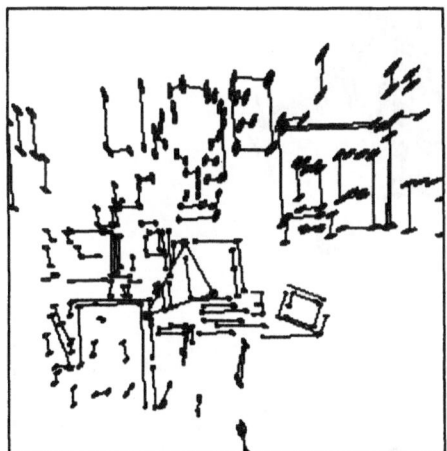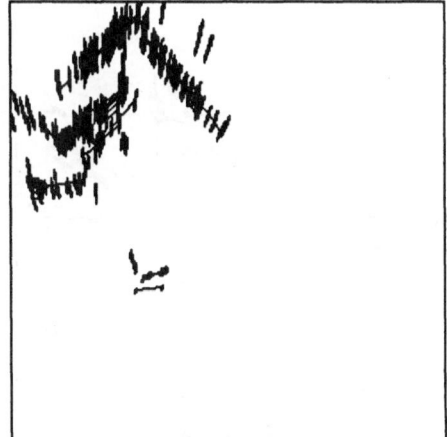

Fig. 5.11. Front and top views of the first 3D scene with ellipse of uncertainty

each direction. The error in translation is almost the same. This is because the relation between measurements and rotation is nonlinear while that between measurements and translation is linear. The EKF methods take into account this nonlinearity, and different measurements contribute, differently but properly, to the final estimation, even they have the same uncertainty distribution.

5.6.2 Results with Real Data

Up to now, the experiments we have carried out used the measurements with the same uncertainty distribution. However, the uncertainty in 3D positions of a line segment obtained by stereo triangulation varies with its distance to the cameras.

Fig. 5.12. Front and top views of the second 3D scene

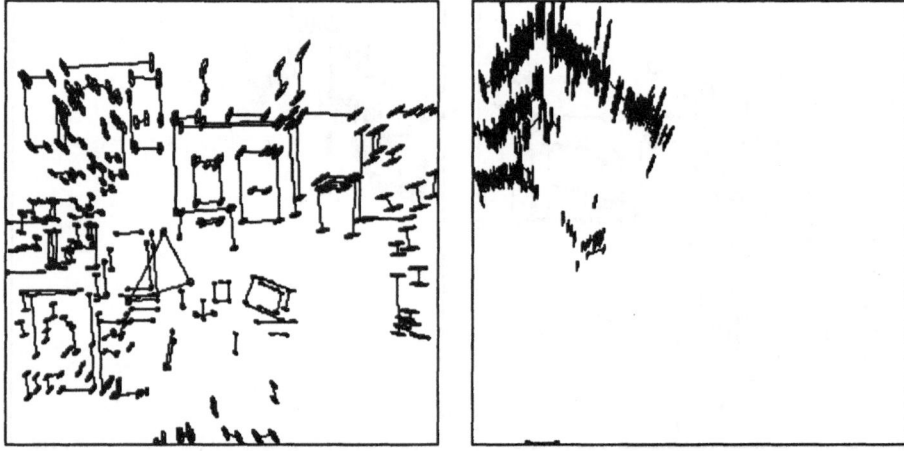

Fig. 5.13. Front and top views of the second 3D scene with ellipse of uncertainty

Generally speaking, it increases with the distance. Furthermore, the uncertainty of a reconstructed 3D point is different in different coordinate directions: a near point has a fairly compact uncertainty; a distant one has a bigger uncertainty in the range than in the horizontal and vertical directions. In this section, we compare the different methods using real data.

Figures 5.10 and 5.12 show two stereo frames reconstructed by our mobile robot in two different positions of a room. In the room, there are walls, tables, a workstation and other furnitures (see Page 176 for the images taken by the three cameras of our trinocular stereo system). The triangle in each frame represents the optical centers of the cameras. Figures 5.11 and 5.13 show the ellipse of uncertainty of the endpoints of the segments. We observe that distant segments

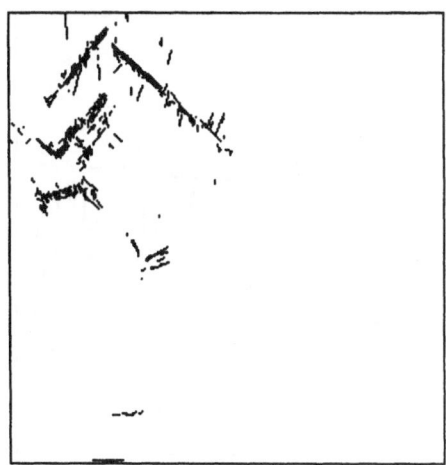

Fig. 5.14. Superposition of two scenes with the result of `EKF-AXIS` using 105 line segment correspondences

have bigger ellipses than nearby ones. There is a large displacement between these two positions (about 10 degrees of rotation and 75 centimeters of translation). We applied the `EKF-AXIS` on these two scenes using 105 correspondences (recovered by the algorithm described in [5.10] and in the following chapters). After two iterations of EKF, the motion parameter is given as $[-5.26\mathrm{e}{-03}, -1.64\mathrm{e}{-01}, 1.45\mathrm{e}{-03}, 3.09\mathrm{e}{+02}, -1.52\mathrm{e}{+00}, 6.89\mathrm{e}{+02}]^T$. The result is shown in Fig. 5.14. We applied the estimated motion to the first scene and superimposed the transformed scene (in dashed lines) on the second scene (in solid lines). The displacement between the two positions can be recognized from the shift of the triangle in Fig. 5.14.

Table 5.5. Comparison using the real data

Methods	Rotation Error	Translation Error
EKF-AXIS	2.25	1.77
EKF-QUAT	2.32	1.37
MIN-AXIS	34.09	22.78
MIN-QUAT	34.09	22.78
EIGEN	53.48	37.74
KIM-AGGA	23.10	37.89

Five arbitrary correspondences are used to compare the different methods. The initial estimate is zero for the four recursive methods. Three iterations of EKF are applied in `EKF-AXIS` and `EKF-QUAT`. The result of `EKF-AXIS` is shown in Fig. 5.15, that of `EKF-QUAT` in Fig. 5.16, that of `MIN-AXIS` in Fig. 5.17, that of

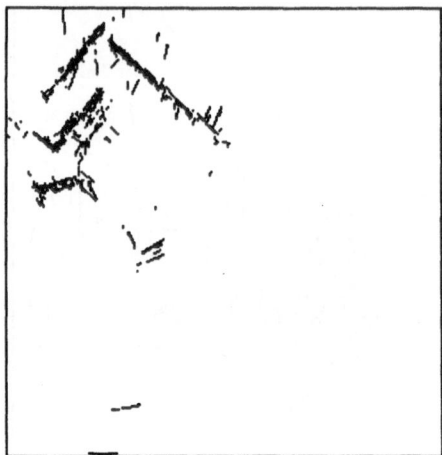

Fig. 5.15. Superposition of two scenes with the result of `EKF-AXIS` using 5 correspondences

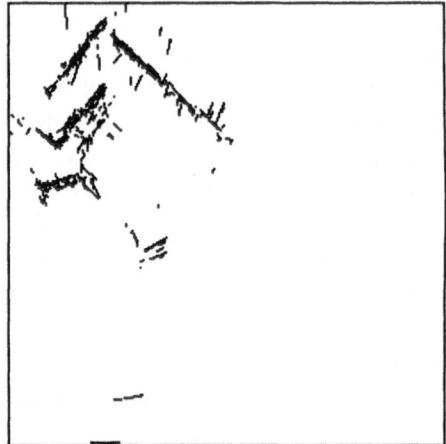

Fig. 5.16. Superposition of two scenes with the result of `EKF-QUAT` using 5 correspondences

`EIGEN` in Fig. 5.18 and that of `KIM-AGGA` in Fig. 5.19. `MIN-QUAT` gives exactly the same result as `MIN-AXIS`, and is thus not shown. All these results are compared with the above one, as displayed in Table 5.5. We conclude that EKF is much better than the others when 3D data come from stereo. The reason is that the EKF takes into account the fact that 3D measurements from stereo have different error distributions, and the others do not.

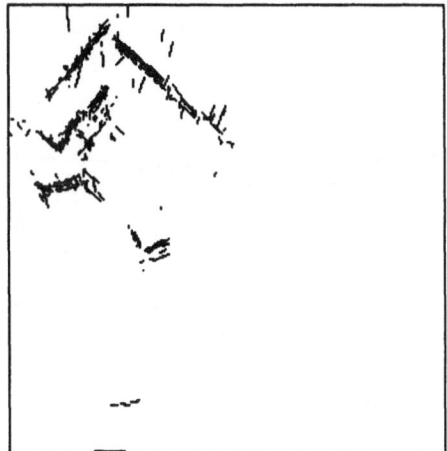

Fig. 5.17. Superposition of two scenes with the result of MIN-AXIS using 5 correspondences

Fig. 5.18. Superposition of two scenes with the result of EIGEN using 5 correspondences

5.7 Summary

In this chapter we have presented a number of methods for determining 3D motion from 3D line segment correspondences. With different representations of line segments and rotation, we showed how to determine motion using the extended Kalman filter, general minimization routines, the singular value decomposition technique and a method proposed by *Kim* and *Aggarwal*. These methods have been compared using both synthetic data and real data obtained by a trinocular stereo.

Fig. 5.19. Superposition of two scenes with the result of KIM-AGGA using 5 correspondences

From our experiments, we observe that:

1. Uncertainty in measurements should be taken into account, especially when measurements have different uncertainty distributions like in a stereo system.
2. When measurements have small uncertainty (less than 2% of segment length), general minimization algorithms give the best results. But when the uncertainty becomes larger, the general minimization algorithms do not give better results than EKF. Furthermore, they are more time consuming.
3. In the general minimization algorithms or the EKF methods, using the quaternion representation is more time consuming and does not give better results than using the rotation axis representation. On the contrary, we observed that using the quaternion representation is less stable.
4. Recursive methods (EKF-AXIS, EKF-QUAT, MIN-AXIS, MIN-QUAT) require an initial guess of the solution. When the initial estimate is far from the true one, recursive methods may give a wrong solution. In our experiments, we observe that the recursive methods can converge to the true solution when the initial estimate varies in a wide range from the true one.
5. Using an iterated extended Kalman filter can reduce the effects of nonlinearity. Even when few correspondences are available, EKF converges to the true estimate after only five or six iterations.
6. Using the quaternion representation of rotation, we can use the singular value decomposition to obtain the analytical solution of motion. KIM-AGGA gives also an analytical solution of motion. Both methods are efficient and do not need an initial motion estimate. EIGEN has one advantage to KIM-AGGA: the matrix A_i can be computed incrementally (see (5.26). However, the results given by both methods are not very significant.
7. EKF can incorporate new measurements incrementally.

We conclude that the extended Kalman filter with the rotation axis representation is preferable, especially when the distribution of errors is not isotropic, like in stereo. Using EKF with the quaternion representation gives also good results, but it is more time consuming and less stable than `EKF-AXIS`.

Since the EKF approach requires an initial guess of motion not very far from the true one, and that `EIGEN` and `KIM-AGGA` give an analytical solution efficiently, we can combine them. That is, we first apply either `EIGEN` or `KIM-AGGA` to obtain a solution and then use it as the initial estimate of EKF. The combined approach integrates the advantages of both analytical and recursive methods. However, if we have only a few (2 or 3) correspondences, the estimate given by `EIGEN` or `KIM-AGGA` is not significant.

We note that all methods presented in this chapter can be directly applied to 3D point data. In that case, at least three (noncollinear) 3D point correspondences are needed. Every three noncollinear points can form two nonparallel segments.

5.8 Appendix: Motion Computation Using the New Line Segment Representation

In this appendix, we address the motion computation problem using the extended Kalman filter and the line segment representation we proposed in Sect. 4.2.2. The method described below is actually used in our motion-estimation algorithm described in the following chapters.

Remember that a 3D line segment S is now represented by ψ and its associated covariance matrix Λ_ψ for the orientation, $\mathbf{m}$ and its associated covariance matrix $\Lambda_\mathbf{m}$ for the location, and l for the length (see Sect. 4.2.2). The l is not used in the motion-computation process. The rotation is represented by its rotation axis $\mathbf{r} = [a,\ b,\ c]^T$ (see Sect. 4.1.2), so the motion is described by a 6-dimensional vector $\mathbf{s} = [\mathbf{r}^T,\ \mathbf{t}^T]^T$, where $\mathbf{t}$ is the translation vector and is assumed preceded by the rotation. The relation between the rotation matrix $\mathbf{R}$ and the rotation axis $\mathbf{r}$ is given by (4.5).

Let $\mathbf{u}$ be the unit direction vector of a segment. For reason of simplicity, we define two nonlinear functions $\mathbf{g}$ and $\mathbf{h}$ to relate ψ and $\mathbf{u}$ (see (4.21) and (4.20)):

$$\psi = \mathbf{g}(\mathbf{u}) \quad \text{and} \quad \mathbf{u} = \mathbf{h}(\psi) . \tag{5.39}$$

If a segment is characterized by $S_1 : [\psi_1^T,\ \mathbf{m}_1^T]^T$ before motion and by $S_2 : [\psi_2^T,\ \mathbf{m}_2^T]^T$ after motion, then we have the following equations:

$$\psi_2 = \mathbf{g}(\mathbf{R}\mathbf{h}(\psi_1)) , \tag{5.40}$$

$$\mathbf{m}_2 = \mathbf{R}\mathbf{m}_1 + \mathbf{t} . \tag{5.41}$$

The above equations say simply that the *transformation* of the segment observed before motion should have the same orientation and location as the segment observed after motion. Here, we use directly (5.41) instead of (5.4), because the uncertainty of a midpoint has been now appropriately modeled, as described in Sect. 4.2.2.

If we define the measurement vector as

$$\mathbf{x}' = [\boldsymbol{\psi}_1^T,\ \mathbf{m}_1^T,\ \boldsymbol{\psi}_2^T,\ \mathbf{m}_2^T]^T\ , \tag{5.42}$$

then we can write down the measurement equation from (5.40) and (5.41) as

$$\mathbf{f}(\mathbf{x}',\mathbf{s}) = \left[\begin{array}{c} \mathbf{g}(\mathbf{R}\mathbf{h}(\boldsymbol{\psi}_1)) - \boldsymbol{\psi}_2 \\ \mathbf{R}\mathbf{m}_1 + \mathbf{t} - \mathbf{m}_2 \end{array} \right] = \mathbf{0}\ . \tag{5.43}$$

This is a 5-dimensional vector equation. In the following, the first two elements in $\mathbf{f}(\mathbf{x}',\mathbf{s})$ are replaced by $\mathbf{f}_1$ and the last three elements by $\mathbf{f}_2$.

In practice, we have a noisy measurement $\mathbf{x}$ instead of $\mathbf{x}'$. The EKF is used to estimate the motion parameters. To this end, it is required to compute the derivatives of $\mathbf{f}(\mathbf{x}',\mathbf{s})$ with respect to $\mathbf{s}$ and $\mathbf{x}'$. In fact,

$$\frac{\partial \mathbf{f}}{\partial \mathbf{s}} = \left[\begin{array}{cc} \dfrac{\partial \mathbf{f}_1}{\partial \mathbf{r}} & \mathbf{0} \\ \dfrac{\partial \mathbf{f}_2}{\partial \mathbf{r}} & \mathbf{I}_3 \end{array} \right], \tag{5.44}$$

$$\frac{\partial \mathbf{f}}{\partial \mathbf{x}'} = \left[\begin{array}{cccc} \dfrac{\partial \mathbf{f}_1}{\partial \boldsymbol{\psi}_1} & \mathbf{0} & -\mathbf{I}_2 & \mathbf{0} \\ \mathbf{0} & \mathbf{R} & \mathbf{0} & -\mathbf{I}_3 \end{array} \right], \tag{5.45}$$

where

$$\frac{\partial \mathbf{f}_1}{\partial \mathbf{r}} = \frac{\partial \mathbf{g}}{\partial \mathbf{u}_1'}\frac{\partial (\mathbf{R}\mathbf{u}_1)}{\partial \mathbf{r}},$$

$$\frac{\partial \mathbf{f}_2}{\partial \mathbf{r}} = \frac{\partial (\mathbf{R}\mathbf{m}_1)}{\partial \mathbf{r}},$$

$$\frac{\partial \mathbf{f}_1}{\partial \boldsymbol{\psi}_1} = \frac{\partial \mathbf{g}}{\partial \mathbf{u}_1'}\mathbf{R}\frac{\partial \mathbf{h}}{\partial \boldsymbol{\psi}_1},$$

with $\mathbf{u}_1 = \mathbf{h}(\boldsymbol{\psi}_1)$ and $\mathbf{u}_1' = \mathbf{R}\mathbf{u}_1$. $\frac{\partial (\mathbf{R}\mathbf{u}_1)}{\partial \mathbf{r}}$ and $\frac{\partial (\mathbf{R}\mathbf{m}_1)}{\partial \mathbf{r}}$ can be computed using (5.11); $\frac{\partial \mathbf{g}}{\partial \mathbf{u}_1'}$ can be computed using (4.24); and $\frac{\partial \mathbf{h}}{\partial \boldsymbol{\psi}_1}$ can be computed using (4.29).

Given an initial estimate of the motion between two views, we can incrementally update it via EKF using the above formulation when new matches are found. See the next two chapters for the matching techniques. Unlike the other formulations presented in this chapter, the above formulation can recover the whole motion parameters even when two line segments are parallel. Of course, the estimate of the translation is more uncertain along the direction of the line segments.

6. Matching and Rigidity Constraints

I n mobile robot navigation, the robot must determine its displacement between successive positions and perhaps find its position relative to a world reference. In traffic surveillance or aerial tracking systems, it is required to find trajectories of objects in successive views. In object manipulation, an object in the model base should be first identified in the scene that the manipulator observes. All of the above examples and many others involve the problem of matching and motion/pose determination. Matching is usually prior to the motion determination process. Under some restricted conditions, for example in [6.1], motion can be computed without registration between two views. In [6.1], the features used are 3D points, and the same set of points is assumed to be observed and to undergo the same motion between successive views. We want to tackle relatively more general problems: multiple moving objects, spurious data, etc.

6.1 Matching as a Search

The matching problem is recognized as a very difficult one. Given two sets of features observed in two views, the task of matching is to establish a correspondence between them. By a correspondence, we mean that the two paired features are the different observations (instances) of a single feature in space undergoing motion. Features in a view can be constructed as a graph, and the matching problem can be considered as searching for subgraph isomorphism.

In [6.2], *Sugihara* proposed an algorithm for determining the congruency of two polyhedra whose complexity is $O(n \log n)$, where n is the number of edges of the polyhedron. He also proposed an algorithm for determining whether a connected part of a polyhedron is congruent with some part of another polyhedron, whose complexity is $O(nm)$ where n and m are the number of edges of the two parts. In his algorithm, *complete* descriptions of polyhedra are assumed available and visibility constraints are introduced. Although we are working in a world of polyhedra, complete descriptions of polyhedra cannot yet be obtained with present capabilities of stereo systems or other range finders.

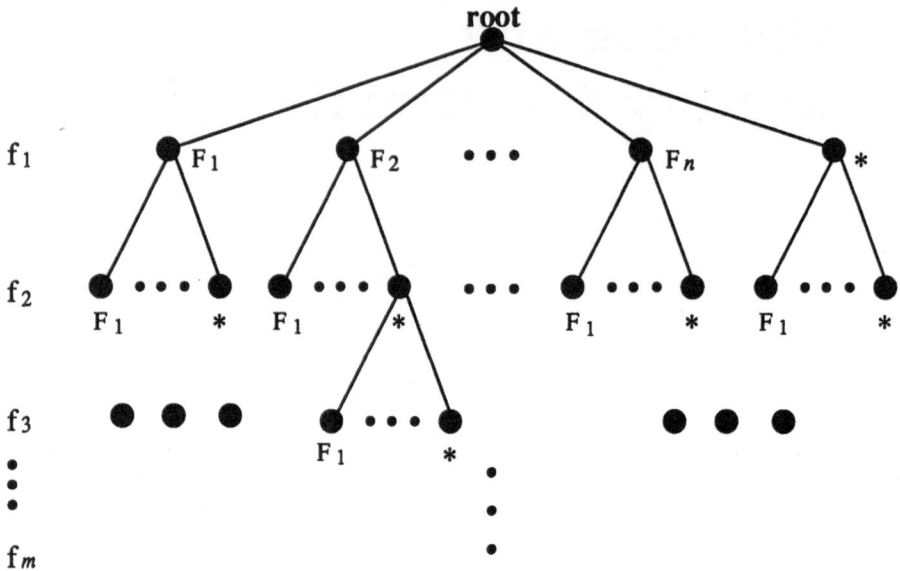

Fig. 6.1. Interpretation tree for matching

The general subgraph isomorphism problem, however, is $\mathcal{NP}$-complete. The problem soon becomes intractable as the number of arcs and nodes increase. To illustrate this, we structure the matching problem as a tree search [6.3–5]. The interpretation tree approach proposed in [6.4] is briefly presented here.

Given m features $\{f_i, i = 1, \ldots, m\}$ in the first frame and n features $\{F_j, j = 1, \ldots, n\}$ in the second frame. Considering that a feature in the first frame may correspond to no feature in the second frame for one reason or another, we introduce a special feature, called a *null character* (denoted by $*$ for distinction), into the second frame (see [6.4]). A pairing between a feature in the first frame and the null character implies that the feature cannot be matched with any real feature in the second frame. This pairing will be excluded from the final interpretation. Now the interpretation tree can be constructed as follows.

Suppose the features in the first frame are ordered in some arbitrary way. At the first level of the tree, each node represents a possible assignment of features in the second frame to the first feature f_1 in the first frame (see Fig. 6.1). All features in the second frame including the null character are possible assignments to f_1, thus at the first level of the interpretation tree we have $n + 1$ nodes. Given each hypothesized assignment, we consider, at the second level of the tree, all possible assignments of features in the second frame to the second feature f_2 in the first frame. So we have again $n + 1$ nodes below each node of the first level of the tree. We can continue in the same manner until all features in the first frame have been considered. The complete interpretation tree has, thus, m levels, each corresponding to a feature in the first frame (see Fig. 6.1). A simple counting shows that there are $(n + 1)^m$ nodes at level m of the interpretation tree.

A node at level l of the interpretation tree describes a partial l-interpretation, in that the nodes lying between the current node and the root of the tree identify an assignment of features in the second frame to the first l features in the first frame. As we can observe, we have actually $(n+1)^m$ m-interpretations. Our goal is to find *consistent* m-interpretations. By a consistent interpretation , we mean there exists a rigid transformation between the two paired feature sets. A straightforward method is to examine all m-interpretations to see whether there exists a rigid transformation bringing each feature f_i in the first frame into alignment with its associated feature from the second frame. This method is evidently unrealistic. In our indoor scenes, n and m are usually of the order of 150. If we take n and m as 100, we have already more than 10^{200} interpretations to examine! In general, most of these interpretations are either inconsistent (i.e., there does not exist any global rigid transformation) or uninteresting (i.e., most of the features of the first frame are matched to the null character). Recognizing all these, we need some constraints to heuristically guide the search without exploring most of nodes while assuring a good interpretation.

In short sequence analysis (bootstrapping mode), the rigidity constraints provide us a powerful tool to deal with the problem. In long sequence analysis (steady-state mode), the constraint on motion smoothness can be used. In the remainder of this chapter, we define the rigidity constraints for 3D line segments which will be used in the next chapter. Discussion of the use of motion constraints in long sequences is deferred to the last part of this monograph.

6.2 Rigidity Constraints

The rigidity assumption implies the conservation of local structure of objects, such as the angle and distance between two line segments, during their motion. Only noise-free data are considered in this section. We return to noisy data in Sect. 6.4.

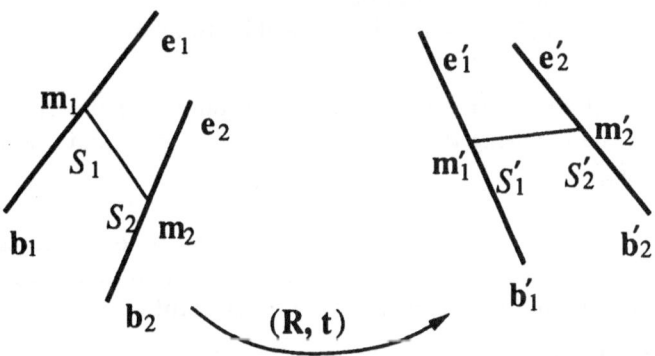

Fig. 6.2. Line segments undergoing a rigid motion

Consider now two line segments undergoing a rigid motion (see Fig. 6.2). The two segments are denoted by S_1 and S_2 before motion and by S_1' and S_2' after motion. Remember that a line segment S is represented by $(\psi, \mathbf{m}, l)$ (see Sect. 4.2). For simplicity, the unit direction vector of S is denoted by $\mathbf{u}$ which can be computed from ψ, and the endpoints are denoted by $\mathbf{b}$ and $\mathbf{e}$ (*Remark:* segments are oriented; see page 56). The rigid motion is represented by a rotation matrix $\mathbf{R}$ and a translation vector $\mathbf{t}$ (see Sect. 4.1). We have the following equations:

- For S_1 and S_1':

$$\mathbf{u}_1' = \mathbf{R}\mathbf{u}_1 , \tag{6.1}$$
$$\mathbf{m}_1' = \mathbf{R}\mathbf{m}_1 + \mathbf{t} , \tag{6.2}$$
$$\mathbf{b}_1' = \mathbf{R}\mathbf{b}_1 + \mathbf{t} , \tag{6.3}$$
$$\mathbf{e}_1' = \mathbf{R}\mathbf{e}_1 + \mathbf{t} . \tag{6.4}$$

- For S_2 and S_2':

$$\mathbf{u}_2' = \mathbf{R}\mathbf{u}_2 , \tag{6.5}$$
$$\mathbf{m}_2' = \mathbf{R}\mathbf{m}_2 + \mathbf{t} , \tag{6.6}$$
$$\mathbf{b}_2' = \mathbf{R}\mathbf{b}_2 + \mathbf{t} , \tag{6.7}$$
$$\mathbf{e}_2' = \mathbf{R}\mathbf{e}_2 + \mathbf{t} . \tag{6.8}$$

Under the assumption of a rigid motion, the geometry of a rigid object remains constant between successive views. In other words, the geometry of the object does not change during motion. We want to derive specific constraints reflecting this invariance. The following requirements should be satisfied:

- The constraints should be independent of the coordinate systems. The relation between the coordinate systems is just what we want to compute.
- The constraints should be as complete as possible, to guarantee the global consistency of the final interpretation.
- The constraints should be as simple as possible, to allow for efficient computation.

From (6.1) to (6.8) and the properties of rotation matrix described in Theorem 4.1, we get the following theorem for the rigidity constraints, which satisfy the above requirements.

Theorem 6.1. *If two segments S_1 and S_2 in the first frame are matched to two segments S_1' and S_2' in the second frame, under the assumption of rigid motion, the following constraints, called the* **rigidity constraints**, *must be satisfied:*

1. **Length Constraints**

$$l_1' = l_1 \qquad \text{and} \qquad l_2' = l_2 ; \tag{6.9}$$

2. **Distance Constraint**

$$\|\mathbf{v}_{12}\| = \|\mathbf{v}_{12}'\| ; \qquad (6.10)$$

3. **Angular Constraints**

(i) $\qquad\qquad\qquad \mathbf{u}_1 \cdot \mathbf{u}_2 = \mathbf{u}_1' \cdot \mathbf{u}_2' ; \qquad (6.11)$

(ii) $\qquad\qquad\qquad \mathbf{u}_1 \cdot \hat{\mathbf{v}}_{12} = \mathbf{u}_1' \cdot \hat{\mathbf{v}}_{12}' ; \qquad (6.12)$

(iii) $\qquad\qquad\qquad \mathbf{u}_2 \cdot \hat{\mathbf{v}}_{12} = \mathbf{u}_2' \cdot \hat{\mathbf{v}}_{12}' ; \qquad (6.13)$

4. **Triple-Product Constraint**

$$<\mathbf{u}_1, \mathbf{u}_2, \hat{\mathbf{v}}_{12}> = <\mathbf{u}_1', \mathbf{u}_2', \hat{\mathbf{v}}_{12}'> . \qquad (6.14)$$

In the above, $<\mathbf{u}_1, \mathbf{u}_2, \mathbf{u}_3>$ denotes the triple product (i.e., $\mathbf{u}_1 \cdot (\mathbf{u}_2 \wedge \mathbf{u}_3)$), $\mathbf{v}_{12}$ denotes the vector joining the midpoints (i.e., $\mathbf{v}_{12} = \mathbf{m}_2 - \mathbf{m}_1$) and $\hat{\mathbf{v}}_{12}$ is a unit length vector parallel to $\mathbf{v}_{12}$ (i.e., $\hat{\mathbf{v}}_{12} = \mathbf{v}_{12}/\|\mathbf{v}_{12}\|$). ∎

Proof. Before examining each constraint, one can easily verify that the following relations hold:

$$\mathbf{v}_{12}' = \mathbf{R}\mathbf{v}_{12} , \qquad (6.15)$$
$$\hat{\mathbf{v}}_{12}' = \mathbf{R}\hat{\mathbf{v}}_{12} . \qquad (6.16)$$

For example,

$$
\begin{aligned}
\mathbf{v}_{12}' &= \mathbf{m}_2' - \mathbf{m}_1' && \text{by definition} \\
&= \mathbf{R}\mathbf{m}_2 + \mathbf{t} - (\mathbf{R}\mathbf{m}_1 + \mathbf{t}) && \text{from (6.2) \& (6.6)} \\
&= \mathbf{R}(\mathbf{m}_2 - \mathbf{m}_1) = \mathbf{R}\mathbf{v}_{12} . && \text{by definition}
\end{aligned}
$$

1. **Length Constraints**: This constraint describes the preservation of the length of a segment during motion.

$$
\begin{aligned}
l_1'' &= \|\mathbf{e}_1' - \mathbf{b}_1'\| && \text{by definition} \\
&= \|\mathbf{R}\mathbf{e}_1 + \mathbf{t} - (\mathbf{R}\mathbf{b}_1 + \mathbf{t})\| && \text{from (6.4) \& (6.3)} \\
&= \|\mathbf{R}(\mathbf{e}_1 - \mathbf{b}_1)\| = \|\mathbf{e}_1 - \mathbf{b}_1\| && \text{from Property 2 of Th. 4.1} \\
&= l_1 . && \text{by definition}
\end{aligned}
$$

We have also $l_2' = l_2$.

2. **Distance Constraint**: This constraint describes the preservation of the distance between the midpoints of two segments during motion. We can immediately deduce this constraint from (6.15).

3. **Angular Constraints**: Three angles are involved (θ, θ_1 and θ_2, see Fig. 6.3). In Fig. 6.3, the segment $//S_2$ is parallel to the segment S_2 and shares a common endpoint with S_1. Instead of angles, we use the cosine of angles. This is because angles are between 0 and π.

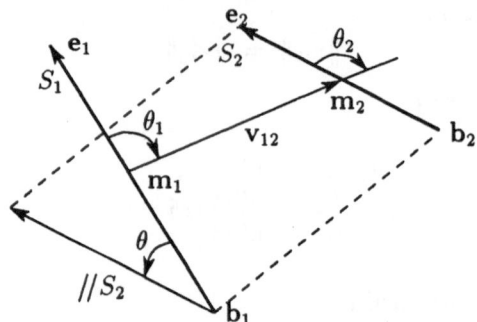

Fig. 6.3. Definition of angles used in rigidity constraints

(i) This subconstraint describes the preservation of the angle θ between two segments during motion.

$$
\begin{aligned}
\cos \theta' &= \mathbf{u}_1' \cdot \mathbf{u}_2' && \text{by definition} \\
&= \mathbf{R}\mathbf{u}_1 \cdot \mathbf{R}\mathbf{u}_2 && \text{from (6.1) \& (6.5)} \\
&= \mathbf{u}_1 \cdot \mathbf{u}_2 && \text{from Property 3 of Th. 4.1} \\
&= \cos \theta . && \text{by definition}
\end{aligned}
$$

(ii) This subconstraint describes the preservation of the angle θ_1 between the first segment S_1 and the segment joining the midpoints $\mathbf{m}_1$ and $\mathbf{m}_2$ during motion.

$$
\begin{aligned}
\cos \theta_1' &= \mathbf{u}_1' \cdot \hat{\mathbf{v}}_{12}' && \text{by definition} \\
&= \mathbf{R}\mathbf{u}_1 \cdot \mathbf{R}\hat{\mathbf{v}}_{12} && \text{from (6.1) \& (6.16)} \\
&= \mathbf{u}_1 \cdot \hat{\mathbf{v}}_{12} && \text{from Property 3 of Th. 4.1} \\
&= \cos \theta_1 . && \text{by definition}
\end{aligned}
$$

(iii) This subconstraint describes the preservation of the angle θ_2 between the second segment S_2 and the segment joining the midpoints $\mathbf{m}_1$ and $\mathbf{m}_2$ during motion. We can verify this subconstraint with similar manipulation to the above.

4. **Triple-Product Constraint:** This constraint is used, as we see later, to resolve the ambiguity due to angle measurements.

$$
\begin{aligned}
<\mathbf{u}_1', \mathbf{u}_2', \hat{\mathbf{v}}_{12}'> &= \mathbf{u}_1' \cdot (\mathbf{u}_2' \wedge \hat{\mathbf{v}}_{12}') && \text{by definition} \\
&= \mathbf{R}\mathbf{u}_1 \cdot (\mathbf{R}\mathbf{u}_2 \wedge \mathbf{R}\hat{\mathbf{v}}_{12}) && \text{from (6.1), (6.5) \& (6.16)} \\
&= \mathbf{R}\mathbf{u}_1 \cdot \mathbf{R}(\mathbf{u}_2 \wedge \hat{\mathbf{v}}_{12}) && \text{from Property 4 of Th. 4.1} \\
&= \mathbf{u}_1 \cdot (\mathbf{u}_2 \wedge \hat{\mathbf{v}}_{12}) && \text{from Property 3 of Th. 4.1} \\
&= <\mathbf{u}_1, \mathbf{u}_2, \hat{\mathbf{v}}_{12}> . && \text{by definition}
\end{aligned}
$$

This ends our derivation of the rigidity constraints.   □

6.3   Completeness of the Rigidity Constraints

The question we want to answer in this section is that of the completeness of the
rigidity constraints:

> Does there exist a *unique* rigid transformation between
> two pairings of segments if they satisfy all the rigidity
> constraints addressed in Theorem 6.1 ?

By linking the endpoints of two noncoplanar line segments, we get a tetrahedron. The answer to the above question will be based on the following definition
on the congruency of two tetrahedra.

Definition 6.1. *Two tetrahedra are said to be* **congruent** *if and only if the
following relations hold:*

1. $\|\overrightarrow{V_iV_j}\| = \|\overrightarrow{V_i'V_j'}\|$, *   for $i,j = 1 \cdots 4$ and $i \neq j$,*

2. $<\overrightarrow{V_1V_2}, \overrightarrow{V_3V_4}, \overrightarrow{V_3V_1}> = <\overrightarrow{V_1'V_2'}, \overrightarrow{V_3'V_4'}, \overrightarrow{V_3'V_1'}>$,

where $\{V_i, (i = 1 \cdots 4)\}$ are the vertices of the first tetrahedron and $\{V_i', (i = 1 \cdots 4)\}$ those of the second. ◇

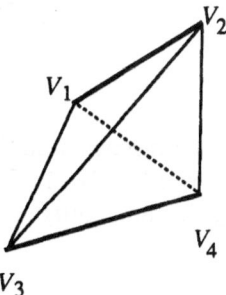 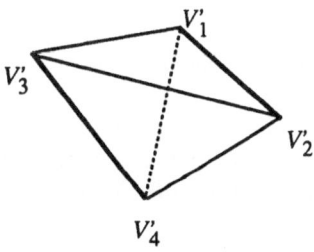

Fig. 6.4. Congruency of two tetrahedra

Figure 6.4 shows two congruent tetrahedra. The first condition of Definition 6.1 guarantees that the two tetrahedra are *isometric*, i.e., two congruent
tetrahedra have the same length in each corresponding side. This condition,
however, is not sufficient, because it cannot disambiguate two polyhedra due to
rotation from those due to *reflection* about the coordinate system. Figure 6.5
gives an example of reflection of two tetrahedra. In this figure, the two tetrahedra
satisfy the first condition of Definition 6.1, but they are certainly not congruent.
For example, if we bring V_1, V_2, V_3 into correspondence with V_1', V_2', V_3' and put
them together, we see that V_4 and V_4' are located on the different sides of the

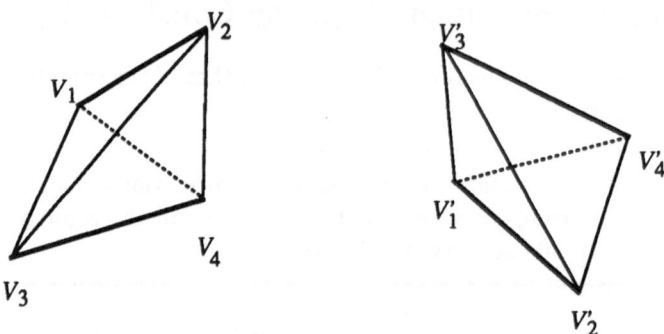

Fig. 6.5. Reflection of two tetrahedra

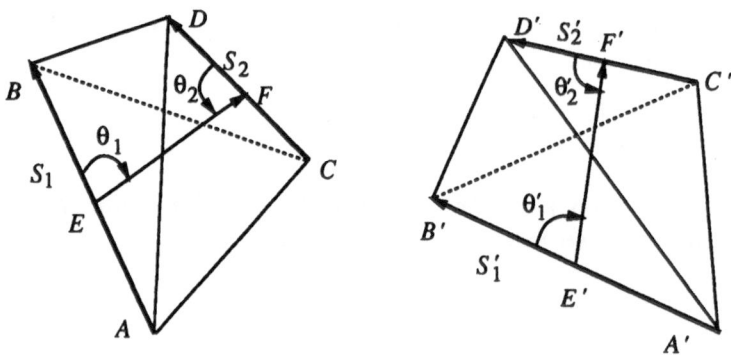

Fig. 6.6. Completeness of the rigidity constraints: noncoplanar case

plane containing V_1, V_2 and V_3. If we compute the transformation from these two tetrahedra, we get a transformation whose determinant is -1. The second condition of Definition 6.1 is just used for resolving this ambiguity.

Theorem 6.2.   *The rigidity constraints described in Theorem 6.1 are complete for 3D line segments, in the sense that they are necessary and sufficient to guarantee the congruency of two sets of 3D line segments.*   ∎

Proof.   Three cases exist due to the configuration of two line segments. We will prove it for each case.

Case 1: noncoplanar In this case, the two line segments form a tetrahedron by linking their endpoints (see Fig. 6.6). We only need to show that if two pairings of segments satisfy all the constraints described in Theorem 6.1, then their resulting tetrahedra are congruent.   The following equations relate the symbols for points in Fig. 6.6 to the parameters of segments:

$$A = b_1 , \quad B = e_1 , \quad C = b_2 , \quad D = e_2 , \quad E = m_1 , \quad F = m_2 ,$$
$$A' = b_1' , \quad B' = e_1' , \quad C' = b_2' , \quad D' = e_2' , \quad E' = m_1' , \quad F' = m_2' .$$
$$(6.17)$$

What we should verify is then (from Definition 6.1) [1]

$$AB = A'B' , \qquad\qquad CD = C'D' , \qquad\qquad (6.18)$$
$$AC = A'C' , \qquad\qquad AD = A'D' , \qquad\qquad (6.19)$$
$$BC = B'C' , \qquad\qquad BD = B'D' , \qquad\qquad (6.20)$$
$$<\overrightarrow{AB}, \overrightarrow{CD}, \overrightarrow{AC}> \ = \ <\overrightarrow{A'B'}, \overrightarrow{C'D'}, \overrightarrow{A'C'}> . \qquad (6.21)$$

From the length constraints given in Theorem 6.1, the relations in (6.18) are satisfied. Now we examine (6.19) and (6.20). Since

$$\overrightarrow{AC} \ = \ b_2 - b_1 = (b_2 - m_2) + (m_2 - m_1) + (m_1 - b_1)$$
$$= \ -l_2 u_2/2 + v_{12} + l_1 u_1/2 ,$$

we have

$$AC^2 \ = \ \overrightarrow{AC} \cdot \overrightarrow{AC} = (-l_2 u_2/2 + v_{12} + l_1 u_1/2) \cdot (-l_2 u_2/2 + v_{12} + l_1 u_1/2)$$
$$= \ \frac{1}{4}(l_1^2 + l_2^2) + (v_{12} \cdot v_{12}) - \frac{1}{2} l_1 l_2 (u_1 \cdot u_2) + l_1 (u_1 \cdot v_{12}) - l_2 (u_2 \cdot v_{12}) .$$

Similarly, we have

$$A'C'^2 = \frac{1}{4}(l_1'^2 + l_2'^2) + (v_{12}' \cdot v_{12}') - \frac{1}{2} l_1' l_2' (u_1' \cdot u_2') + l_1' (u_1' \cdot v_{12}') - l_2' (u_2' \cdot v_{12}') .$$

From the length, distance and angular constraints given in Theorem 6.1, we get

$$AC^2 = A'C'^2 .$$

In the same manner, we can show that the other relations in (6.19) and (6.20) are also true.

It remains to verify (6.21). In fact, we have

$$<\overrightarrow{AB}, \overrightarrow{CD}, \overrightarrow{AC}> \ = \ \overrightarrow{AB} \cdot (\overrightarrow{CD} \wedge \overrightarrow{AC})$$
$$= \ l_1 u_1 \cdot (l_2 u_2 \wedge (-l_2 u_2/2 + v_{12} + l_1 u_1/2))$$
$$= \ l_1 l_2 u_1 \cdot (u_2 \wedge v_{12}) = l_1 l_2 \|v_{12}\| <u_1, u_2, \hat{v}_{12}> ,$$

and

$$<\overrightarrow{A'B'}, \overrightarrow{C'D'}, \overrightarrow{A'C'}> \ = \ l_1' l_2' \|v_{12}'\| <u_1', u_2', \hat{v}_{12}'> .$$

From the length, distance and triple product constraints given in Theorem 6.1, we see that Equation (6.21) is true.

---

[1]In the proof, AB and $\|\overrightarrow{AB}\|$ are synonymous.

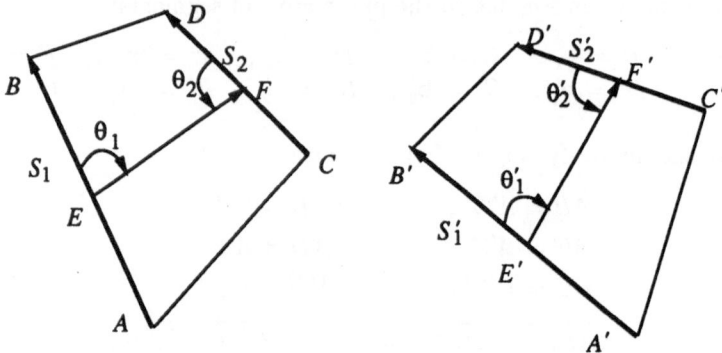

Fig. 6.7. The tetrahedra in Fig. 6.6 degenerate to the quadrilaterals in the case where two segments are coplanar

Case 2: coplanar but noncollinear In this case, the tetrahedron degenerates to a quadrilateral (see Fig. 6.7). The necessary-sufficient conditions of congruency of two quadrilaterals are

$$AB \;=\; A'B'\,, \qquad CD \;=\; C'D'\,, \qquad AC \;=\; A'C'\,, \qquad BD = B'D'\,,$$
$$\overrightarrow{AB} \cdot \overrightarrow{AC} \;=\; \overrightarrow{A'B'} \cdot \overrightarrow{A'C'}\,.$$

(6.22)

One can easily verify that these conditions are satisfied given the rigidity constraints in Theorem 6.1.

Case 3: collinear In this special case, the local shape is determined by the lengths of two segments, the separation between them and their relative position. We can show that the rigidity constraints guarantee the preservation of local shape. Since it does not allow us to recover fully the rigid motion, this special configuration is not considered furthermore.

We have thus completely proved this theorem.   □

Now we return to the question asked in the beginning of the present section.

Theorem 6.3. *The motion computed from two congruent tetrahedra or quadrilaterals is rigid and unique.*   ■

Proof.   Once two segments are noncollinear (see Figs. 6.6 and 6.7), we can choose two vectors $\mathbf{u}_1$, $\hat{\mathbf{v}}_{12}$ to construct a matrix M as

$$M = [\mathbf{u}_1, \; \hat{\mathbf{v}}_{12}, \; \mathbf{u}_1 \wedge \hat{\mathbf{v}}_{12}]\,,$$

(6.23)

and another matrix M' as

$$M' = [\mathbf{u}_1', \; \hat{\mathbf{v}}_{12}', \; \mathbf{u}_1' \wedge \hat{\mathbf{v}}_{12}']\,.$$

(6.24)

The following equation relates M and M':

$$M' = \mathbf{R}M , \tag{6.25}$$

because $\mathbf{u}'_1 = \mathbf{R}\mathbf{u}_1$, $\hat{\mathbf{v}}'_{12} = \mathbf{R}\hat{\mathbf{v}}_{12}$ and Property 4 of $\mathbf{R}$ in Theorem 4.1. Since the three columns in M are linearly independent, M is invertible. $\mathbf{R}$ is then computed as

$$\mathbf{R} = M'M^{-1} . \tag{6.26}$$

It can be verified that $\mathbf{R}\mathbf{R}^T = \mathbf{R}^T\mathbf{R} = \mathbf{I}_3$ and $|\mathbf{R}| = 1$, i.e., $\mathbf{R}$ is orthonormal.

We can show that $\mathbf{R}$ remains the same whichever two vectors we choose, i.e., $\mathbf{R}$ is unique. For example, choosing $\mathbf{u}_2$, $\hat{\mathbf{v}}_{12}$, we can construct a matrix M_2 as

$$M_2 = [\mathbf{u}_2, \ \hat{\mathbf{v}}_{12}, \ \mathbf{u}_2 \wedge \hat{\mathbf{v}}_{12}] , \tag{6.27}$$

and another matrix M'_2 as

$$M'_2 = [\mathbf{u}'_2, \ \hat{\mathbf{v}}'_{12}, \ \mathbf{u}'_2 \wedge \hat{\mathbf{v}}'_{12}] . \tag{6.28}$$

The matrices M_2 and M'_2 are not singular. Assume the rotation computed from M_2 and M'_2 is $\mathbf{R}_2$, i.e.,

$$M'_2 = \mathbf{R}_2 M_2 . \tag{6.29}$$

From the rigidity constraints, we have

$$\begin{aligned}
\mathbf{u}_1 \cdot \mathbf{u}_2 &= \mathbf{u}'_1 \cdot \mathbf{u}'_2 &= a , \\
\hat{\mathbf{v}}_{12} \cdot \mathbf{u}_2 &= \hat{\mathbf{v}}'_{12} \cdot \mathbf{u}'_2 &= b , \\
(\mathbf{u}_1 \wedge \hat{\mathbf{v}}_{12}) \cdot \mathbf{u}_2 &= (\mathbf{u}'_1 \wedge \hat{\mathbf{v}}'_{12}) \cdot \mathbf{u}'_2 &= c .
\end{aligned}$$

Thus we have

$$\mathbf{u}_2 = \begin{bmatrix} \mathbf{u}_1^T \\ \hat{\mathbf{v}}_{12}^T \\ (\mathbf{u}_1 \wedge \hat{\mathbf{v}}_{12})^T \end{bmatrix}^{-1} \begin{bmatrix} a \\ b \\ c \end{bmatrix} = (M^T)^{-1} \begin{bmatrix} a \\ b \\ c \end{bmatrix} ,$$

and

$$\mathbf{u}'_2 = \begin{bmatrix} \mathbf{u}'^T_1 \\ \hat{\mathbf{v}}'^t_{12} \\ (\mathbf{u}'_1 \wedge \hat{\mathbf{v}}'_{12})^T \end{bmatrix}^{-1} \begin{bmatrix} a \\ b \\ c \end{bmatrix} = (M'^T)^{-1} \begin{bmatrix} a \\ b \\ c \end{bmatrix} .$$

From (6.25), we have

$$\mathbf{u}'_2 = (M'^T)^{-1} \begin{bmatrix} a \\ b \\ c \end{bmatrix} = (M^T\mathbf{R}^T)^{-1} \begin{bmatrix} a \\ b \\ c \end{bmatrix} = (\mathbf{R}^T)^{-1}(M^T)^{-1} \begin{bmatrix} a \\ b \\ c \end{bmatrix} = \mathbf{R}\mathbf{u}_2 . \tag{6.30}$$

Here we have used $(\mathbf{R}^T)^{-1} = \mathbf{R}$ and $(M\mathbf{R})^{-1} = \mathbf{R}^{-1}M^{-1}$. Since $\hat{\mathbf{v}}'_{12} = \mathbf{R}\hat{\mathbf{v}}_{12}$, we have

$$\mathbf{u}'_2 \wedge \hat{\mathbf{v}}'_{12} = \mathbf{R}\mathbf{u}_2 \wedge \mathbf{R}\hat{\mathbf{v}}_{12} = \mathbf{R}(\mathbf{u}_2 \wedge \hat{\mathbf{v}}_{12}) . \tag{6.31}$$

From the above equations, we thus have

$$M'_2 = \mathbf{R}M_2 .$$

(6.32)

From (6.29) and (6.32), we have

$$(\mathbf{R} - \mathbf{R}_2)M_2 = 0 ,$$

(6.33)

where 0 is the 3×3 zero matrix. Since M_2 is not singular, we have

$$\mathbf{R}_2 = \mathbf{R} .$$

(6.34)

We can choose a point, for example, $\mathbf{m}_1$, to compute the translation vector $\mathbf{t}$

$$\mathbf{t} = \mathbf{m}'_1 - \mathbf{R}\mathbf{m}_1 .$$

(6.35)

If we choose another point, for example, $\mathbf{b}_1$, then the translation vector $\mathbf{t}'$ is given by

$$\mathbf{t}' = \mathbf{b}'_1 - \mathbf{R}\mathbf{b}_1 = (\mathbf{m}'_1 - \frac{l''_1}{2}\mathbf{u}'_1) - \mathbf{R}(\mathbf{m}_1 - \frac{l_1}{2}\mathbf{u}_1) = (\mathbf{m}'_1 - \mathbf{R}\mathbf{m}_1) - \frac{l''_1}{2}(\mathbf{u}'_1 - \mathbf{R}\mathbf{u}_1) .$$

Since $\mathbf{u}'_1 = \mathbf{R}\mathbf{u}_1$, we have

$$\mathbf{t}' = \mathbf{m}'_1 - \mathbf{R}\mathbf{m}_1 = \mathbf{t} .$$

(6.36)

In fact, $\mathbf{t}$ is independent of the point chosen, i.e., $\mathbf{t}$ is unique.   □

The above method for computing a 3D motion is only applicable to noise-free data, or in the limit, to data with very little noise. In practice, we use the extended Kalman filtering approach described in the last chapter. For other methods to determine 3D motion, the reader is also referred to the last chapter.

6.4   Error Measurements in the Constraints

In the preceding sections, we have proposed a number of rigidity constraints for 3D line segments, and shown that they are complete (necessary and sufficient conditions for the existence of a rigid displacement). We have also shown that we can compute a unique rigid motion from two pairings of segments satisfying the rigid constraints. However, all derivations were based on the noise-free assumption. The data we have are always corrupted with noise, due to uncertainties in polygonal approximation, stereo calibration and 3D reconstruction. The equalities in Theorem 6.1 are not true anymore. In this section, we reformalize the rigidity constraints by explicitly taking into account the uncertainty of measurements. The idea is to compute *dynamically* a threshold for each constraint from the uncertainty in 3D data given by the stereo.

Examining Theorem 6.1, we find that the rigidity constraints take three forms:

(*i*)	**Norm Constraint**	$\|\mathbf{v}\| = \|\mathbf{v}'\|$,	(6.37)
(*ii*)	**Dot-Product Constraint**	$\mathbf{u} \cdot \mathbf{v} = \mathbf{u}' \cdot \mathbf{v}'$,	(6.38)
(*iii*)	**Triple-Product Constraint**	$< \mathbf{u}_1, \mathbf{u}_2, \mathbf{u}_3 > = < \mathbf{u}'_1, \mathbf{u}'_2, \mathbf{u}'_3 >$.	(6.39)

6.4.1 Norm Constraint

The norm constraint says that the difference between the norms of two vectors should be zero. For convenience, we use the squared norm of a vector instead of its norm. Taking into account the uncertainty of measurements, we can formalize the constraint as follows:

$$| \, \|\mathbf{v}\|^2 - \|\mathbf{v}'\|^2 \, | < \varepsilon_n \, , \tag{6.40}$$

where ε_n is the threshold of the norm constraint to be determined.

Replacing $\|\mathbf{v}\|^2 - \|\mathbf{v}'\|^2$ by d_n, we have

$$d_n = \mathbf{v} \cdot \mathbf{v} - \mathbf{v}' \cdot \mathbf{v}' = \mathbf{v}^T \mathbf{v} - \mathbf{v}'^T \mathbf{v}' \, . \tag{6.41}$$

Given the covariance matrix $\Lambda_{\mathbf{v}}$ of $\mathbf{v}$ and the covariance matrix $\Lambda_{\mathbf{v}'}$ of $\mathbf{v}'$, we now compute the variance Λ_{d_n} of d_n. Under the first order approximation (see Theorem 2.3), we have

$$\Lambda_{d_n} = J_{\mathbf{v}}^{d_n} \Lambda_{\mathbf{v}} J_{\mathbf{v}}^{d_n T} + J_{\mathbf{v}'}^{d_n} \Lambda_{\mathbf{v}'} J_{\mathbf{v}'}^{d_n T} \, , \tag{6.42}$$

where $J_{\mathbf{v}}^{d_n}$ is the Jacobian matrix of d_n with respect to $\mathbf{v}$ and $J_{\mathbf{v}'}^{d_n}$, the one with respect to $\mathbf{v}'$. Here we assume that $\mathbf{v}$ and $\mathbf{v}'$ are two independent Gaussian random variables.

The Jacobian $J_{\mathbf{v}}^{d_n}$ is given by

$$J_{\mathbf{v}}^{d_n} = \frac{\partial d_n}{\partial \mathbf{v}} = \frac{\partial}{\partial \mathbf{v}} (\mathbf{v}^T \mathbf{v} - \mathbf{v}'^T \mathbf{v}') = 2\mathbf{v}^T \, . \tag{6.43}$$

Similarly, we have

$$J_{\mathbf{v}'}^{d_n} = -2\mathbf{v}'^T \, . \tag{6.44}$$

Now Equation (6.42) can be rewritten down as

$$\Lambda_{d_n} = 4(\mathbf{v}^T \Lambda_{\mathbf{v}} \mathbf{v} + \mathbf{v}'^T \Lambda_{\mathbf{v}'} \mathbf{v}') \, . \tag{6.45}$$

Therefore, the norm constraint (6.37) is finally expressed in the real case as

$$\frac{d_n^2}{\Lambda_{d_n}} < \kappa_n \, , \tag{6.46}$$

where d_n is computed by (6.41), Λ_{d_n} is computed by (6.45), and κ_n is a coefficient.

In fact, $\frac{d_n^2}{\Lambda_{d_n}}$ can be considered, up to the first order approximation that we have used, as a random variable following a χ^2 distribution with 1 degree of freedom. Looking up the χ^2 table, we can choose an appropriate κ_n. This gives us the probability that d_n^2 falls into the interval $[0, \kappa_n \Lambda_{d_n}]$. For example, we can take $\kappa_n = 3.84$ for a probability of 95% when we consider the lengths of segments, and $\kappa_n = 1.32$ for a probability of 75% when we consider the distance between the midpoints of the two segments. That is, we impose a stricter constraint on the distance between midpoints than on the lengths of segments.

6.4.2 Dot-Product Constraint

The dot-product constraint says that the difference between the cosines of angles between two vectors should be zero. Suppose we are given two vectors $\mathbf{u}$ and $\mathbf{v}$, and their covariance matrices $\Lambda_\mathbf{u}$ and $\Lambda_\mathbf{v}$. Here, $\mathbf{u}$ and $\mathbf{v}$ are assumed unit vectors. Denote the difference between the cosines of angles as d_c, i.e.,

$$d_c = \mathbf{u} \cdot \mathbf{v} - \mathbf{u}' \cdot \mathbf{v}' = \mathbf{u}^T\mathbf{v} - \mathbf{u}'^T\mathbf{v}' . \tag{6.47}$$

We now compute the variance Λ_{d_c} of d_c. Under the first order approximation (see Theorem 2.3), we have

$$\Lambda_{d_c} = J_\mathbf{u}^{d_c}\Lambda_\mathbf{u}J_\mathbf{u}^{d_c T} + J_\mathbf{v}^{d_c}\Lambda_\mathbf{v}J_\mathbf{v}^{d_c T} + J_{\mathbf{u}'}^{d_c}\Lambda_{\mathbf{u}'}J_{\mathbf{u}'}^{d_c T} + J_{\mathbf{v}'}^{d_c}\Lambda_{\mathbf{v}'}J_{\mathbf{v}'}^{d_c T} , \tag{6.48}$$

where $J_\mathbf{u}^{d_c}$, $J_\mathbf{v}^{d_c}$, $J_{\mathbf{u}'}^{d_c}$ and $J_{\mathbf{v}'}^{d_c}$ are the Jacobian matrices of d_c with respect to $\mathbf{u}$, $\mathbf{v}$, $\mathbf{u}'$ and $\mathbf{v}'$, respectively. We assume that $\mathbf{u}$, $\mathbf{v}$, $\mathbf{u}'$ and $\mathbf{v}'$ are four independent Gaussian random variables.

The Jacobian $J_\mathbf{u}^{d_c}$ is given by

$$J_\mathbf{u}^{d_c} = \frac{\partial d_c}{\partial \mathbf{u}} = \frac{\partial}{\partial \mathbf{u}}(\mathbf{u}^T\mathbf{v} - \mathbf{u}'^T\mathbf{v}') = \mathbf{v}^T . \tag{6.49}$$

Similarly, we have

$$J_\mathbf{v}^{d_c} = \mathbf{u}^T , \quad J_{\mathbf{u}'}^{d_c} = -\mathbf{v}'^T \quad \text{and} \quad J_{\mathbf{v}'}^{d_c} = -\mathbf{u}'^T . \tag{6.50}$$

Now Equation (6.48) can be rewritten down as

$$\Lambda_{d_c} = \mathbf{v}^T\Lambda_\mathbf{u}\mathbf{v} + \mathbf{u}^T\Lambda_\mathbf{v}\mathbf{u} + \mathbf{v}'^T\Lambda_{\mathbf{u}'}\mathbf{v}' + \mathbf{u}'^T\Lambda_{\mathbf{v}'}\mathbf{u}' . \tag{6.51}$$

Therefore, the dot-product constraint (6.38) is finally expressed in the real case as

$$\frac{d_c^2}{\Lambda_{d_c}} < \kappa_c , \tag{6.52}$$

where d_c is computed by (6.47), Λ_{d_c} is computed by (6.51), and κ_c is a coefficient. We can choose $\kappa_c = 1.32$ for a probability of 75%.

6.4.3 Triple-Product Constraint

The same manipulation can be done for the triple-product constraint. Recall first the notation "$\tilde{}$" introduced in (4.2). If $\mathbf{v} = [x, y, z]^T$, then $\tilde{\mathbf{v}}$ is defined as

$$\tilde{\mathbf{v}} = \begin{bmatrix} 0 & -z & y \\ z & 0 & -x \\ -y & x & 0 \end{bmatrix} .$$

For any three-dimensional vectors $\mathbf{u}$ and $\mathbf{v}$, we then have $\mathbf{u} \wedge \mathbf{v} = \tilde{\mathbf{u}}\mathbf{v}$.

Given 6 unit vectors: $\mathbf{u}_1$, $\mathbf{u}_2$, $\mathbf{u}_3$, $\mathbf{u}'_1$, $\mathbf{u}'_2$, $\mathbf{u}'_3$ and their covariance matrices. Denote the difference between the two triple-products by d_t, i.e.,

$$
\begin{aligned}
d_t &= <\mathbf{u}_1, \mathbf{u}_2, \mathbf{u}_3> - <\mathbf{u}'_1, \mathbf{u}'_2, \mathbf{u}'_3> \\
&= \mathbf{u}_1 \cdot (\mathbf{u}_2 \wedge \mathbf{u}_3) - \mathbf{u}'_1 \cdot (\mathbf{u}'_2 \wedge \mathbf{u}'_3) = \mathbf{u}_1^T (\tilde{\mathbf{u}}_2 \mathbf{u}_3) - \mathbf{u}'^T_1 (\tilde{\mathbf{u}}'_2 \mathbf{u}'_3) \, . \quad (6.53)
\end{aligned}
$$

The variance Λ_{d_t} of d_t is given, under the first order approximation, by

$$
\begin{aligned}
\Lambda_{d_t} &= J^{d_t}_{\mathbf{u}_1} \Lambda_{\mathbf{u}_1} J^{d_t}_{\mathbf{u}_1}{}^T + J^{d_t}_{\mathbf{u}_2} \Lambda_{\mathbf{u}_2} J^{d_t}_{\mathbf{u}_2}{}^T + J^{d_t}_{\mathbf{u}_3} \Lambda_{\mathbf{u}_3} J^{d_t}_{\mathbf{u}_3}{}^T \\
&\quad + J^{d_t}_{\mathbf{u}'_1} \Lambda_{\mathbf{u}'_1} J^{d_t}_{\mathbf{u}'_1}{}^T + J^{d_t}_{\mathbf{u}'_2} \Lambda_{\mathbf{u}'_2} J^{d_t}_{\mathbf{u}'_2}{}^T + J^{d_t}_{\mathbf{u}'_3} \Lambda_{\mathbf{u}'_3} J^{d_t}_{\mathbf{u}'_3}{}^T \, , \quad (6.54)
\end{aligned}
$$

where $J^{d_t}_{\mathbf{x}}$ is the Jacobian matrix of d_t with respect to $\mathbf{x}$. We assume that the six vectors are independent Gaussian random variables.

The Jacobian $J^{d_t}_{\mathbf{u}_2}$, for example, can be computed as follows

$$
\begin{aligned}
J^{d_t}_{\mathbf{u}_2} &= \frac{\partial d_t}{\partial \mathbf{u}_2} = \frac{\partial}{\partial \mathbf{u}_2} [\mathbf{u}_1^T (\tilde{\mathbf{u}}_2 \mathbf{u}_3) - \mathbf{u}'^T_1 (\tilde{\mathbf{u}'_2} \mathbf{u}'_3)] \\
&= \frac{\partial}{\partial \mathbf{u}_2} [-\mathbf{u}_1^T (\tilde{\mathbf{u}}_3 \mathbf{u}_2)] = -\mathbf{u}_1^T \tilde{\mathbf{u}}_3 \, .
\end{aligned}
$$

Similarly we can compute the other Jacobian matrices. Thus the variance Λ_{d_t} can be computed by (6.54).

Therefore, the triple-product constraint (6.39) is finally expressed in the real case as

$$
\frac{d_t^2}{\Lambda_{d_t}} < \kappa_t \, , \quad (6.55)
$$

where κ_t is a coefficient.

In our implementation, however, we do not compute the variance of the d_t, because the computation is relatively expensive. Instead, we give a predefined threshold ε_t. If $|d_t| < \varepsilon_t$, then the triple-product constraint is considered satisfied; otherwise, not satisfied. $|d_t|$ reaches a maximum of 2 when $\mathbf{u}_1$, $\mathbf{u}_2$, $\mathbf{u}_3$ are perpendicular to each other, and $\mathbf{u}'_1$, $\mathbf{u}'_2$, $\mathbf{u}'_3$ are perpendicular to each other. In that case, they are not congruent (i.e., reflection). $|d_t|$ reaches a minimum of 0 when they are congruent. We choose $\varepsilon_t = 0.5$ in our implementation to account for noise.

6.5 Other Formalisms of Rigidity Constraints

In this section, we present several formalisms proposed in the literature.

Grimson and *Lozano-Pérez* [6.6] formalized the rigidity constraints in a different way. The length constraint and the constraint on the angle between two segments are similar to ours (see (6.9) and (6.11)). The two other angular constraints are not used in their approach. The distance constraint is different. They state the distance constraint as the preservation of the *range* of distances (i.e., the minimum and maximum distances) between two line segments. The maximum distance is the maximum of the four distances between endpoints of segments (the dashed thin lines in Fig. 6.8). The minimum distance[2] is the minimum of the following three parts (see Fig. 6.8):

1. distances between the endpoints of segments (in dashed thin lines),
2. all possible distances between an endpoint of one segment and its projection on the other segment if the projection is on that segment (in solid thin lines),
3. distance between the supporting lines if the common perpendicular intersects both segments (in dashed thick line).

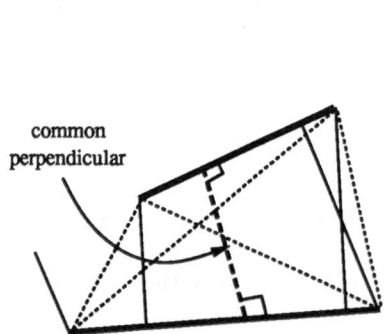

Fig. 6.8. Distance constraints   **Fig. 6.9.** Component constraints

They define a new constraint called *component constraint*. Given two 3D segments S_1 and S_2, we can construct a 3D coordinate system by $\mathbf{u}_1$, $\mathbf{u}_2$ and $\mathbf{u}_1 \wedge \mathbf{u}_2$ if those two segments are not parallel (see Fig. 6.9). Consider a vector $\mathbf{v}$ which starts from an arbitrary point on S_1 and ends at an arbitrary point on S_2. Projecting $\mathbf{v}$ on the axes $\mathbf{u}_1$, $\mathbf{u}_2$ and $\mathbf{u}_1 \wedge \mathbf{u}_2$, we get three components. For all possible $\mathbf{v}$'s, we get a range of values for each component. The component constraint says that those three ranges of components must be preserved during motion.

Pollard et al. [6.7] used the constraints on the segment length and the angle between two segments, as well as the following two constraints (see Fig. 6.10):

1. minimum separation between *lines* supporting segments. If two segments are not parallel, then the distance is given by

$$d = (\mathbf{m}_2 - \mathbf{m}_1) \cdot \mathbf{u}_1 \wedge \mathbf{u}_2 / \|\mathbf{u}_1 \wedge \mathbf{u}_2\| . \qquad (6.56)$$

If they are parallel, $d = \|(\mathbf{m}_2 - \mathbf{m}_1) - [(\mathbf{m}_2 - \mathbf{m}_1) \cdot \mathbf{u}_1]\mathbf{u}_1\|$.

---

[2]In the modelization of the 2D distance constraint, they *implicitly* exclude the case where two segments intersect. The minimum distance is zero in this case, but they do not consider it in their formulation.

2. distances between the endpoints of each segment and the point of minimum separation (i.e., the intersection of that segment with the common perpendicular). It is applicable only to nonparallel line segments. In Fig. 6.10, a_1 and c_1 are those distances for the first segment, and a_2 and c_2, those for the second.

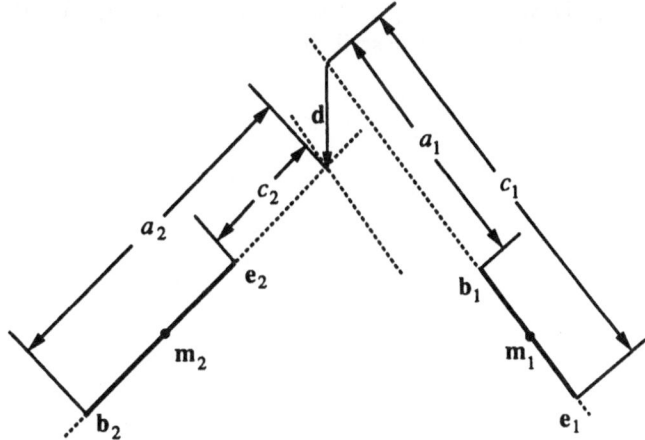

Fig. 6.10. Constraints used by *Pollard* et al.

Chen and *Huang* [6.8] used only the constraint on the angle between line segments in their matching algorithm. This constraint allows to obtain several hypotheses of the rotation between two sets of line segments which are later pruned by a Hough-like procedure to recover the translation.

Note that unlike our approach, all above approaches use some predefined error ranges in the measurements. As we have pointed out earlier, the errors of measurements given by a stereo system have different distributions in different positions. One cannot handle such phenomenon by some prefixed values. Our approach described in the previous section is expected to handle it better.

6.6 Summary

In this chapter, we have first structured the matching problem as a standard tree search approach. As we have seen, the tree search approach is combinatorially explosive. The problem soon becomes intractable as the number of features increases. This shows that we need some constraints to guide the search heuristically without exploring most of the nodes in the tree while assuring a good interpretation.

To this end, we have proposed a set of rigidity constraints for 3D line segments. We have shown that the rigidity constraints derived are complete in the sense that they are necessary and sufficient to guarantee the congruency of two pairings of 3D line segments. We have also shown that we can compute a unique rigid motion from two pairings of line segments satisfying the rigidity constraint.

In the real case, the equalities in the rigidity constraints are not true anymore due to inherent noise in the measurements. We have then reformalized the rigidity constraints to determine dynamically the thresholds from the error measurements of 3D data given by a stereo system. The last formalism will be used in the matching algorithms described in the next chapter.

7. Hypothesize-and-Verify Method for Two 3D View Motion Analysis

We present in this chapter a method based on the *hypothesize-and-verify* paradigm to register two sets of 3D line segments obtained from stereo and to compute the transformation (motion) between them. We assume that the environment is static and that it is only the stereo rig that has moved. The multiple-object-motions problem is dealt with in Chap. 9.

7.1 General Presentation

The rigidity assumption about the environment and objects is used in most matching algorithms, although the implementation differs from one to another in the details. Basically, we can divide existing algorithms into two approaches: global and local. The *global* approach consists in finding directly a global rigid transformation/motion which can best bring features from one frame into another. The *local* approach consists in first identifying a best set of matches based on the conservation of local geometrical relationships (such as the rigidity constraints described in the previous chapter). The transformation between frames is then computed from identified matches. In both approaches, tree-searching strategies are exploited. The interpretation-tree-like approaches presented in Sect. 6.1 are the local ones.

7.1.1 Search in the Transformation Space

The global approach exploits the tree search in the Transformation Space (TS). A point in this space represents a possible transformation between two adjacent frames. A straightforward method is to sample points in the transformation space at some spacing (which should correspond to some predetermined precision). For each sampled point, we can use the associated transformation to bring the first frame to the second. By defining a goodness measure, we can assign a score to the transformation. The goodness measure may be defined as the number of matches

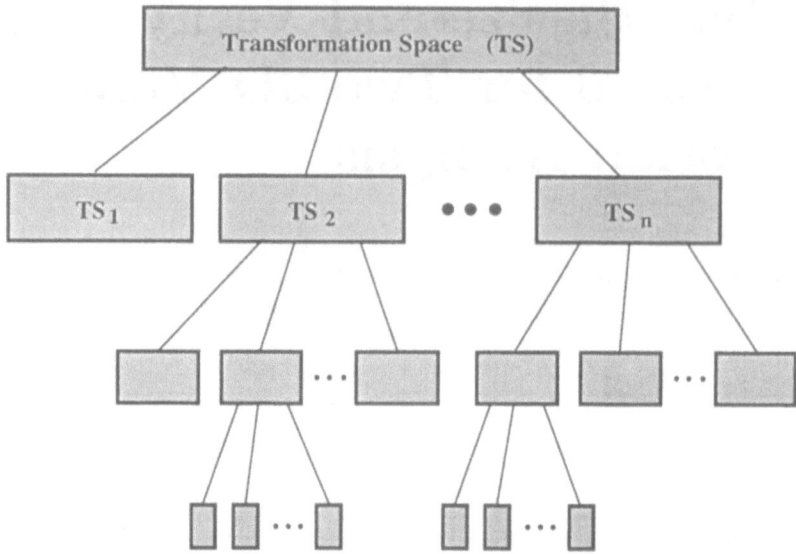

Fig. 7.1. Search method in the transformation space

by the transformation. After all sampled points are evaluated, we can simply choose the transformation with the highest score as the optimal transformation between the two frames. Since the transformation space is six-dimensional, this method tends either to require unrealistic large amount of computation, or to yield results with poor precision.

A more intelligent method may exploit a coarse-to-fine strategy. The coarse-to-fine strategy has been successfully used for solving many problems in Computer Vision such as edge detection and pattern recognition. Figure 7.1 illustrates how the strategy is used to solve our motion analysis problem. The transformation space is represented by a rectangle. We first sample the large transformation space into several (100, for example) smaller subspaces. As described in the last paragraph, we can evaluate the goodness measure for each subspace. For each subspace whose score is bigger than some threshold, we resample it into still smaller subspaces and evaluate them. We can iterate the same procedure until enough accuracy in the transformation is achieved. We should note that the goodness measure must be appropriately defined at each level according to the uncertainty of the transformations.

The above method is simple and has a big potentiality for parallel implementation. Although we have not implemented it, it suffers *a priori* from some problems:

1. To get a reasonable accuracy, we should evaluate a large amount of transformations. For example, if we assume the rotation angle in each axis is in the range $[-\pi, \pi]$ and the translation in each axis is in the range $[-1000, 1000]$

(in millimeters), we should sample 40 points in each axis to obtain an initial accuracy in rotation to within 9 degrees and in translation to within 50 millimeters in each axis. This implies that 40^6 or 56,000,000 different transformations must be evaluated at the first level. It is difficult to obtain a solution in a reasonable time even on parallel machines, not mentioning on SUN workstations.

2. The goodness measure of a transformation is difficult to define precisely, especially when the transformation is very uncertain (at the first levels of the transformation tree).

7.1.2 Hypothesize-and-Verify Method

We now extend the above idea to develop a hypothesize-and-verify method for solving effectively the motion analysis problem. The hypothesize-and-verify paradigms is one of the most popular paradigms to deal with the matching problem [7.1–7]. Several of those approaches start with first heuristically determining a small number of focus features, which allow a considerable reduction of the search space.

Our idea is simple. Instead of evaluating all possible transformations, we use the rigidity constraints on the interrelationships between features to generate some hypothetical feature correspondences between two successive frames. We compute an initial estimate of the transformation for each hypothesis. We then evaluate the validity of these hypothetical transformations. Due to the exploitation of the rigidity constraints (described in the previous chapter), the number of hypothetical transformations is usually very small, and computational efficiency is achieved. Our method can be considered as a hybrid approach in that it starts with a search in the interpretation tree described in the last chapter and follows by a search in the transformation space described above.

Figure 7.2 illustrates diagrammatically the principle of our hypothesize-and-verify method. In the following, we explain in detail the method and provide some experimental results.

7.2 Generating Hypotheses

The first stage of the algorithm is called the *hypothesis generation*. We first give the definition of a hypothesis.

7.2.1 Definition and Primary Algorithm

Definition 7.1.   *Two pairings of 3D line segments (S_1, S_2 in Frame 1, S_1', S_2' in Frame 2) form a **hypothesis** if and only if they satisfy all rigidity constraints described in Theorem 6.1, up to the thresholds defined in Section 6.4. This hypothesis is denoted by $\{S_1, S_1'; S_2, S_2'\}$.*   ◇

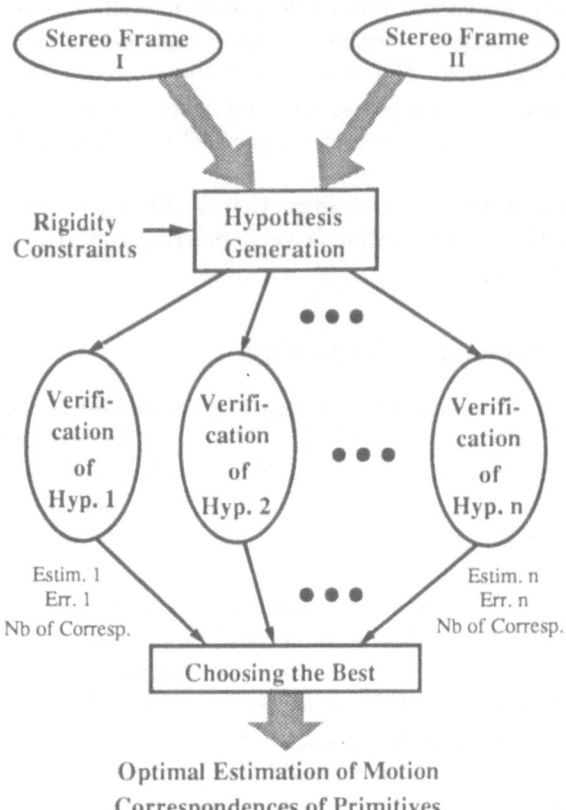

Fig. 7.2. Diagram of the motion-analysis algorithm based on the hypothesize-and-verify paradigm

Since we apply the rigidity constraints to any pair of segments, if we explore all possible pairs, the number of operations required is

$$\binom{m}{2}\binom{n}{2}2! = \tfrac{1}{2}mn(m-1)(n-1) \ ,$$

where m is the number of segments in the first frame and n that in the second frame. Therefore, the complexity is $O(m^2n^2)$. The primary algorithm is given by Algorithm 7.1.

7.2.2 Control Strategies in Hypothesis Generation

The above algorithm is certainly very slow. Since at this stage we do not want to recover all matches between two frames, but to recover all potential motions between them, we do not need to explore the search space exhaustively. We

---

Algorithm 7.1: Primary Algorithm of Hypothesis Generation

- Set *list_hyps* to NIL   /* initialization of the list of hypotheses */
- **for** each two segments S_1 and S_2 in Frame 1
    - → **for** each two segments S_1' and S_2' in Frame 2
        - ↪examine the rigidity of $\{S_1, S_1';\ S_2, S_2'\}$
        - ↪**if** they satisfy the rigidity constraints
            - **then** retain $\{S_1, S_1';\ S_2, S_2'\}$ in *list_hyps*
        - ↪**endif**
    - ← **endfor**
- ⇑ **endfor**

---

reduce the complexity of the hypothesis generation phase by using a number of heuristics:

Sort the segments Sort all segments in each frame in decreasing length order, so that we can easily find, by binary search, the segments in the second frame which are compatible in length with the segments in the first one.

Control the search depth Rather than finding all possible pairings compatible with a given pairing, we stop if we have found a sufficient number of compatible pairings (5, for instance).

Avoid redundant hypotheses If a pairing has already been retained as a potential match in some early hypothesis, it is not considered further, because it does not give us new information about the motion between the two frames.

Reduce the search width Consider segments of the first frame only in the central part of the frame, because segments on the sides are likely to move out of the view field in the next frame.

Reduce the number of segments Choose only the longest segments in the first scene, for instance, the m/q longest ($q = 2$ or 3) . This reduces also the search width.

7.2.3 Additional Constraints

To make full use of 3D information obtained by our passive stereo system and to reduce the number of incorrect hypotheses, we can derive an additional constraint on the change of the orientation of a segment.

Theorem 7.1. *Suppose an oriented segment undergoes a rigid displacement between two successive frames with a rotation angle between 0 and π, then the change of orientation of the segment is less than or equal to the rotation angle between the two frames.* ∎

Proof. Let $\mathbf{u} = [p, \ q, \ r]^T$ be the unit orientation vector of the segment in the first frame, and $\mathbf{u}'$ be that in the second frame, then we have $\mathbf{u}' = \mathbf{R}\mathbf{u}$ where $\mathbf{R}$ is the rotation matrix.

Let the vector $\mathbf{r} = [a, \ b, \ c]^T$, which gives the direction of the axis of rotation and its norm is the rotation angle around this axis. If we choose the rotation axis representation for rotation (see Chap. 5), then the rotation matrix $\mathbf{R}$ is given by (4.4):

$$\mathbf{R} = \mathbf{I} + \mathrm{f}(\theta)\tilde{\mathbf{r}} + \mathrm{g}(\theta)\tilde{\mathbf{r}}^2 \ ,$$

where $\theta = \sqrt{a^2 + b^2 + c^2}$ is the rotation angle, $\mathrm{f}(\theta) = \frac{\sin\theta}{\theta}$ and $\mathrm{g}(\theta) = \frac{1-\cos\theta}{\theta^2}$.

Now it is easy to compute the angle between $\mathbf{u}$ and $\mathbf{u}'$, denoted by $\widehat{\mathbf{u} \ \mathbf{u}'}$, the cosine of which is given by

$$\cos(\widehat{\mathbf{u} \ \mathbf{u}'}) = \cos(\widehat{\mathbf{u} \ \mathbf{R}\mathbf{u}}) = \frac{1 - \cos\theta}{\theta^2}(ap + bq + cr)^2 + \cos\theta \ .$$

Because $\frac{1-\cos\theta}{\theta^2}(ap + bq + cr)^2 \geq 0$, we have $\cos(\widehat{\mathbf{u} \ \mathbf{u}'}) \geq \cos\theta$, i.e., $\widehat{\mathbf{u} \ \mathbf{u}'} \leq \theta$ for $0 \leq \theta < \pi$.

Note that when $\theta \neq 0$ (not pure translation), $\widehat{\mathbf{u} \ \mathbf{u}'} = \theta$ if and only if $ap + bq + cr = 0$, that is, if a segment undergoes a movement whose rotation axis is perpendicular to the segment, then the change of the segment orientation is equal to the rotation angle. If not, it is less than the rotation angle. $\square$

In general, the rotation angle between two successive frames does not go beyond 60 degrees, so we can impose that the orientation difference of a pairing of segments to be matched must be less than 60 degrees.

Other constraints can be integrated in the algorithm to speed up even more the hypothesis generation process. In the indoor mobile robot navigation example, the robot and the objects usually move horizontally (in the ground plane). In our stereo coordinate system, the ground plane is parallel to the plane $y = 0$. Thus we can impose another unitary constraint that if two segments (S in Frame 1, S' in Frame 2) can be matched to each other, they must have almost the same y coordinates.

7.2.4 Algorithm of Hypothesis Generation

Algorithm 7.2 shows the algorithm of the hypothesis generation incorporated with most of the above heuristics. In the algorithm, **continue** and **break** have the same definitions as in the C programming language.

In order to obtain more reliable hypotheses, we have actually implemented a slightly modified algorithm. We use triplets of segments. Let $\{S_1, S_1'; \ S_2, S_2'\}$ be two compatible pairings (in the sense that they satisfy the rigidity constraints), if another pairing S_3 and S_3' is also compatible with S_1 and S_1', then we verify whether this pairing is compatible with S_2 and S_2'. If so, S_1, S_2, S_3 and S_1', S_2', S_3' can form a hypothesis. This means that each hypothesis has propagated to at least one segment. Therefore, we have a more reliable hypothesis, on one hand;

Algorithm 7.2: Algorithm of Hypothesis Generation

- Set *list_hyps* to NIL   /* initialization of the list of hypotheses */
- Sort segments in Frame 1: $I_1[1..m]$ contains their indices sorted
- Sort segments in Frame 2: $I_2[1..n]$ contains their indices sorted
- **for** a segment S_1 in $I_1[1..m/q]$
  - → Find, by binary search, segments in Frame 2 compatible in length with S_1: $I_2[b_1..e_1]$
  - → **for** a segment S_1' in $I_2[b_1..e_1]$
    - ↪**if** S_1 and S_1' are not congruent in orientation
      or their y coordinates are very different
      then continue
    - ↪**endif**
    - ↪Set *number* = 0
    - ↪**for** a segment S_2 in $I_1[1..m]$ and $S_1 \neq S_2$
      - ⇒ Find, by binary search, segments in Frame 2 compatible in length with S_2: $I_2[b_2..e_2]$
      - ⇒ **for** a segment S_2' in $I_2[b_2..e_2]$
        - ⊢ **if** $S_1' = S_2'$ or S_1 and S_1' are not congruent in orientation
          or their y coordinates are very different
          then continue
        - ⊣ **endif**
        - ⊢ **if** $\{S_1, S_1'; S_2, S_2'\}$ satisfy the rigidity constraints
          then ⊤ Retain $\{S_1, S_1'; S_2, S_2'\}$ in *list_hyps*
          ⊤ Increment *number* by 1
          ⊤ **if** *number* > 5 **then break endif**
        - ⊣ **endif**
      - ⇐ **endfor**
      - ⇒ **If** *number* > 5 **then break endif**
    - ↩ **endfor**
  - ← **endfor**
- ⇑ **endfor**

---

on the other hand, we increase the complexity of the algorithm. The slowdown in hypothesis generation because of this modification is not significant because the third pairing is only searched among those compatible pairings, and will be compensated for in hypothesis verification because we now have less hypotheses but more reliable.

7.3 Verifying Hypotheses

The second stage of the algorithm is called the *hypothesis verification*. We propagate each hypothesis generated above to the whole frame and try to match more segments. This stage can be done in parallel for each hypothesis. Then we can choose the best ones, best being defined later. The process is performed as follows.

7.3.1 Estimating the Initial Rigid Motion

From Theorem 6.3, we know that a unique rigid motion can be computed for each hypothesis. We compute an initial estimation of the rotation and translation via the extended Kalman filter described in Sect. 5.8. As input to the filter, an initial estimate of the state vector (representing the rigid motion) is required. In our implementation, the state vector s for the motion is simply initialized to zero (i.e. $\hat{s}_0 = 0$) but with a big covariance matrix. The covariance matrix $\Lambda_{\hat{s}_0}$ is initialized as follows: $\Lambda_{\hat{s}_0}[i][i] = 2.0$ for $i = 0, 1, 2$ (rotation components), $\Lambda_{\hat{s}_0}[i][i] = 1.0\mathrm{e}{+06}$ for $i = 3, 4, 5$ (translation components) and $\Lambda_{\hat{s}_0}[i][j] = 0$ for $i \neq j$. This is equivalent to saying that we assume a standard deviation of 114 degrees of rotation along each axis and a standard deviation of 1 meter of translation along each axis.

As stated in Chap. 2, because we use a first order approximation, if the initial estimate is not very good, the final estimate given by the filter may be different from the true value. In order to reduce the effect of nonlinearities, we can apply iteratively the extended Kalman filter. Usually, we obtain a good estimate after a few iterations (typically 3 or 4 for three matches) (see Sect. 5.6).

7.3.2 Propagating Hypotheses

We now have an initial estimate of the motion for each hypothesis. We apply this estimate to the first frame and compare the transformed frame with the second one. (See the appendix to this chapter for the details of how to transform a 3D line segment). If a transformed segment from the first frame is near enough to some segment in the second frame (see Sect. 7.4), then this pair is considered as being matched, and again, the extended Kalman filter is used to update the motion estimate. After all segments have been processed, we obtain an estimate of the motion, its uncertainty, and also the correspondences between segments.

We said that we transformed all segments of the first frame once, using the initial estimate of the motion. This is the one-shot approach. The matching results will depend heavily upon the initial estimate. Since we may have a poor initial estimate, we may not get a satisfactory matching result. To overcome this problem, we transform only one unmatched segment of the first frame at a time. This is the many-shots approach. If we find a match for it, we update the motion estimate and we transform another unmatched segment of the first

frame using the (new) updated motion estimate (not the initial estimate anymore). If we cannot find a match, we process another segment of the first frame.
After all segments have been processed, we obtain an estimate of the motion, its
uncertainty, and also the correspondences between segments.

We have implemented the two approaches and found that the second one gives
much better performances. Starting from two slightly different initial estimates,
the first approach yields two different (not dramatically different, of course)
results of matchings and motion estimates, while the second converges to almost
the same result.

In order to obtain a precise estimation and because of the nonlinearities, we
iterate the above procedure twice.

7.3.3 Choosing the Best Hypothesis

We now discuss how to choose the best hypothesis using the error given by the
Kalman filter. The criterion must be a function of the number of segments
actually matched and of the error made in approximating the match by a rigid
displacement. We use the following criterion:

$$C = \sum_{i=1}^{p} E_i + (N - p)E_{min} , \qquad (7.1)$$

where E_i is the error of the i-th match (the sum of the distances given by (7.10)
and (7.11), see Sect. 7.4), $N = \min(m, n)$ is the smallest number of segments
in the two frames, p is the number of matched segments and E_{min} is the error
corresponding to the threshold determining whether two segments are matched
or not (which is equal to $\kappa_\psi + \kappa_m$ given in Sect. 7.4). Then, the hypothesis with
a minimal C is chosen as the best one. From (7.1), we see that the more matches
we have, usually the smaller the error is. Note that from our experiences, if we
simply define the score C' of a hypothesis as p (the number of matches) and
choose the hypothesis with a maximal C' as the best one, we obtain almost the
same result.

7.3.4 Algorithm of Hypothesis Verification

Algorithm 7.3 shows the algorithm for the hypothesis verification phase. This
algorithm is applied to each hypothesis. i_{max} is the fixed number of iterations
to obtain a more precise estimation (see Sect. 7.3.2). In our case, $i_{max} = 2$.
list_matches is the list containing all recovered matches, and *list_tmp_matches*
is a local list containing matches with the current segment S' in Frame 2. Our
algorithm allows multiple matches, that is, a segment in Frame 1 can be matched
to several segments in Frame 2, and *vice versa*. The variable *admit_multiple* is a
boolean taking value of **true** or **false**. If *admit_multiple* is equal to **true**, then
the algorithm allows multiple matches; otherwise not. In our implementation,
admit_multiple is set to **false** in the first iteration and to **true** in the second
iteration.

Algorithm 7.3: Algorithm of Hypothesis Verification

- Set $s = 0$ and Λ_s very big /* initialization of motion parameters */
- Compute an initial motion estimate for the current hypothesis: s, Λ_s
- **for** i from 1 to i_{max} /* i_{max} is the number of iterations */
  → Set *list_matches* to NIL /* global list of matches */
  → **for** a segment S in $I_1[1..m]$
    ↳ Transform S to $\hat{S}$ using s, Λ_s
    ↳ Set *list_tmp_matches* to NIL /* local list of matches for S */
    ↳ Find, by binary search, segments in Frame 2
      compatible in length with S: $I_2[b..e]$
    ↳ **for** a segment S' in $I_2[b..e]$
      ⇒ **if** $\hat{S}$ and S' are near enough
        then retain (S, S') in *list_tmp_matches*
      ⇐ **endif**
    ↪ **endfor**
    ↳ **if** the length of *list_tmp_matches* = 1: $\{\mathcal{M}\}$
      then ⇒ Update s, Λ_s via EKF using $\mathcal{M}$
           ⇒ Retain $\mathcal{M}$ in *list_matches*
      else if the length of *list_tmp_matches* > 1 **and** *admit_multiple*
      then ⇒ **for** each match $\mathcal{M}$ in *list_tmp_matches*
           ⊢ Update s, Λ_s via EKF using $\mathcal{M}$
           ⊢ Retain $\mathcal{M}$ in *list_matches*
         ⇐ **endfor**
    ↪ **endif**
  ← **endfor**
⇑ **endfor**

We note that the complexity of the algorithm in the worst case is $O(mn)$ for each hypothesis. The speed of the algorithm depends essentially on the ability to get access quickly to the segments of Frame 2 which are close to the transformed segment $\hat{S}$. We have used several approaches to achieve this. One approach is to use binary search to discard segments of Frame 2 which are not compatible in length with S. Another efficient method is to use the *bucketing techniques*, which allow us to obtain a list of segments in the neighborhood of some segment. The preprocessing necessary to sort segments into buckets can be done very quickly (the complexity is linear in the number of segments). We will explain this technique in Sect. 15.1.

7.4 Matching Noisy Segments

The question we want to answer in this section is the following:

> Given a transformed segment $\hat{S}$ from Frame 1, which segment in Frame 2 can be matched to $\hat{S}$?

Our matching technique has improved twice. We now explain how the different versions work. See the appendix to this chapter for the details of how to transform a 3D line segment.

7.4.1 Version 1

In the first version, we measure the Mahalanobis distance between the endpoints of segments. The transformed segment $\hat{S}$ from Frame 1 is represented by its endpoints M_1, M_2 and their covariance matrices Λ_{M_1}, Λ_{M_2}. A segment S' in the Frame 2 is represented by its endpoints M_1', M_2' and their covariance matrices $\Lambda_{M_1'}$, $\Lambda_{M_2'}$. The Mahalanobis distance between the first endpoints is given by

$$d_1 = (M_1 - M_1')^T (\Lambda_{M_1} + \Lambda_{M_1'})^{-1} (M_1 - M_1') \,, \tag{7.2}$$

and that between the second endpoints is given by

$$d_2 = (M_2 - M_2')^T (\Lambda_{M_2} + \Lambda_{M_2'})^{-1} (M_2 - M_2') \,. \tag{7.3}$$

In order for the two segments to be matched, the Mahalanobis distances d_1 and d_2 must be less than some threshold ε_d. Looking up the χ^2 table with three degrees of freedom, we can choose an appropriate ε_d, for example, 7.81 for a probability of 95%.

As we can see later, a number of segments cannot be matched using this technique. The main problem is that the endpoints are not reliable, because a segment of an object can be segmented differently in successive views, i.e., the endpoints of two matched segments do not correspond to the same physical point in space. Of course, we can increase the threshold ε_d, but this will result in many false matches.

7.4.2 Version 2

In this version, segments are still characterized by their endpoints, as in Version 1. On the other hand, instead of measuring directly the distance between the endpoints, we measure the Mahalanobis distance between an endpoint of one segment and its orthogonal projection on the other segment, as detailed below.

Two segments which can be matched must be in one of the configurations[1] shown in Fig. 7.3, where M_1^P and M_2^P are the orthogonal projections of the endpoints M_1 and M_2 of segment $\hat{S}$ on segment S'. That is, the projection of segment $\hat{S}$ on segment S' should have some common part with segment S'. The comparison goes in two steps.

---

[1]The two segments need not be coplanar.

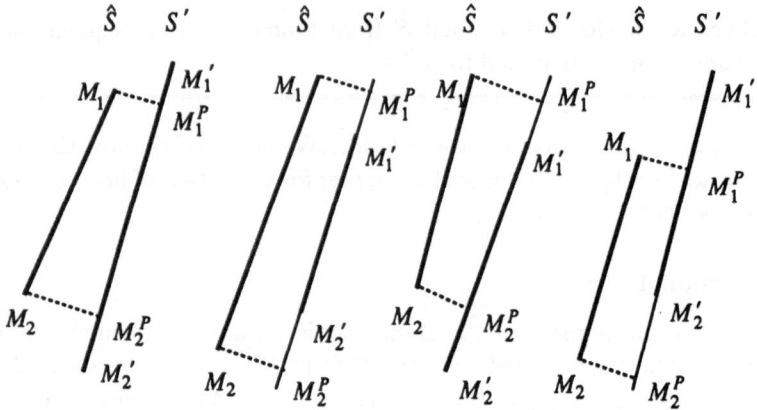

Fig. 7.3. Configurations of matched segments

Step 1: Examine the similarity in orientation. Let **u** be the unit direction vector of S and $\mathbf{u}'$ that of S'. Their covariance matrices are $\Lambda_\mathbf{u}$ and $\Lambda_{\mathbf{u}'}$, respectively. In the noise-free case, we should have $\mathbf{u} \cdot \mathbf{u}' = 1$. In the noisy case, we expect $1 - \mathbf{u} \cdot \mathbf{u}'$ to be less than some threshold ε_o. Using a similar derivation to that in Chap. 6 for the rigidity constraints, we can compute the variance Λ_o of $\mathbf{u} \cdot \mathbf{u}'$ (see (6.51)):

$$\Lambda_o = \mathbf{u}'^T \Lambda_\mathbf{u} \mathbf{u}' + \mathbf{u}^T \Lambda_{\mathbf{u}'} \mathbf{u} \ . \tag{7.4}$$

The condition for the similarity in orientation is formalized as follows:

$$(1 - \mathbf{u} \cdot \mathbf{u}')^2 / \Lambda_o < \kappa_o \ , \tag{7.5}$$

where κ_o is a coefficient. We can choose $\kappa_o = 3.8$ for a probability of 95%.

Step 2: Examine the distance between two segments. We must first examine whether the configuration of two segments to be matched is among the above four configurations, that is, whether they share some common part. The following condition must be satisfied:

$$(M_{1x}^P \geq M_{2x}' \quad \text{and} \quad M_{2x}^P \leq M_{1x}') \quad \text{or} \quad (M_{1x}^P \leq M_{2x}' \quad \text{and} \quad M_{2x}^P \geq M_{1x}') \tag{7.6}$$

where M_x is the x coordinate of point M. Of course, we can use the y or z coordinates, without any modification of the condition.

Now, we measure the Mahalanobis distance between M_1 and M_1^P, d_1 and that between M_2 and M_2^P, d_2, which are given by

$$d_1 = \overrightarrow{M_1 M_1^P}^T \Lambda_1^{-1} \overrightarrow{M_1 M_1^P} \ , \tag{7.7}$$

$$d_2 = \overrightarrow{M_2 M_2^P}^T \Lambda_2^{-1} \overrightarrow{M_2 M_2^P} \ . \tag{7.8}$$

If d_1 and d_2 are less than some threshold κ_d, we then consider these two segments matched. We can choose $\kappa_d = 7.8$, for example, for a probability of 95%. In (7.7) and (7.8), $\Lambda_i = \Lambda_{M_i} + \Lambda_{M_i^P}$. The covariance matrix $\Lambda_{M_i^P}$ of point M_i^P can be approximated as follows:

$$\Lambda_{M_i^P} = \frac{1}{(M'_{2x} - M'_{1x})^2}[(M'_{2x} - M^P_{ix})^2 \Lambda_{M'_1} + (M^P_{ix} - M'_{1x})^2 \Lambda_{M'_2}] , \qquad (7.9)$$

because by interpolation we have the following relation

$$M_i^P = \frac{1}{M'_{2x} - M'_{1x}}[(M'_{2x} - M^P_{ix})M'_1 + (M^P_{ix} - M'_{1x})M'_2] .$$

Here we assume that S' is not perpendicular to the x axis. If S' is perpendicular to the x axis, then y or z coordinate should be used.

Using this new technique and choosing appropriately κ_o and κ_d, we can recover almost all possible matches between successive frames, even if the segments are very noisy. Note that theoretically the condition (7.5) is not needed since we have the conditions (7.7) and (7.8), and those two conditions already constrain the orientation. In practice, we can use the condition (7.5) to reject a large portion of the candidates before we enter the second step, whose computation is more expensive. Furthermore, we can replace the right hand side of (7.5) by a constant. If Θ (for instance, 30 degrees) is the given tolerance in orientation, we can replace (7.5) by the following:

$$\mathbf{u} \cdot \mathbf{u'} > \cos \Theta .$$

7.4.3 Version 3

This version uses the new representation of 3D line segments described in Sect. 4.2. Although Version 2 gives satisfactory results, its computation is a little bit expensive. Using the new representation, matching can be done very efficiently.

The segment $\widehat{S}$ is now represented by $(\widehat{\psi}, \widehat{m}, \widehat{l})$ and its covariance matrix $\Lambda_{\widehat{\psi}}$, $\Lambda_{\widehat{m}}$. The segment S' is now represented by (ψ', m', l') and its covariance matrix is $\Lambda_{\psi'}$, $\Lambda_{m'}$. The length and its variance are not used. As in Version 2, the first step is to examine the similarity in the orientations. The Mahalanobis distance between the orientations is given by

$$d_\psi = (\widehat{\psi} - \psi')^T (\Lambda_{\widehat{\psi}} + \Lambda_{\psi'})^{-1} (\widehat{\psi} - \psi') . \qquad (7.10)$$

If d_ψ is less than some threshold κ_ψ, we then go to the second step: examine the distance between the two segments. Since we have well characterized the uncertainty of a segment by that of its midpoint, we need simply to examine the distance between the midpoints of the two segments. The Mahalanobis distance between the midpoints is given by

$$d_\mathbf{m} = (\widehat{m} - m')^T (\Lambda_{\widehat{m}} + \Lambda_{m'})^{-1} (\widehat{m} - m') . \qquad (7.11)$$

If $d_{\mathbf{m}}$ is still less than some threshold $\kappa_{\mathbf{m}}$, we then consider the two segments as being matched.

Before computing the Mahalanobis distance between ψ's (7.10), special care is taken to cope with the discontinuity of ϕ when a segment is nearly parallel to the plane $y = 0$ (see Sect. 4.2). The treatment is very simple: if $\phi < \pi/2$ and $\phi' > 3\pi/2$, then set ϕ' to be $\phi' - 2\pi$; else if $\phi > 3\pi/2$ and $\phi' < \pi/2$, then set ϕ to be $\phi - 2\pi$; else do nothing. Note that adding a constant to a random variable does not affect its covariance matrix.

By choosing appropriately the threshold κ_{ψ} and $\kappa_{\mathbf{m}}$, we can recover almost all possible matches, but the computation, as we can remark, is much reduced comparing with in Version 2. Looking up the χ^2 distribution table, we take $\kappa_{\psi} = 6.0$ for a probability of 95% with 2 degrees of freedom, and $\kappa_{\mathbf{m}} = 7.8$ for a probability of 95% with 3 degrees of freedom.

7.5   Experimental Results

The algorithm presented in this chapter has been tested for more than two years using real stereo data acquired by our trinocular stereo system. At least two hundred pairs of stereo frames (i.e., 400 individual stereo frames) have been used. The algorithm has succeeded in correctly computing 3D motion and registering two views in almost all cases. The few cases of failure are due to too little common intersection (less than 20%) between two stereo frames and because the reconstructed frames are very noisy.

In this section, we provide three experimental examples to demonstrate the matching process. In each figure shown, if there are four pictures, the upper left one is the top view (projection on the ground plane), the lower left one is the front view (projection on the plane in front of the stereo system and perpendicular to the ground plane), the lower right one is a side view (projection on the plane which is perpendicular to the two previous planes) and the upper right one is the view from the first camera (perspective projection using the calibrated camera transformation matrix). If there are only two pictures in each figure, then the left one is the front view and the right one is the top view.

7.5.1   Indoor Scenes with a Large Common Part

Figures 7.4 and 7.5 show two stereo frames reconstructed by our mobile robot in two different positions. The triangle in each picture represents the optical centers of the cameras of our trinocular stereo system. The 2D images taken by the cameras can be found in Sect. 11.4. We have 261 segments in the first frame and 250 segments in the second. Note that there is a large displacement between these two positions (about 10 degrees of rotation and 75 centimeters of translation) which can be noticed by superposing the two frames (see Fig. 7.6). The maximum shift in 2D images is about 95 pixels (the image resolution is 512×512). However, the common part between the two frames is relatively big.

Fig. 7.4. Different views of stereo frame 1 (uniform scale)

Applying our hypothesis generation procedure to these two frames, we obtain 12 hypotheses. All these hypotheses are propagated to the whole frame in order to match more segments and to update the motion estimate. In the end, 9 hypotheses correctly give the estimate of the displacement. The one which matches the largest number of segments and gives the minimal matching error is kept as the best one. To determine how good this estimate is, we apply it to the first frame and superimpose the transformed one on the second, i.e., in the coordinate system of the second frame (see Fig. 7.8). The shift of the triangle in this figure displays in fact the displacement of the cameras which are rigidly attached to the robot. Figure 7.7 shows only the superposition of the matched segments obtained by using the third version of the matching technique. One

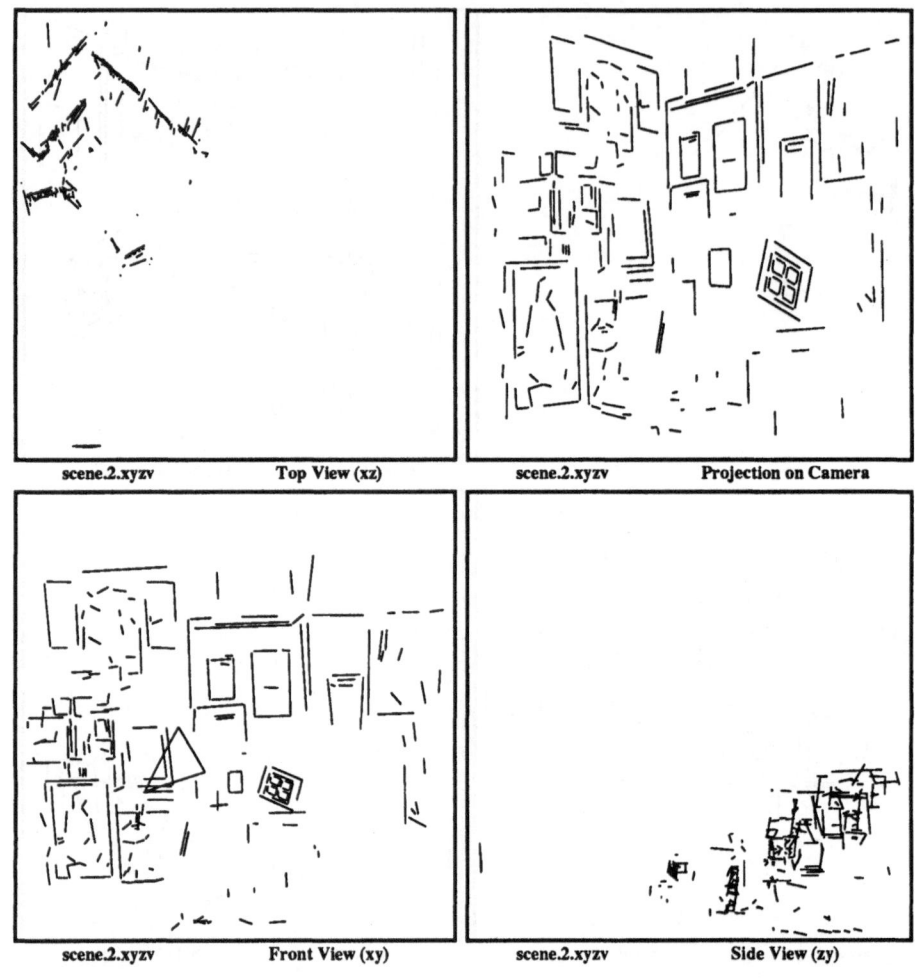

Fig. 7.5. Different views of stereo frame 2 (uniform scale)

can observe the very good accuracy of this estimate. It takes about 3 minutes on a SUN 3 workstation to process these two frames.

Several remarks can be made at this point:

- Segments in the foreground (near the observer) are better superimposed than others. This is reasonable since segments close to the observer are more precisely reconstructed by the stereo system than distant ones (see Figs. 5.11 and 5.13).
- There is a better agreement in the lateral coordinates of the segments than in the range, because the range component is usually much more uncertain than the other components (see Figs. 5.11 and 5.13).
- Those two remarks justify the use of the weighted least-squares property of the Kalman filtering approach (see Sect. 2.2).

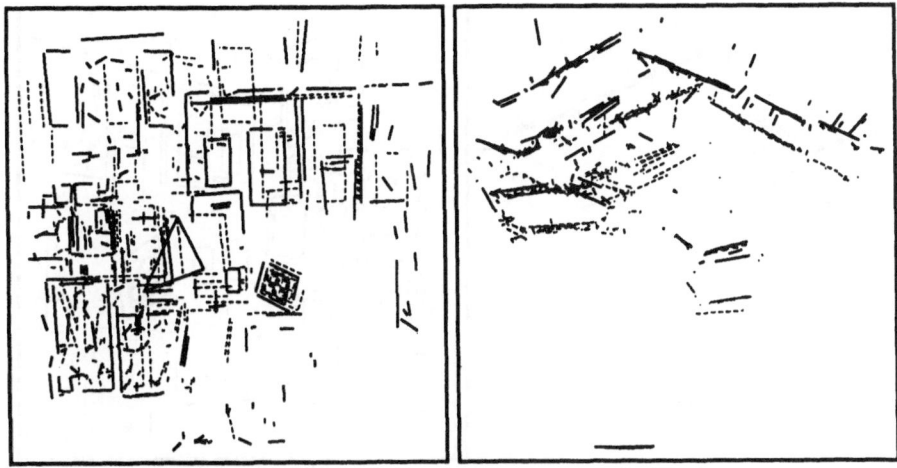

Fig. 7.6. Superposition of the two original frames: segments of Frame 1 are represented in dashed lines and those of Frame 2 in solid lines

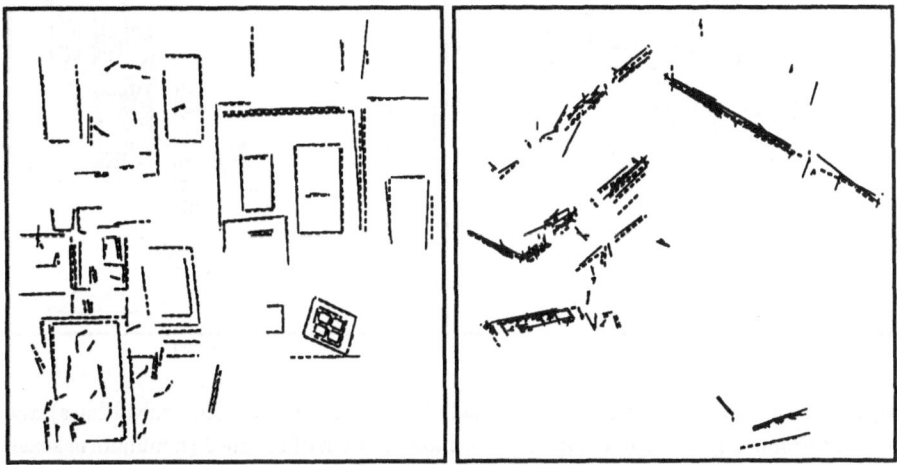

Fig. 7.7. Superposition of the matched segments after applying the computed motion to Frame 1: segments of Frame 1 are represented in dashed lines and those of Frame 2 in solid lines

Fig. 7.8. Superposition of the transformed segments of Frame 1 (in dashed lines) and those of Frame 2 (in solid lines) in the coordinate system of Frame 2 (nonuniform scale)

In the propagation phase, we also recover the matches between the two frames. If we use the criterion of the Mahalanobis distance between endpoints of segments as in our early version of the algorithm, we recover only 79 matches (see Figs. 7.9 and 7.10). If we use the technique of Version 2, we recover 157 matched segments (60% of the total segments) (see Figs. 7.11 and 7.12). A remarkable improvement can be observed. Figures 7.13 and 7.14 show the matched segments by using the technique of Version 3, i.e., based on the new representation of 3D line segments. Using that technique, we have found 170 matches. Their superposition is displayed in Fig. 7.19.

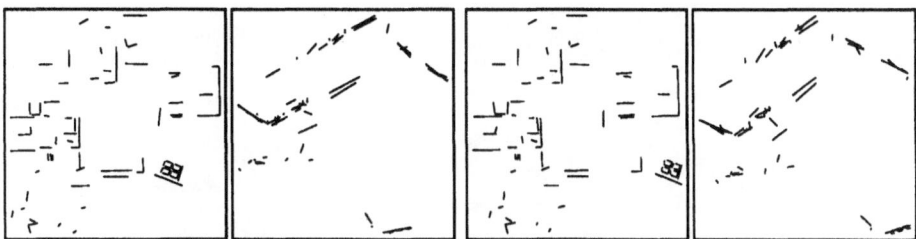

Fig. 7.9. Matched segments of Frame 1 based on the distances between endpoints (Version 1 of the matching technique)

Fig. 7.10. Matched segments of Frame 2 based on the distances between endpoints (Version 1 of the matching technique)

Fig. 7.11. Matched segments of Frame 1 based on the distances between an endpoint and its projection (Version 2 of the matching technique)

Fig. 7.12. Matched segments of Frame 2 based on the distances between an endpoint and its projection (Version 2 of the matching technique)

7.5.2 Indoor Scenes with a Small Common Part

The second example shows an experiment with a very small common part between two successive frames. Theoretically, our algorithm works if there are three common segments in two frames. In pratice, as 3D line segments reconstructed are rather noisy, we need about a dozen of segments visible to both views.

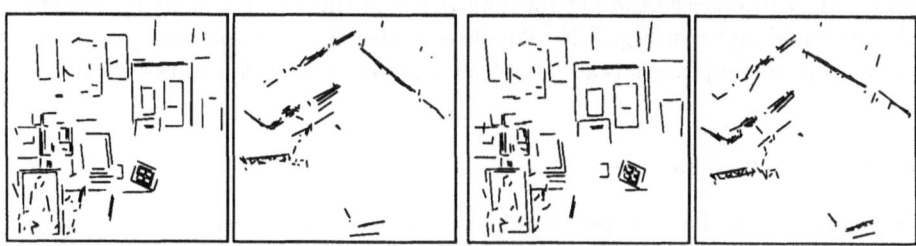

Fig. 7.13. Matched segments of Frame 1 using the new line segment representation (Version 3 of the matching technique)

Fig. 7.14. Matched segments of Frame 2 using the new line segment representation (Version 3 of the matching technique)

Fig. 7.15. Images of the first camera: the left one is at t_1 and the right one is at t_2

Figure 7.15 shows the images taken by the first camera of the stereo rig at two instants. As can be observed, there is only a small common part between the two successive frames. In fact, their is a big rotation between the two views. Comparing the two images, we see that the boxes on the table have a displacement of about 200 pixels in the image plane (resolution: 512×512).

Figures 7.16 and 7.17 show the pair of stereo frames. We have 79 segments in the first frame and 121 segments in the second. There exists a big rotation (about 16.5 degrees) between the two positions, which can be noticed by superimposing the two frames (see Fig. 7.18).

Applying the motion-estimation algorithm to these two frames, we obtain 5 hypotheses. All these hypotheses are evaluated. In the end, 4 hypotheses yield the correct estimate of the displacement. 37 matches are recovered. To determine how good this estimate is, as in the last example, we apply the computed estimate to the first frame and superimpose the transformed one on the second. Figure 7.19 shows such superposition of the matched segments and Figure 7.20 shows such superposition of the whole frames. The displacement estimate is very good. We can observe the big rotation between the two positions by the shift of the two triangles in Fig. 7.20. We observe also that the common part is very small. The whole process takes about 50 seconds on a SUN 3 workstation.

7.5.3 Rock Scenes

As pointed out earlier, our program is developed in the context of visual navigation of a mobile robot in an indoor scene. We expect to be able to describe most of the objects in such an environment by line segments. In this subsection, we describe an example to show that our algorithm also works in a cluttered scene

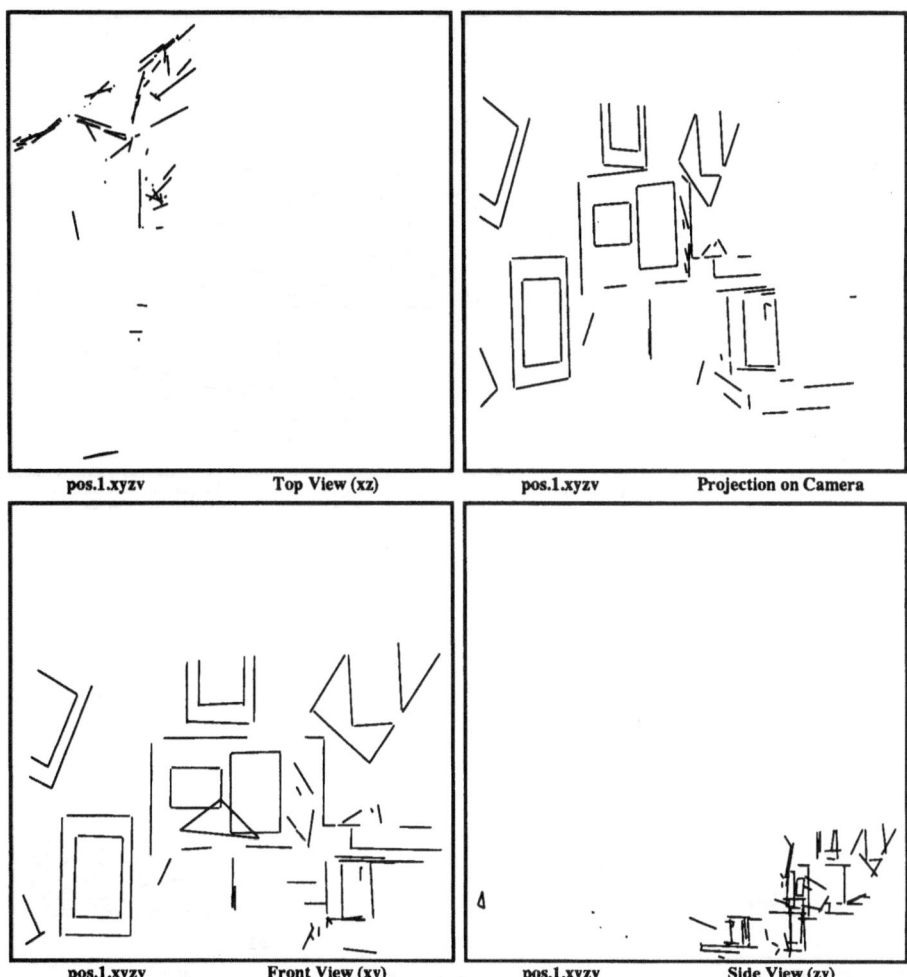

| pos.1.xyzv | Top View (xz) | pos.1.xyzv | Projection on Camera |

| pos.1.xyzv | Front View (xy) | pos.1.xyzv | Side View (zy) |

Fig. 7.16. Different views of stereo frame 1 (uniform scale)

containing rocks. Figure 7.21 shows two images of such a rock scene taken by the first camera of the stereo rig at two instants.

Two 3D frames are reconstructed by our trinocular stereo system which are displayed in Figs. 7.22 and 7.23. We have obtained 211 line segments in the first frame and 208 in the second. Two remarks can be made: the first is that the segments reconstructed are very noisy and even spurious (for example, some are under the ground, as can be observed in the front views); the second is that they are usually very short. As we can observe, there is a rather big difference between the two frames (about 11 degrees of rotation and 34 centimeters of translation).

In our program, a parameter l (for *level*), ranging from –5 to 5, is used to control the thresholds in the rigidity constraints. A change of one for l corresponds to a change of 0.2 for the thresholds. For example, the threshold for the dot-

pos.2.xyzv Top View (xz)	pos.2.xyzv Projection on Camera
pos.2.xyzv Front View (xy)	pos.2.xyzv Side View (zy)

Fig. 7.17. Different views of stereo frame 2 (uniform scale)

product constraint $\kappa_c = 1.32+0.2l$. The parameter l is set to zero in the program, which corresponds to the thresholds indicated in Sect. 6.4. When l decreases, the constraints are imposed more strictly during the hypothesis generation process. When l increases, the constraints are imposed less strictly. For indoor scenes, we do not need to change l, i.e., the thresholds are image-independent (they are certainly stereo-system-dependent). However, as the 3D frames reconstructed from the rock scene are very noisy, the motion program generates many false hypotheses although the final result is good. If we set l to –3, i.e., we impose more strict rigidity constraints, we obtain five hypotheses. All these hypotheses are propagated to the whole frame to match more segments and to update the motion estimate. In the end, 4 hypotheses give the correct estimate of the displacement.

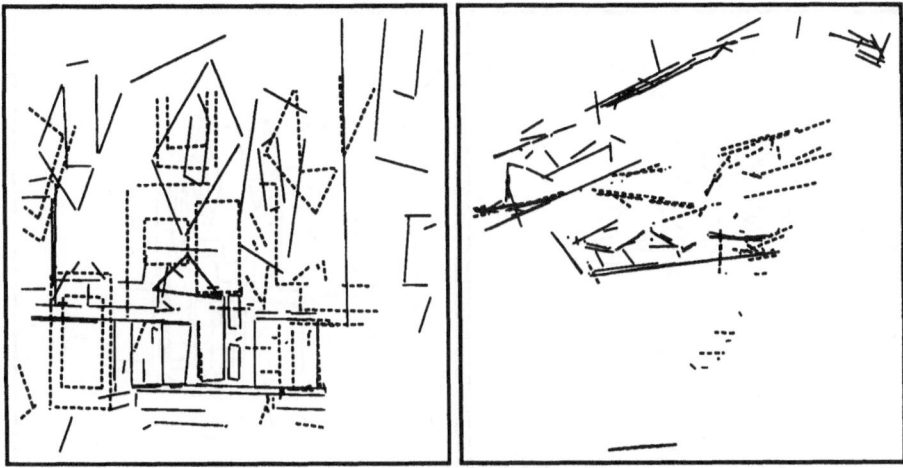

Fig. 7.18. Superposition of the two original frames: segments of Frame 1 are represented in dashed lines and those of Frame 2 in solid lines

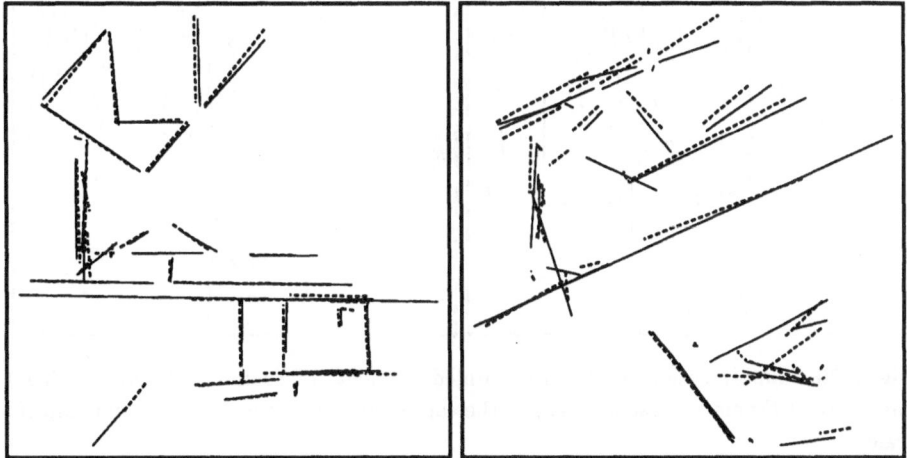

Fig. 7.19. Superposition of the matched segments after applying the computed motion to Frame 1: segments of Frame 1 are represented in dashed lines and those of Frame 2 in solid lines

Fig. 7.20. Superposition of the transformed segments of Frame 1 (in dashed lines) and those of Frame 2 (in solid lines) in the coordinate system of Frame 2 (nonuniform scale)

93 matches are recovered. To determine how good this estimate is, we apply the computed estimate to the first frame and superimpose the transformed one on the second, which is displayed in Fig. 7.24. We can find that the motion estimate is still very good even for such complicated scene. The displacement of the robot is shown by the shift of the two triangles in the top view of the superposition. The fifth hypothesis gives a suboptimal solution (21 matches are found by this hypothesis). The whole process takes about 50 seconds on a SUN 4 workstation: 40 seconds in generating the hypotheses and 10 seconds in verifying them.

Fig. 7.21. Two images of a rock scene taken by the first camera

7.6 Summary

We have described in detail an algorithm to deal with the analysis of two 3D views. The algorithm is based on the *hypothesize-and-verify* paradigm. The rigidity constraints described in the previous chapter are heavily used to generate hypotheses of matches between two stereo frames. Other constraints such as the constraint on the similarity in segment orientations are used to speed up further the generation process. The extended Kalman filter is used to compute an initial estimate of the motion for each hypothesis and is also used in the verification phase to incrementally update the estimate. Several techniques to match noisy segments have been described and uncertainty in measurements is again taken into account. The proposed algorithm is completely automatic. We have tested it in our laboratory for more than two years and it provides us with very satisfactory results. Three experimental results have been shown and excellent results demonstrated. As a final point, it should be noted that the algorithm has very strong potentialities for parallel implementation.

7.7 Appendix: Transforming a 3D Line Segment

We now describe how to transform a 3D line segment from one coordinate system to another. More precisely, the question we want to answer is:

Given a 3D line segment S with the parametrization (ψ, $\mathbf{m}$, l) and its covariance matrices (Λ_ψ, $\Lambda_\mathbf{m}$), what is the parametrization of the transformed segment $\widehat{S}$ after undergoing a motion with $\mathbf{s} = [\mathbf{r}^T, \mathbf{t}^T]^T$ and its covariance matrix $\Lambda_\mathbf{s}$?

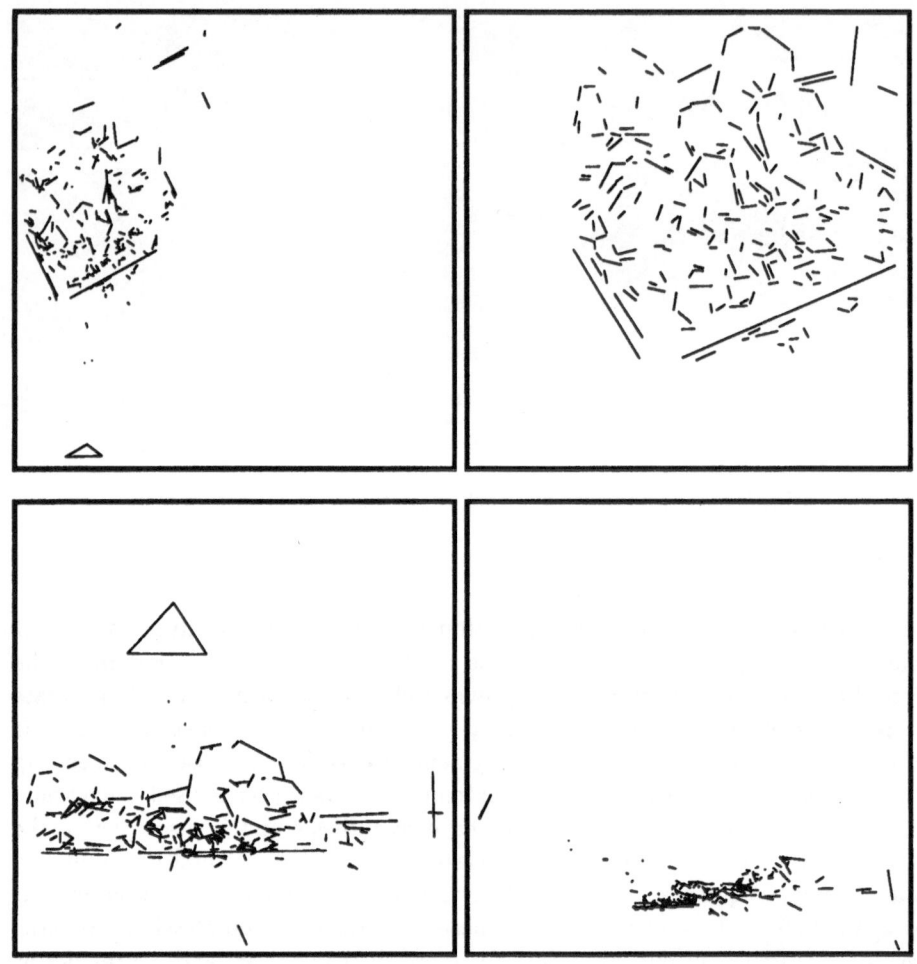

Fig. 7.22. Different views of the 3D frame reconstructed at t_1 (uniform scale)

Let the parametrization of the transformed segment $\widehat{S}$ be $(\widehat{\psi}, \widehat{m}, \widehat{l})$ and its covariance matrices $(\Lambda_{\widehat{\psi}}, \Lambda_{\widehat{m}})$. For simplicity, we define two nonlinear functions $\mathbf{g}$ and $\mathbf{h}$ to relate ψ and $\mathbf{u}$ (see (4.21) and (4.20)):

$$\psi = \mathbf{g}(\mathbf{u}) \quad \text{and} \quad \mathbf{u} = \mathbf{h}(\psi) . \tag{7.12}$$

Clearly, we have

$$\widehat{\psi} = \mathbf{g}(\mathbf{R}\mathbf{h}(\psi)) , \tag{7.13}$$
$$\widehat{m} = \mathbf{R}\mathbf{m} + \mathbf{t} , \tag{7.14}$$
$$\widehat{l} = l . \tag{7.15}$$

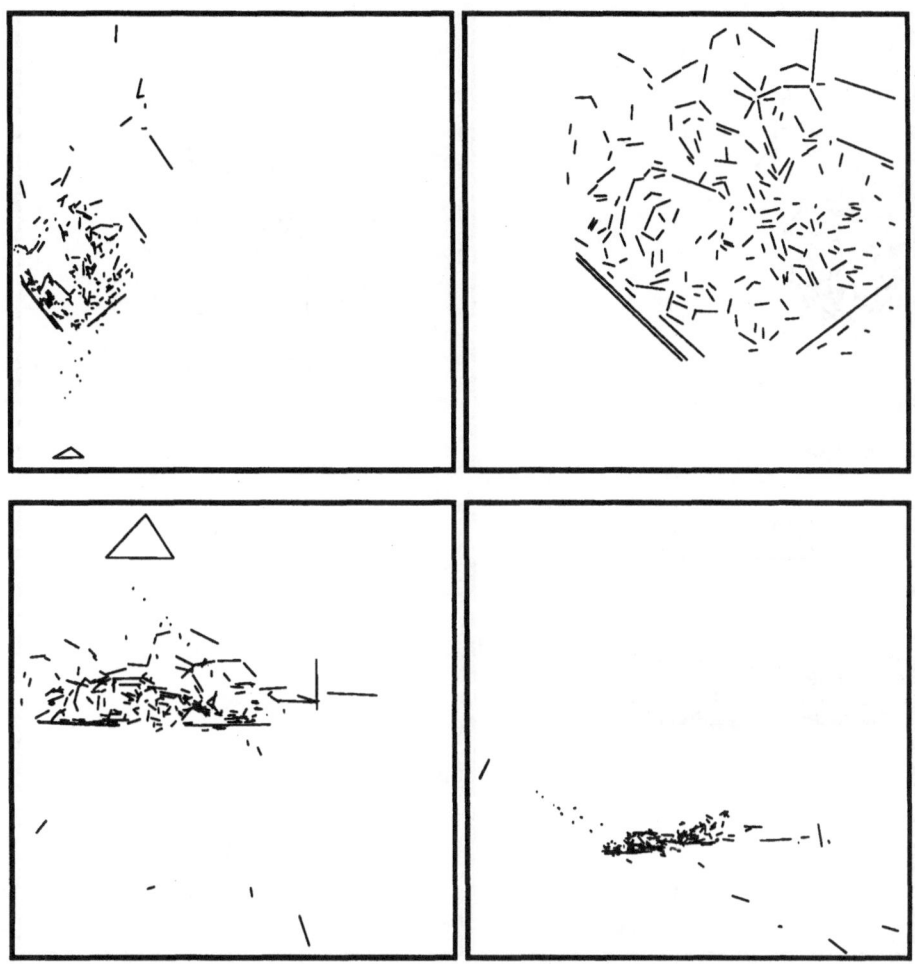

Fig. 7.23. Different views of the 3D frame reconstructed at t_2 (uniform scale)

Based on Theorem 2.3, we can compute the covariance matrix $\Lambda_{\widehat{\psi}}$, under the first order approximation, as

$$\Lambda_{\widehat{\psi}} = J_{\psi}^{\widehat{\psi}} \Lambda_{\psi} J_{\psi}^{\widehat{\psi}^T} + J_{s}^{\widehat{\psi}} \Lambda_{s} J_{s}^{\widehat{\psi}^T} \, , \tag{7.16}$$

where the Jacobian matrices are given by

$$J_{\psi}^{\widehat{\psi}} = \frac{\partial \mathbf{g}(\hat{\mathbf{u}})}{\partial \mathbf{u}} \mathbf{R} \frac{\partial \mathbf{h}(\boldsymbol{\psi})}{\partial \boldsymbol{\psi}} \, ,$$

$$J_{s}^{\widehat{\psi}} = \left[\frac{\partial \mathbf{g}(\hat{\mathbf{u}})}{\partial \mathbf{u}} \frac{\partial (\mathbf{R}\mathbf{u})}{\partial \mathbf{r}} \quad \mathbf{0}_3 \right] \, ,$$

where $\mathbf{u} = \mathbf{h}(\boldsymbol{\psi})$, $\hat{\mathbf{u}} = \mathbf{R}\mathbf{u}$ and $\mathbf{0}_3$ is the 3×3 null matrix.

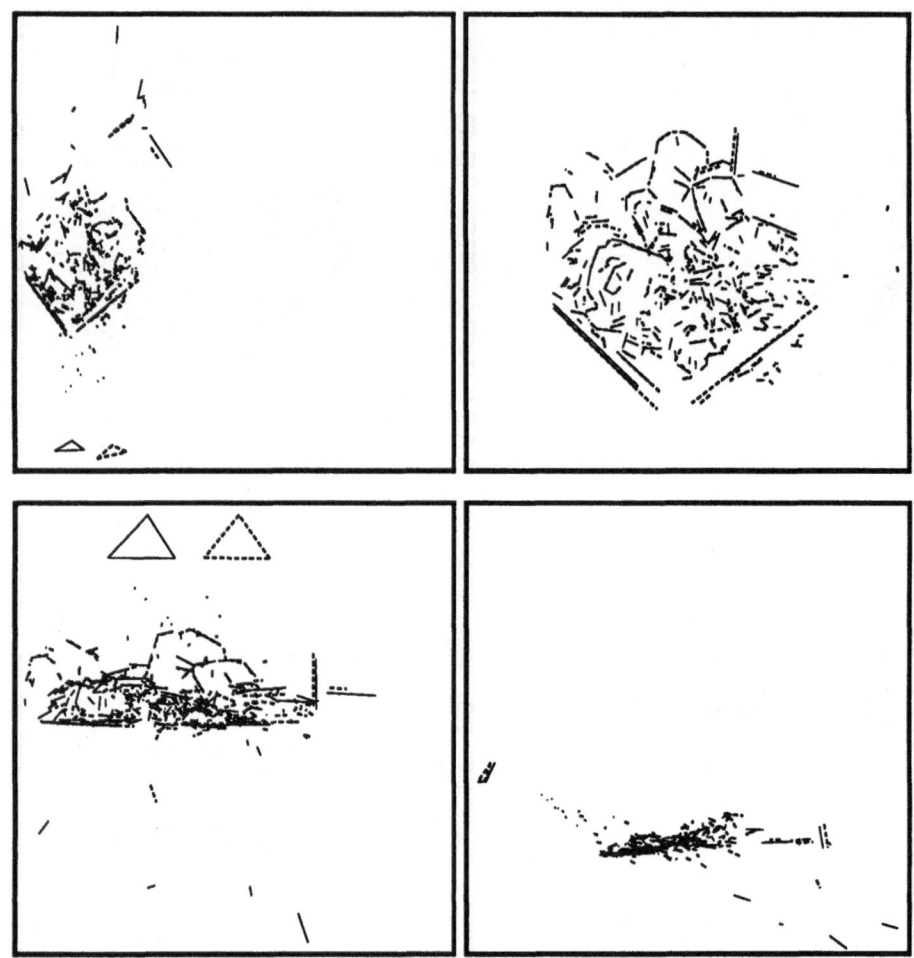

Fig. 7.24. Superposition of the transformed segments of Frame 1 (in dashed lines) and those of Frame 2 (in solid lines) (uniform scale)

Similarly, the covariance matrix $\Lambda_{\widehat{\mathbf{m}}}$ of the midpoint $\widehat{\mathbf{m}}$ is given by

$$\Lambda_{\widehat{\mathbf{m}}} = \mathbf{R}\Lambda_{\mathbf{m}}\mathbf{R}^T + J_{\mathbf{s}}^{\widehat{\mathbf{m}}}\Lambda_{\mathbf{s}}J_{\mathbf{s}}^{\widehat{\mathbf{m}}^T} , \qquad (7.17)$$

where the Jacobian matrix of $\widehat{\mathbf{m}}$ with respect to $\mathbf{s}$ is

$$J_{\mathbf{s}}^{\widehat{\mathbf{m}}} = \left[\frac{\partial(\mathbf{Rm})}{\partial\mathbf{r}} \quad \mathbf{I}_3 \right] . \qquad (7.18)$$

In the above, $\mathbf{u} = \mathbf{h}(\boldsymbol{\psi})$ and $\widehat{\mathbf{u}} = \mathbf{Ru}$; $\partial(\mathbf{Ru})/\partial\mathbf{r}$ and $\partial(\mathbf{Rm})/\partial\mathbf{r}$ are computed using (5.11); $\partial\mathbf{g}/\partial\widehat{\mathbf{u}}$ is computed using (4.24); $\partial\mathbf{h}/\partial\boldsymbol{\psi}$ is computed using (4.29).

8. Further Considerations on Reducing Complexity

W e have described in the last chapter a method based on the hypothesize-and-verify paradigm for registering two 3D frames obtained from stereo and computing their motion (transformation). Like almost all practical matching algorithms, we can use some heuristics to guide the search. Several have already been addressed in the last chapter, and we present in this chapter several more.

8.1 Sorting Data Features

A key consideration in all matching algorithms is how to restrict the search. Sorting data features is very important in this regard, because by sorting data features we can find correct correspondences with less search.

Data features can be sorted by saliency or significance such as length of line segments. This has two fundamental effects. First, since data features are sorted, we can find very efficiently a subset of features satisfying the length constraint by, for example, binary search. Thus we reduce significantly the search width. Second, salient features usually provide more information about the environment, in the sense that they are likely to be observed in successive views. Although a long segment may be broken into small ones in the edge detection and polygonal approximation processes, a part can at least be observed. Short segments are usually more noisy and may correspond to nothing in space. Thus by choosing the longest segments, we reduce the search involved with incorrect but temporarily consistent initial matches. The idea has already been used in the last chapter, and is similar to the Local-Feature-Focus (LFF) method [8.1–3] for object recognition. It selects a set of focus features (regions, edges, corners, etc.) based on a function of several factors. Some of these factors are the uniqueness of the feature, its expected contribution, the cost of detecting it, and the likelihood of detection. Each focus feature is used to initialize the search and predict a few nearby features. Once possible matches for nearby features have been found, the LFF method constructs a graph of possible data-model pairings using geometric constraints together with a graph matching algorithm. A maximal clique in the graph is finally sought as the best set of mutually consistent pairings of data and model features. The system HYPER [8.4] developed in our laboratory for the recognition and positioning of two-dimensional objects uses a similar idea. 2D line segments are used as primitives. The system associates with a model a set of focal segments, typically the 10 longest ones. It considers all possible pairings of a focal segment with a data segment to generate a pose hypothesis. Each generated pose hypothesis is propagated to the whole model in order to find more matches which support the pose. The hypothesis verification stage is carried out in a similar manner to that described in the last chapter.

The data features can be organized based on neighborhood relations, for example, by a bucketing technique. Suppose we have matched a segment in one frame to one segment in another. To extend this match, we can select segments in the neighborhood of the currently matched segments as candidates, which equivalently restricts the search width. Note that this approach exploits implicitly, although not exactly, the binary constraint on distance. In the case of multiple moving objects, the above strategy reflects the fact that features close to each other are more likely to have come from a single object than distant features.

8.2 "Good-Enough" Method

Due to the difference between two successive views, one cannot expect to assign each feature of one frame with a feature of another frame. A registration between two views is never perfect. Once a registration is found, we do not know whether there exists a better one, so we continue searching. Such exhaustive search method produces all possible registrations. By a simple comparison of those registrations, one can find that they can be easily classified into two categories: with high score and with low score. The *score* of a registration can be quantified by its number of matches. The difference between the high and low scores is usually very big. The low score is usually in the order of 5 percent of the total number of primitives in a frame. The high score is usually in the order of 30 percent or even higher depending on the difference between two views. A further remark is that the best registrations are almost the same. All these imply that exhaustive searching is not necessary. If we define a threshold to measure the goodness of a registration (the score described earlier is an example of the goodness measure), we can terminate the search once we obtain a registration whose goodness measure exceeds the predefined threshold. The registration is considered good enough. This method usually allows us to reduce considerably the computational expense, at the cost of possibly missing a better registration. This method has been used in a number of object-recognition systems like HYPER.

The key problem of this method is to choose the predefined threshold of the score. The threshold should be large enough so that there are no expected false registrations, but at the same time, it should be small enough such that a good registration is retained as early as possible and the search work is reduced as much as possible. As the number of matches depends tightly on the displacement between two views, ideally the threshold should be a function of displacement. However, displacements are not always known. In our implementation, the "good-enough" method is only used in real-time demonstrations where the threshold is set to 30 percent of the total number of primitives in a frame.

8.3 Speeding Up the Hypothesis Generation Process Through Grouping

If we can divide segments in each frame into several groups such that segments in each group are likely to have originated from a single object, we can considerably speed up the hypothesis generation process. This can be seen from the following. Assume we have divided the two consecutive frames into g_1 and g_2 groups. For simplicity, assume each group in a frame contains the same number of segments. Thus a group in the first frame, G_{1i}, has m/g_1 segments, and a group in the second frame, G_{2i}, has n/g_2 segments, where m and n are the total numbers of segments in the first and second frames. Possible matches of the segments in G_{1i} $(i = 1, \ldots, g_1)$ are restricted in one of the G_{2j}'s $(j = 1, \ldots, g_2)$. That is, it is not possible that one segment in G_{1i} is matched to one segment in G_{2j} and another segment in the same G_{1i} is matched to one segment in G_{2k} $(k \neq j)$. We call this condition the *completeness of grouping*. We show in Fig. 8.1 a noncomplete grouping in the second frame. The completeness of a grouping implies that we need only apply the hypothesis generation process (Algorithm 7.2) to each pairing of groups. As we have $g_1 \times g_2$ such pairings and that the complexity for each paring is $O((\frac{m}{g_1})^2(\frac{n}{g_2})^2)$, the total complexity is now $O(\frac{m^2 n^2}{g_1 g_2})$. Thus we have a speedup of $O(g_1 g_2)$.

Take a concrete example (see Fig. 8.2). We have 6 and 12 segments in Frame 1 and Frame 2, respectively. If we directly apply the hypothesis generation algorithm to these two frames, we need $6 \times 5 \times 12 \times 11/2 = 1980$ operations in the worst case. If the first frame is divided into 2 groups and the second into 3, we have 6 pairings of groups. Applying the hypothesis generation algorithm to each pairing requires $3 \times 2 \times 4 \times 3/2 = 36$ operations. The total number of operations is $6 \times 36 = 216$, and we have a speedup of 9. We should remember that the $O(g_1 g_2)$ speedup is achieved at the cost of a prior grouping process. Whether the speedup is significant depends upon whether the grouping process is efficient.

As each group must contain at least 2 segments, the maximal number of groups in Frame 1 is $m/2$ and that in Frame 2, $n/2$. This means that the maximal speedup which can be expected is $O(mn/4)$. In other words, with the assistance of grouping, the run time of the hypothesis generation process (after grouping) may be reduced from $O(m^2 n^2)$ to $O(mn)$. One possible grouping yielding this maximal speedup is to follow an (maybe arbitrary) order and consider each two consecutive segments as a group. However, such grouping does not satisfy the condition of the above analysis: completeness of grouping. The violation of

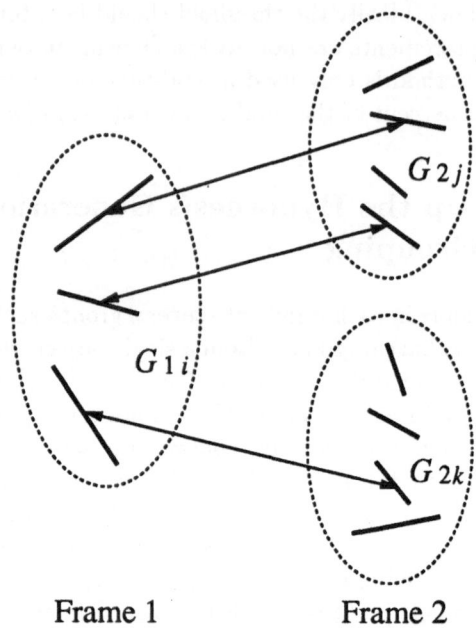

Fig. 8.1. A grouping which is not complete

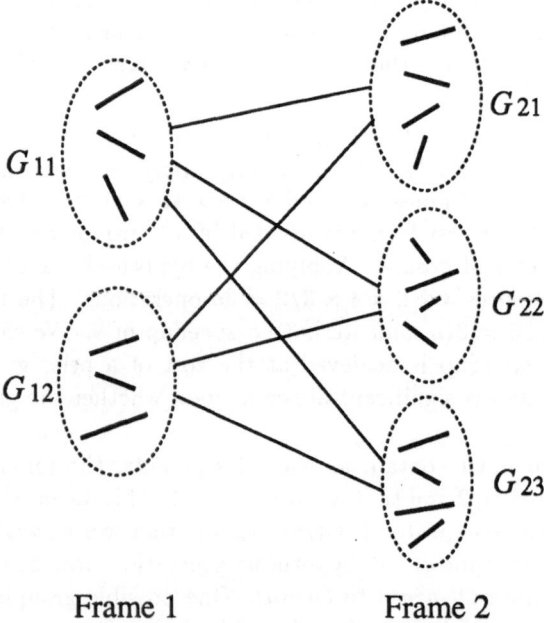

Fig. 8.2. Illustration of how grouping speeds up the hypothesis generation process

completeness may be tolerable if there is only one moving object in the scene like the problem addressed in this chapter (only the robot moves). This may result in a loss of important motion hypotheses when several moving objects exist in the scene like the problem addressed in Chap. 9.

One of the most influential grouping techniques is called the *perceptual grouping*, pioneered by *Low* [8.5]. He suggests organizing features (points or lines) by detecting properties such as symmetry, collinearity, curvilinearity, parallelism, connectivity, clustering and repetitive textures in an image. To demonstrate the importance of grouping, he implemented in his system SCERPO, which is concerned with the recognition of three-dimensional objects from a single two-dimensional image, some of the 2D perceptual organizations. The system detects, in particular, the occurrence of collinearity, proximity of endpoints, and parallelism of straight line edge segments. These are used together with information about object models to hypothesize a pose of the corresponding 3D object, which is then verified by searching for additional evidence. The use of perceptual grouping can effectively reduce the search.

In our algorithm, grouping is performed based on proximity and coplanarity of 3D line segments. Use of proximity allows us to roughly divide the scene into clusters, each constituting a geometrically compact entity. As we mainly deal with indoor environment, many polyhedral objects can be expected. Use of coplanarity allows us to further divide each cluster into semantically meaningful entities, namely planar facets.

8.4 Finding Clusters Based on Proximity

Two segments are said to be proximally connected if one segment is in the neighborhood of the other one. There are many possible definitions of a neighborhood. Our definition is: the neighborhood of a segment S is a cylindrical space C with radius r whose axis is coinciding with the segment S. This is shown in Fig. 8.3. The top and bottom of the cylinder are chosen such that the cylinder C contains completely the segment S. S intersects the two planes at A and B. The segment AB is called the *extended segment* of S. The distance from one endpoint of S to the top or bottom of the cylinder is b. Thus the volume V of the neighborhood of S is equal to $\pi r^2(l + 2b)$, where l is the length of S. We choose $b = r$. The volume of the neighborhood is then determined by r.

A segment S_i is said in the neighborhood of S if S_i intersects the cylindrical space C. From this definition, S_i intersects C if either of the following condition is satisfied:

1. At least one of the endpoints of S_i is in C.
2. The distance between the supporting lines of S and S_i is less than r and the common perpendicular intersects both S_i and the *extended* segment of S.

The first condition is obvious. The second is best explained in Fig. 8.4, where the left picture shows an example that two segments are proximally connected while the right one shows a contrary example.

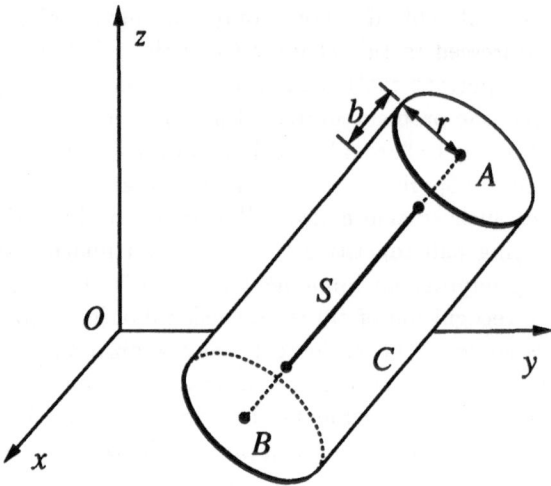

Fig. 8.3. Definition of a segment's neighborhood

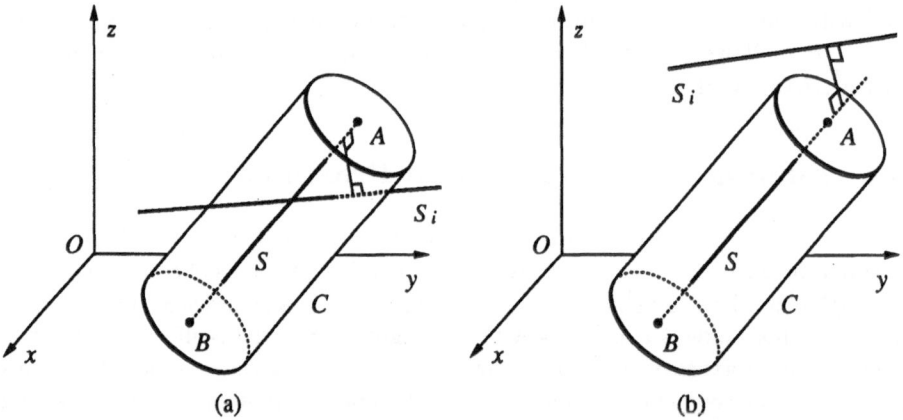

(a)   (b)

Fig. 8.4. Illustration of the connectivity of two segments

We define a *cluster* as a group of segments, every two of which are proximally connected in the sense defined above either directly or through one or more segments in the same group. A simple implementation to find clusters by testing the above conditions leads to a complexity of $O(n^2)$ in the worst case, where n is the number of segments in a frame. As the operation to examine whether a line segment intersects a cylinder is not a simple operation, the computation of finding clusters may be too expensive for the grouping to be useful to the motion-determination process. In the following, we present a method based on a *bucketing technique* to find clusters. First, the minima and maxima of the x,

y and z coordinates are computed, denoted by $x_{\min}$, $y_{\min}$, $z_{\min}$ and $x_{\max}$, $y_{\max}$, $z_{\max}$. Then the parallelepiped formed by the minima and maxima is partitioned into m^3 buckets W_{ijk} ($m = 16$ in our implementation). Fig. 8.5 illustrates the situation for $m = 4$. To each bucket W_{ijk} we attach the list of segments L_{ijk} intersecting it. The key idea of bucketing techniques is that on the average the number of segments intersecting a bucket is much smaller than the total number of segments in the frame. The computation of attaching segments to buckets can be performed very fast by an algorithm whose complexity is linear in the number of segments. Finally, a recursive search is performed to find clusters.

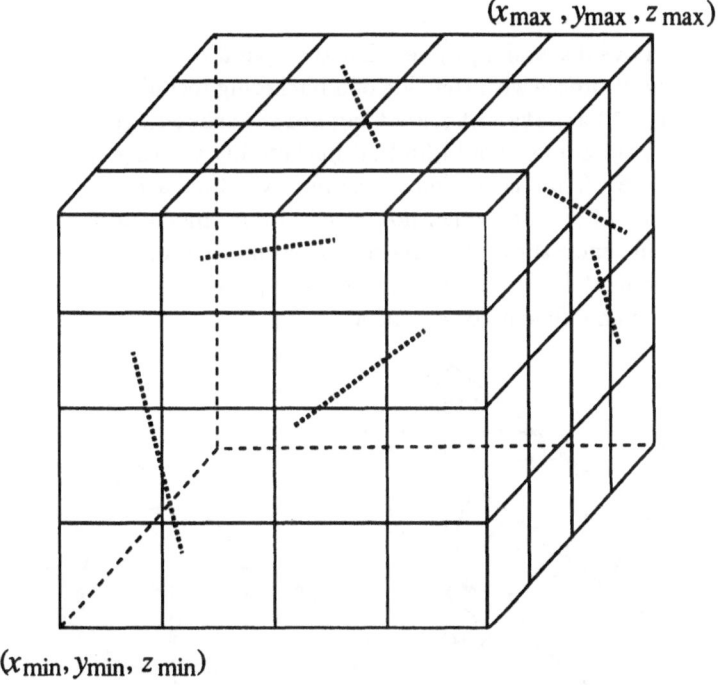

$(x_{\max}, y_{\max}, z_{\max})$

$(x_{\min}, y_{\min}, z_{\min})$

Fig. 8.5. Illustration of a bucketing technique

We can write an algorithm to find a cluster containing segment S in pseudo C codes as:

```
List find_cluster(S)
Segment S ;
{
    List cluster = NULL ;
    if is_visited(S) return NULL ;
    mark_segment_visited(S) ;
    list_buckets = find_buckets_intersecting_neighborhood_of(S) ;
    list_segments = union_of_all_segments_in(list_buckets) ;
```

```
for (S_i = each_segment_in(list_segments))
    cluster = cluster ∪ {S_i} ∪ find_cluster(S_i) ;
return cluster ;
}
```

where List is a structure storing a list of segments, defined as

```
struct cell {
    Segment seg ;
    struct cell *next ;
} *List ;
```

From the above discussion, we see that the operations required to find a cluster are really very simple with the aide of a bucketing technique, except probably the function find_buckets_intersecting_neighborhood_of(S). This function, as indicated by its name, should find all buckets intersecting the neighborhood of the segment S. That is, we must examine whether a bucket intersects the cylindrical space C or not, which however is by no means a simple operation. The gain in efficiency through using a bucketing technique may become non-significant. Fortunately, we have a very good approximation as described below which allows for an efficient computation.

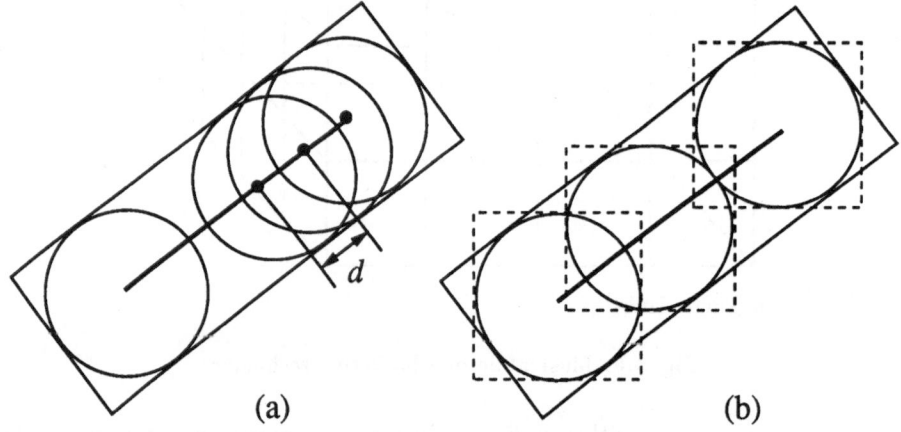

(a)   (b)

Fig. 8.6. Approximation of a neighborhood of a segment

Let us fill the cylindrical space C with many spheres, each just fitting into the cylinder, i.e., whose radius is equal to r. We allow intersection between spheres. Fig. 8.6a illustrates the situation using only a section passing through the segment S. The union of those spheres gives an approximation of C. When the distance d between successive sphere centers approaches to zero (i.e., $d \to 0$), the approximation is almost perfect, except in the top and bottom of C. The error of the approximation in this case is $\frac{1}{3}\pi r^3$. This part is not very important because it is the farthest to the segment. Although the operation to examine the intersectionness between a bucket and a sphere is simpler than between a bucket and a cylinder, it is not beneficial if we use too many spheres. What we do is further, as illustrated in Fig. 8.6b. Spheres are not allowed to intersect with each other (i.e., $d = 2r$) with the exception of the last sphere. The center of the last sphere is always at the endpoint of S, so it may intersects with the previous sphere. The number of spheres required is equal to $\lceil \frac{l}{2r} + 1 \rceil$, where $\lceil a \rceil$ denotes the smallest integer greater than or equal to a. It is obvious that the union of these spheres is always smaller than the cylindrical space C. Now we replace each sphere by a cube circumscribing it and aligned with the coordinate axes (represented by a dashed rectangle in Fig. 8.6b). Each cube has a side length of $2r$. It is now almost trivial to find which buckets intersect a cube. Let the center of the cube be $[x, y, z]^T$. Let

$$i_{min} = \max[0, \lfloor (x - x_{min} - r)/dx \rfloor], \quad i_{max} = \min[m - 1, \lceil (x - x_{min} + r)/dx - 1 \rceil],$$
$$j_{min} = \max[0, \lfloor (y - y_{min} - r)/dy \rfloor], \quad j_{max} = \min[m - 1, \lceil (y - y_{min} + r)/dy - 1 \rceil],$$
$$k_{min} = \max[0, \lfloor (z - z_{min} - r)/dz \rfloor], \quad k_{max} = \min[m - 1, \lceil (z - z_{min} + r)/dz - 1 \rceil],$$

where $\lfloor a \rfloor$ denotes the greatest integer less than or equal to a, m is the dimension of the buckets, $dx = (x_{max} - x_{min})/m$, $dy = (y_{max} - y_{min})/m$, and $dz = (z_{max} - z_{min})/m$. The buckets intersecting the cube is simply $\{W_{ijk} \mid i = i_{min}, \ldots, i_{max}, j = j_{min}, \ldots, j_{max}, k = k_{min}, \ldots, k_{max}\}$.

If a segment is parallel to either x, y or z axis, it can be shown that the union of the cubes is bigger than the cylindrical space C. However, if a segment is near 45°(or 135°) to an axis (as shown in Fig. 8.6b), there are some gaps in approximating C. These gaps are usually filled by buckets, because a whole bucket intersecting one cube is now considered as part of the neighborhood of S.

8.5 Finding Planes

Several methods have been proposed in the literature to find planes. One common method is to directly use data from range finders [8.6, 7]. *Faugeras* and *Lustman* [8.8, 9] propose a method to find planes from two monocular views of 3D segments. The expert system developed by *Thonnat* [8.10] for scene interpretation is capable of finding planes from 3D line segments obtained from stereo. The system being developed by *Grossmann* [8.11, 12] aims at extracting, also from 3D line segments, visible surfaces including planar, cylindrical, conical and spherical ones. In the latter system, each coplanar crossing pair of 3D line segments (i.e., they are neither collinear nor parallel) forms a candidate plane. Each candidate plane is tested for compatibility with the already existing planes. If compatibility is established, the existing plane is updated by the candidate plane. In this section, we present a new method to find planes from 3D line segments.

As we do not know how many and where are the planes, we first try to find two coplanar line segments which can define a plane, and then try to find more line segments which lie in this hypothetical plane until all segments are processed. The segments in this plane are marked visited. For those unvisited segments, we repeat the above process to find new planes.

Let a segment be represented by its midpoint $\mathbf{m}$, its unit direction vector $\mathbf{u}$ and its length l. Because segments reconstructed from stereo are always corrupted by noise, we attach to each segment an uncertainty measure (covariance matrix) of $\mathbf{m}$ and $\mathbf{u}$, denoted by $\Lambda_{\mathbf{m}}$ and $\Lambda_{\mathbf{u}}$. The uncertainty measure of l is not required. For two noncollinear segments: $(\mathbf{m}_1, \mathbf{u}_1)$ with $(\Lambda_{\mathbf{m}_1}, \Lambda_{\mathbf{u}_1})$ and $(\mathbf{m}_2, \mathbf{u}_2)$ with $(\Lambda_{\mathbf{m}_2}, \Lambda_{\mathbf{u}_2})$, the coplanarity condition is

$$c \overset{\triangle}{=} (\mathbf{m}_2 - \mathbf{m}_1) \cdot (\mathbf{u}_1 \wedge \mathbf{u}_2) = 0 , \tag{8.1}$$

where $\cdot$ and $\wedge$ denote the dot product and the cross product of two vectors, respectively. In reality, the condition (8.1) is unlikely met. Instead, we impose that $|c|$ is less than some threshold. We determine the threshold in a dynamic manner by relating it to the uncertainty measures. The variance of c, denoted by Λ_c, is computed from the covariance matrices of the two segments by

$$\begin{aligned}\Lambda_c = &(\mathbf{u}_1 \wedge \mathbf{u}_2)^T (\Lambda_{\mathbf{m}_1} + \Lambda_{\mathbf{m}_2})(\mathbf{u}_1 \wedge \mathbf{u}_2) + [\mathbf{u}_1 \wedge (\mathbf{m}_2 - \mathbf{m}_1)]^T \Lambda_{\mathbf{u}_2}[\mathbf{u}_1 \wedge (\mathbf{m}_2 - \mathbf{m}_1)] \\ &+ [\mathbf{u}_2 \wedge (\mathbf{m}_2 - \mathbf{m}_1)]^T \Lambda_{\mathbf{u}_1}[\mathbf{u}_2 \wedge (\mathbf{m}_2 - \mathbf{m}_1)] .\end{aligned}$$

Here we assume there is no correlation between the two segments. Since c^2/Λ_c follows a χ^2 distribution with one degree of freedom, two segments are said coplanar if

$$c^2/\Lambda_c \leq \kappa , \tag{8.2}$$

where κ can be chosen by looking up the χ^2 table such that $\Pr(\chi^2 \leq \kappa) = \alpha$. In our implementation, we set $\alpha = 50\%$, or $\kappa = 0.5$.

As discussed in the last paragraph, the two segments used must not be collinear. Two segments are collinear if and only if the following two conditions are satisfied:

$$\begin{aligned}\mathbf{u}_1 \pm \mathbf{u}_2 &= 0 , \\ \mathbf{u}_1 \wedge (\mathbf{m}_2 - \mathbf{m}_1) &= 0 .\end{aligned} \tag{8.3}$$

The first says that two collinear segments should have the same direction (Remark: segments are oriented in our stereo system). The second says that the midpoint of the second segment lies on the first segment. In reality, of course, these conditions are rarely satisfied. A treatment similar to the coplanarity should be performed.

Once two segments are identified to lie in a single plane, we estimate the parameters of the plane. A plane is described by

$$ux + vy + wz + d = 0 , \tag{8.4}$$

where $\mathbf{n} = [u, v, w]^T$ is parallel to the normal of the plane, and $|d|/\|\mathbf{n}\|$ is the distance of the origin to the plane. It is clear that for an arbitrary scalar $\lambda \neq 0$,

$\lambda[u, v, w, d]^T$ describes the same plane as $[u, v, w, d]^T$. Thus the minimal representation of a plane has only three parameters. One possible minimal representation is to set $w = 1$, which gives

$$ux + vy + z + d = 0 \ .$$

However, it cannot represent planes parallel to the z-axis. To represent all planes in 3D space, we should use three maps [8.13]:

Map 1: $ux + vy + z + d = 0$ for planes nonparallel to the z-axis,   (8.5)
Map 2: $x + vy + wz + d = 0$ for planes nonparallel to the x-axis,   (8.6)
Map 3: $ux + y + wz + d = 0$ for planes nonparallel to the y-axis,   (8.7)

In order to determine which map to be used, we first compute an initial estimate of the plane normal $n_0 = u_1 \wedge u_2$. If the two segments are parallel, $n_0 = u_1 \wedge (m_2 - m_1)$. If the z component of n has a maximal absolute value, Map 1 (8.5) will be used; if the x component has a maximal absolute value, Map 2 (8.6) will be used; otherwise, Map 3 (8.7) will be used. An initial estimate of d can then be computed using the midpoint of a segment. In the sequel, we use Map 1 for explanation. The derivations are easily extended to the other maps.

We use an extended Kalman filter (see Sect. 2.2) to estimate the plane parameters. Let the state vector be $s = [u, v, d]^T$. We have an initial estimate s_0 available, as described just above. Since this estimate is not good, we set the diagonal elements of the initial covariance matrix Λ_{s_0} to a very big number and the off-diagonal elements to zero. A 3D segment with parameters (u, m) is identified as lying in the plane. Define the measurement vector as $x = [u^T, m^T]^T$. We have two equations relating s to x:

$$f(x, s) = \left\{ \begin{array}{l} n^T u \\ n^T m + d \end{array} \right. = 0 \ , \tag{8.8}$$

where $n = [u, v, 1]^T$. The first equation says that the segment is perpendicular to the plane normal, and the second says that the midpoint of the segment is in the plane. In order to apply the Kalman filter, we must linearize the above equation (see Sect. 2.2.2), which gives

$$y = Ms + \xi \ , \tag{8.9}$$

where y is the new measurement vector, M is the observation matrix, and ξ is the noise disturbance in y, and they are given by

$$M = \frac{\partial f(x, s)}{\partial s} = \left[\begin{array}{ccc} u_x & u_y & 0 \\ m_x & m_y & 1 \end{array} \right] \ , \tag{8.10}$$

$$y = -f(x, s) + Ms = \left[\begin{array}{c} -u_x \\ -m_z \end{array} \right] \ , \tag{8.11}$$

$$\xi = \frac{\partial f(x, s)}{\partial x} \delta_x = \left[\begin{array}{cc} n^T & 0^T \\ 0^T & n^T \end{array} \right] \delta_x \ , \tag{8.12}$$

where $\mathbf{0}$ is the 3D zero vector and $\boldsymbol{\delta_x}$ is the noise disturbance in $\mathbf{x}$. Now that we have two segments which have been identified to be coplanar and an initial estimate of the plane parameters, we can apply the extended Kalman filter based on the above formulation to obtain a better estimate of $\mathbf{s}$ and its error covariance matrix $\Lambda_{\mathbf{s}}$.

Once we have estimated the parameters of the plane, we try to find more evidences of the plane, i.e., more segments in the same plane. If a 3D segment $\mathbf{x} = [\mathbf{u}^T, \mathbf{m}^T]^T$ with $(\Lambda_{\mathbf{u}}, \Lambda_{\mathbf{m}})$ lies in the plane, it must satisfy (8.8). Since data are noisy, we do not expect to find a segment having exactly $\mathbf{p} \stackrel{\triangle}{=} \mathbf{f}(\mathbf{x}, \mathbf{s}) = \mathbf{0}$. Instead, we compute the covariance matrix $\Lambda_{\mathbf{p}}$ of $\mathbf{p}$ as follows:

$$\Lambda_{\mathbf{p}} = \frac{\partial \mathbf{f}(\mathbf{x}, \mathbf{s})}{\partial \mathbf{s}} \Lambda_{\mathbf{s}} \frac{\partial \mathbf{f}(\mathbf{x}, \mathbf{s})}{\partial \mathbf{s}}^T + \frac{\partial \mathbf{f}(\mathbf{x}, \mathbf{s})}{\partial \mathbf{x}} \Lambda_{\mathbf{x}} \frac{\partial \mathbf{f}(\mathbf{x}, \mathbf{s})}{\partial \mathbf{x}}^T \quad , \tag{8.13}$$

where $\frac{\partial \mathbf{f}(\mathbf{x}, \mathbf{s})}{\partial \mathbf{s}}$ and $\frac{\partial \mathbf{f}(\mathbf{x}, \mathbf{s})}{\partial \mathbf{x}}$ are computed by (8.10) and (8.12), respectively. If

$$\mathbf{p}^T \Lambda_{\mathbf{p}}^{-1} \mathbf{p} \leq \kappa_p \quad , \tag{8.14}$$

then the segment is considered as lying in the plane. Since $\mathbf{p}^T \Lambda_{\mathbf{p}}^{-1} \mathbf{p}$ follows a χ^2 distribution with 2 degrees of freedom, we can choose an appropriate κ_p by looking up the χ^2 table such that $\Pr(\chi^2 \leq \kappa_p) = \alpha_p$. We choose $\alpha_p = 50\%$, or $\kappa_p = 1.4$. Each time we find a new segment in the plane, we update the plane parameters $\mathbf{s}$ and $\Lambda_{\mathbf{s}}$ and try to find still more. Finally, we obtain a set of segments supporting the plane and also an estimate of the plane parameters accounting for all these segments.

8.6 Experimental Results

8.6.1 Grouping Results

In this section we show the grouping results using an indoor scene. A stereo rig takes three images, one of which is displayed in Fig. 8.7. After performing edge detection, edge linking and linear segment approximation, the three images are supplied to a trinocular stereo system, which reconstructs a 3D frame consisting of 137 3D line segments. Figure 8.8 shows the front view (projection on the plane in front of the stereo system and perpendicular to the ground plane) and the top view (projection on the ground plane) of the reconstructed 3D frame.

We then apply the bucketing technique to this 3D frame to sort segments into buckets, which takes about 0.02 seconds of user time on a Sun 4/60 workstation. The algorithm described in Sect. 8.4 is then applied, which takes again 0.02 seconds of user time to find two clusters. They are respectively shown in Figs. 8.9 and 8.10. Comparing these with Fig. 8.8, we observe that the two clusters do correspond to two geometrically distinct entities.

Finally we apply the algorithm described in Sect. 8.5 to each cluster, and it takes 0.35 seconds of user time to find in total 11 planes. The four largest

Fig. 8.7. Image taken by the first camera

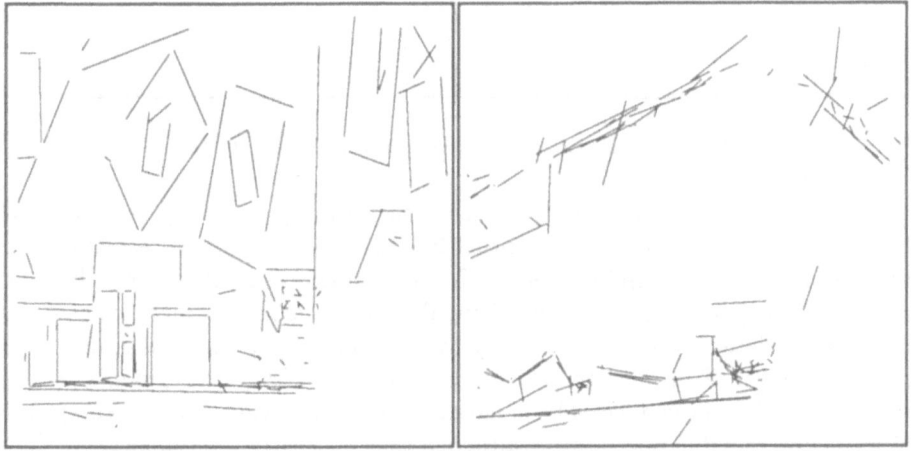

Fig. 8.8. Front and top views of the reconstructed 3D frame

Fig. 8.9. Front and top views of the first cluster

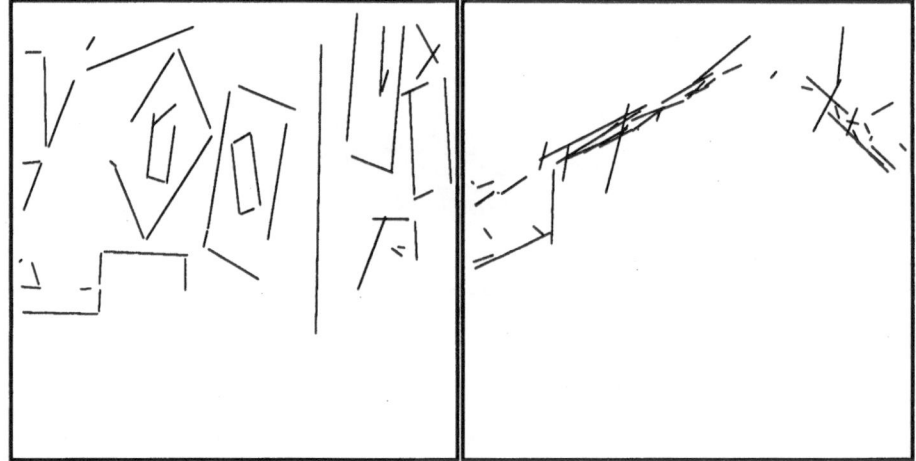

Fig. 8.10. Front and top views of the second cluster

planes contain 17, 10, 25 and 13 segments, respectively, and they are shown in Figs. 8.11 to 8.14. Other planes contain less than 7 segments, corresponding to the box faces, the table and the terminal. From these results, we observe that our algorithm can reliably detect planes from 3D line segments obtained from stereo, but a plane detected does not necessarily correspond to a physical plane. The planes shown in Figs. 8.12 to 8.14 correspond respectively to segments on the table, the wall and the door. The plane shown in Fig. 8.11, however, is composed of segments from different objects, although they do satisfy the coplanarity. This is because in our current implementation any segment in a *cluster* satisfying the coplanarity is retained as a support of the plane. One possible solution to this

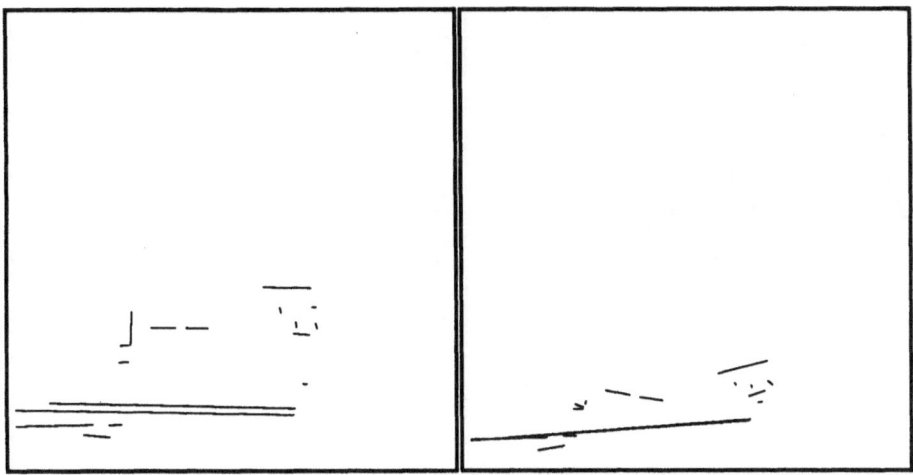

Fig. 8.11. Front and top views of the first plane

Fig. 8.12. Front and top views of the second plane

problem is to grow the plane by look for segments in the *neighborhood* of the segments already retained as supports of the plane.

8.6.2 Motion Results

Now we show the overall performance of the motion-determination algorithm. The 3D frame just described is reconstructed in the first position. After the stereo rig makes a displacement, we take another image triplet. The image taken by the first camera is shown in Fig. 8.15, which should be compared with Fig. 8.7. A large displacement can be observed. The shift in the image plane is about 190 pixels (the image resolution is 512×512).

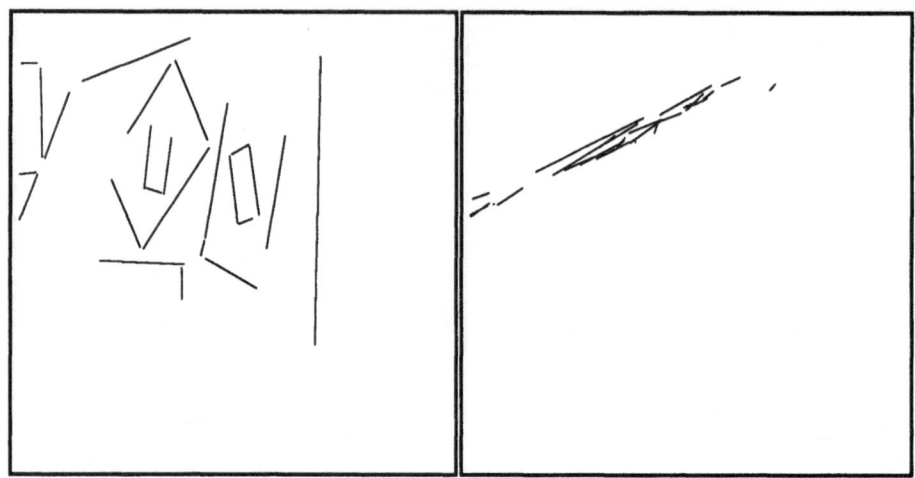

Fig. 8.13. Front and top views of the third plane

Fig. 8.14. Front and top views of the fourth plane

The trinocular stereo reconstructs also 137 3D segments (by chance, the same number as in the first frame). Applying the grouping technique to this 3D frame, We find two clusters and nine planes. Note that there is a large displacement between the two positions, which can be noticed by superimposing the two original frames (see Fig. 8.16). Applying the hypotheses generation process to these two frames, we obtain nine hypotheses. All these hypotheses are evaluated, and the hypothesis yielding maximal matches and minimal matching errors is chosen as the final displacement estimate. 48 matches are found. The final motion estimate is: 15.5 degrees in rotation and 32 centimeters in translation. The whole

Fig. 8.15. Image taken by the first camera in the second position

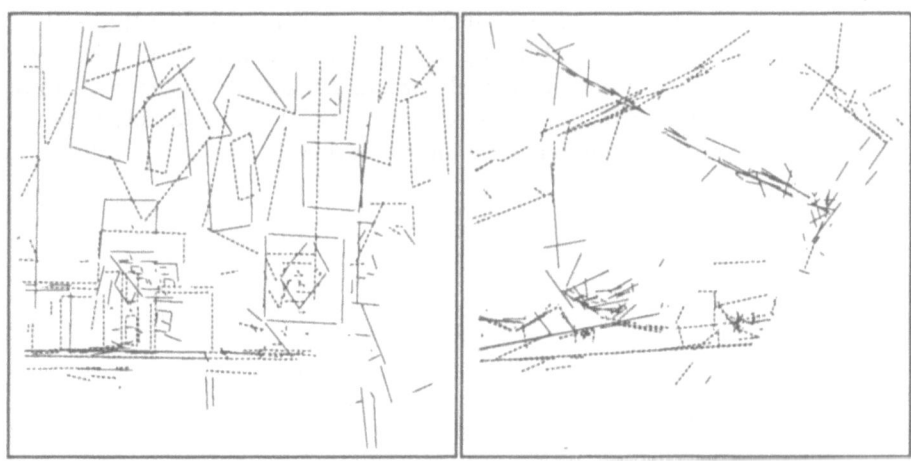

Fig. 8.16. Front and top views of the superposition of the two original 3D frames

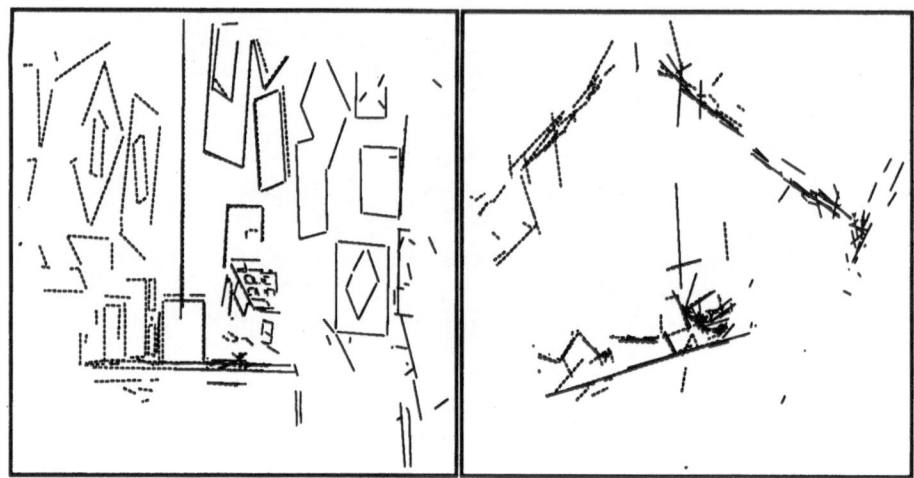

Fig. 8.17. Front and top views of the superposition of the two 3D frames after applying the estimated motion to the first frame

process takes about 11 seconds. To see whether this estimate is good or not, we apply it to the first frame and superimpose the transformed frame on the second frame, which is shown in Fig. 8.17.

8.7 Conclusion

In this chapter, we have presented several methods to reduce the complexity of the motion-determination algorithm described in the last chapter. These methods include sorting data features, "good-enough" method and grouping. In particular, we have described in great detail how we can speed up the motion-determination algorithm through grouping. A formal analysis has been done. Two criteria have been proposed, namely proximity to find clusters which are geometrically compact and coplanarity to find planes. We should note that the two procedures are also useful for scene interpretation.

9. Multiple Object Motions

U p to now, we have been interested in the motion analysis of two frames in which there is only one moving object. In other words, we can solve the following two problems:

1. The observer navigates in a static environment where there are no other moving objects.
2. The observer is static and only one moving object is seen.

A preprocessing is required to solve the second problem: The two observed frames should be subtracted from each other in order to eliminate data from the background.

The above context is, of course, very restrictive. Usually, in a static environment (indoor scene, for example), the observer moves and at the same time there exist other moving objects, or the observer is fixed but there exist more than one moving object (surveillance applications, for example). We call this the **multiple-object-motions** problem.

9.1 Multiple Object Motions

From a single view, it is difficult to segment a scene into objects. However, using the temporal information contained in an image sequence, the task becomes much easier. We are interested here in segmenting scenes into objects that move differently in space. That is, objects undergoing the same motion are considered as a single entity. All static objects are considered as a single one. If the separation of such objects is necessary, other knowledge such as the proximity criterion should be used.

Two approaches are proposed in the literature. The first is based on the computation of the *motion field*, which describes the two-dimensional projection of the three-dimensional motion of scene points relative to the camera [9.1]. However, what we compute from the image differential measurements is the local apparent motion, known as the *optical flow*. It does not necessarily correspond to the 2D motion field, as pointed out by *Verri* and *Poggio* [9.2]. Furthermore, there is a problem known as the *aperture problem*: only the component of the optical flow perpendicular to the local image edge can be recovered from local dif-

ferential information [9.3]. Many researchers have recently followed a qualitative
approach. For example, *Nelson* and *Aloimonos* [9.4] use certain measures of flow
field divergence as a qualitative motion cur for obstacle avoidance. *Nagel* and
his colleagues [9.5, 6] propose an approach to automatically characterize vehicle
trajectories from image sequences in terms of motion verbs in natural languages.

Another approach is feature-based, which consists in identifying corresponding features for each object through, for example, maximal matching. *Chen* and
Huang propose in [9.7] an algorithm for matching 3D points of multiple rigid
objects. The algorithm starts a matching with initiating a pairing of a triplet
of noncollinear points in one frame with a triplet of points in another. It then
grows the matching by searching for more pairings of points, one at a time, using
distance and angular constraints. Each of nonoverlapping matchings corresponds
to a rigid object. They also propose in [9.8] a two-stage algorithm for matching two sets of 3D line segments corresponding to a scene containing multiple
moving objects. In the first stage, a tree-search procedure based on the angular
constraints on pairwise line segments is used to establish potential matches. For
each hypothetical matching a rotation can be computed. In the second stage, a
Hough clustering technique based on the position of line segments is applied to
verify potential matches. For each verified matching, a translation estimate is
also available. Different object motions are then detected in the clustering.

The algorithm developed in Chap. 7 is directly applicable to analyze the
multiple-object-motions problem. In the hypothesis generation phase, the algorithm tries to find all pairs of pairings of segments between two frames which
satisfy the rigidity constraints described in Chap. 6. Every two pairings constitute a hypothesis of the potential motion between two frames. If two segments
in a hypothesis are from a single object, the hypothesis will give the motion
of that object. If an object has more than two segments and two of them are
precisely reconstructed by the stereo system, then at least one hypothesis among
all generated by the algorithm belongs to the object. "Precisely" is, of course,
related to the thresholds in the rigidity constraints.

Algorithm 9.1 explains how to extend the algorithm developed in Chap. 7
to solve the multiple-object-motions problem. We first apply Algorithm 7.2 to
the two observed frames. In order not to miss any potential motions between
them, one modification should be made. Remember that in order to speed up
the generation process, we choose only the m/q ($q = 2$ or 3) longest segments
in the first frame. It is very likely that for a small object, none of its segments
is among the m/q longest, thus the motion of such object may not be detected.
In order to avoid this, we use all segments in the first frame, i.e., $q = 1$. Once
the hypotheses are generated, we apply Algorithm 7.3 to each of them to update
the motion estimate and to find segment correspondences. We then use the
criterion described in (7.1) to sort those hypotheses. The best one is chosen as
the motion of Object 1 and the corresponding segments in the two frames are
labeled as belonging to that object. For each of the other hypotheses, once it is
not compatible with any previous one, we retain it as representing a new motion.

Algorithm 9.1: Multiple-Object-Motions Algorithm

- Generate all hypotheses of motion between two frames
- Evaluate all generated hypotheses
- Sort the hypotheses in the order of decreasing score:
  $\{\mathcal{H}_1, \cdots, \mathcal{H}_n\}$
- Choose the first one as the motion of Object 1
- **for** i from 2 to n
  $\rightarrow$ **if** $\mathcal{H}_i$ is not compatible with any one of the recovered objects
     then retain it as a new object
  $\leftarrow$ **endif**
$\Uparrow$ **endfor**

In the end, all motions are recovered and the scene is segmented into objects.

Two methods can be used to test the compatibility of two hypotheses. The first one is to compute the similarity of the corresponding motion estimates. If the first hypothesis yields a motion estimate s_1 with its covariance matrix Λ_1, and the second one yields a motion estimate s_2 with its covariance matrix Λ_2, then the Mahalanobis distance between them can be computed as

$$d_s = (s_1 - s_2)^T (\Lambda_1 + \Lambda_2)^{-1} (s_1 - s_2) \ . \tag{9.1}$$

The distance d_s follows a χ^2 distribution with 6 degrees of freedom. We can choose $\kappa_s = 12.6$ for a probability of 95%. If the distance between two hypotheses $d_s \leq \kappa_s$, they are considered compatible. Another approach is to look at the segment correspondences recovered by the hypotheses. For example, if half of the correspondences recovered by a hypothesis are among those recovered by another hypothesis, then the two hypotheses are considered to be compatible.

9.2   Influence of Egomotion on Observed Object Motion

If there are some moving objects in the environment and the robot also moves, we can use Algorithm 9.1 to find multiple possible motions: egomotion (motion of the robot) and motions corresponding to objects.

Suppose now there is only one moving object besides the robot (the following results can be easily extended to the case of multiple objects). Using Algorithm 9.1, we can recover two different motions: $\mathbf{R}_E, \mathbf{t}_E$, the inverse of the robot motion and $\mathbf{R}'_O, \mathbf{t}'_O$, the object motion as observed by the robot (in the coordinate system of the second frame). However, $\mathbf{R}'_O, \mathbf{t}'_O$ is not the motion of the object with respect to the world coordinate system. Indeed, it is influenced by the egomotion and we have following relations:

$$\begin{aligned}
\mathbf{R}_O &= \mathbf{R}'_O \mathbf{R}_E^T , \\
\mathbf{t}_O &= \mathbf{t}'_O - \mathbf{R}_O \mathbf{t}_E ,
\end{aligned} \tag{9.2}$$

where $\mathbf{R}_O, \mathbf{t}_O$ is the object motion with respect to the world coordinate system.

Equations (9.2) can be easily verified (see Fig. 9.1). If the object is static, it should be observed by the robot at O'_{b2} in the second frame. $O'_{b2} = \mathbf{R}_E O_{b1} + \mathbf{t}_E$, where O_{b1} is the object position in the first frame. In reality, the object is observed at O_{b2} in the second frame, so the difference between O'_{b2} and O_{b2} is the real motion of the object (in the coordinate system of the second frame), that is:

$$O_{b2} = \mathbf{R}_O O'_{b2} + \mathbf{t}_O .$$

But since

$$O_{b2} = \mathbf{R}'_O O_{b1} + \mathbf{t}'_O$$

and

$$O'_{b2} = \mathbf{R}_E O_{b1} + \mathbf{t}_E ,$$

after simple calculation, we obtain (9.2).

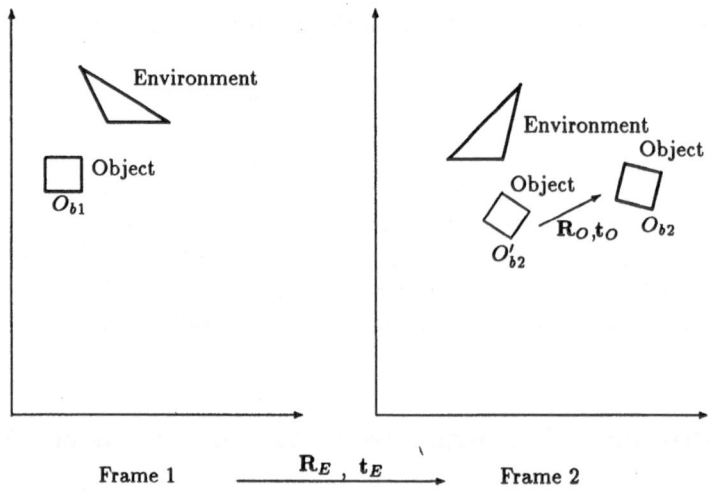

Fig. 9.1. Influence of egomotion on observed object motion

From the foregoing discussion, we have at least two approaches to correctly recover the object motions.

1. Recover first all possible motions between two frames, then determine the egomotion and use the above equation (9.2) to compute the motions of objects with respect to the world coordinate system.

2. If the segments corresponding to the environment are more numerous than those of objects, we need not explore all possible matches. We can choose only half of the segments in the first frame and take as the egomotion the motion which can bring into correspondence the largest number of segments. Once the egomotion is recovered, we can apply it to the first frame. Applying again the matching process between the first frame after transformation and the second one, we now recover the real motions of objects using all remaining segments *unmatched* by the egomotion, since we work in the same coordinate system (that of the second frame).

In the second approach, it is reasonable to choose only half of the segments, because we only need to match three segments belonging to the environment to compute the egomotion.

There remains a problem, however. How to select the egomotion among all possible motions? Two possible solutions exist. If we know that the environment contains more segments than objects, we can then select the motion estimate that matches the largest number of segments. If we know an *a priori* estimate of the robot motion (for example, an estimate given by the odometric system of the robot), we can select the motion estimate which is nearest to the *a priori* estimate.

9.3   Experimental Results

9.3.1   Real Scene with Synthetic Moving Objects

Considering that the static environment contains much more segments than the moving objects, we adopt the second approach described in Sect. 9.2 to determine the motions of multiple objects. Figures 9.2 and 9.3 display the reconstructed segments of an indoor scene observed by the mobile vehicle in two different positions. On the left is the projection on a vertical plane (front view); On the right is the projection on a horizontal plane (top view). The two positions differ only by a camera rotation of 5 degrees around an almost vertical axis. We have added to the scene two synthetic objects, a small house (containing 10 segments) and a chair (containing 12 segments), which have two different motions and whose segments are corrupted by noises. The simulated motion of the house is a rotation of 30 degrees around a vertical axis and a translation of about 80 cms. The chair has been rotated by –40 degrees about a vertical axis and translated also by about 80 cms. There are 147 segments in the first frame and 161 in the second.

In order to speed up the process of hypotheses generation, we choose only the first 73 longest segments in the first frame. Eight hypotheses are generated among which, five correspond to the static environment, one corresponds to a moving object, and two are wrong hypotheses (after verification, we find that these yield very few, less than five, segments matched). The hypothesis that matches the largest number of segments is chosen as representing the egomotion. Figure 9.4

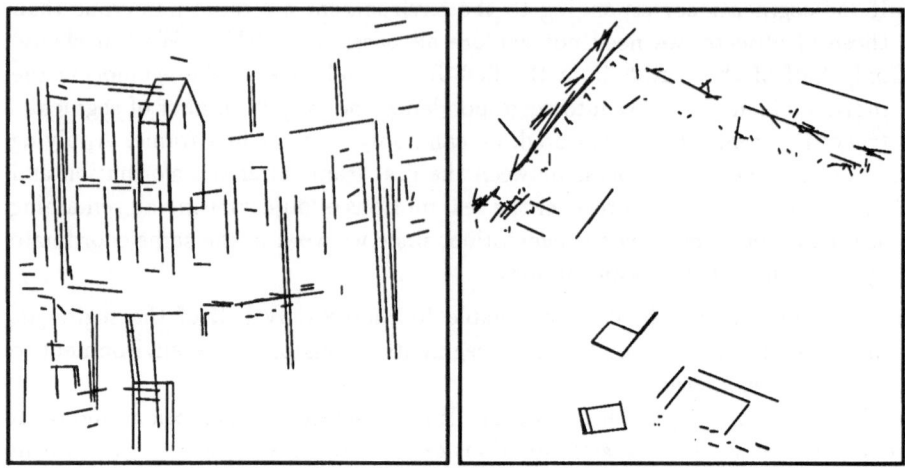

Fig. 9.2. Frame 1 of the first example

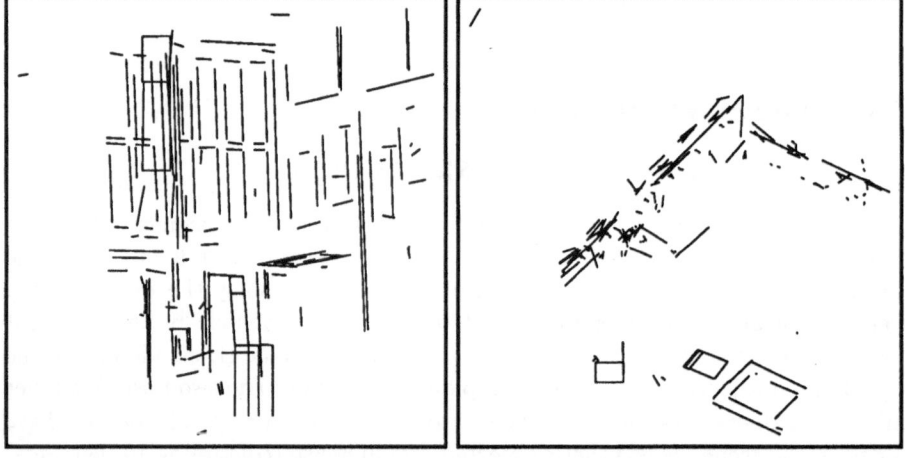

Fig. 9.3. Frame 2 of the first example

shows the result of the estimated egomotion. The transformed segments in Frame 1 are displayed in dotted lines overlaying Frame 2 where segments are drawn in continuous lines. Clearly, most of the background (68 segments) have been matched, while those of the moving objects have not.

After determining the motion of the robot, we have matched a number of segments of the static environment between the two 3D frames. But there still exist some segments of the environment and of the moving objects which have not been matched. This is shown in Figs. 9.5 and 9.6. These segments have not been matched for several reasons:

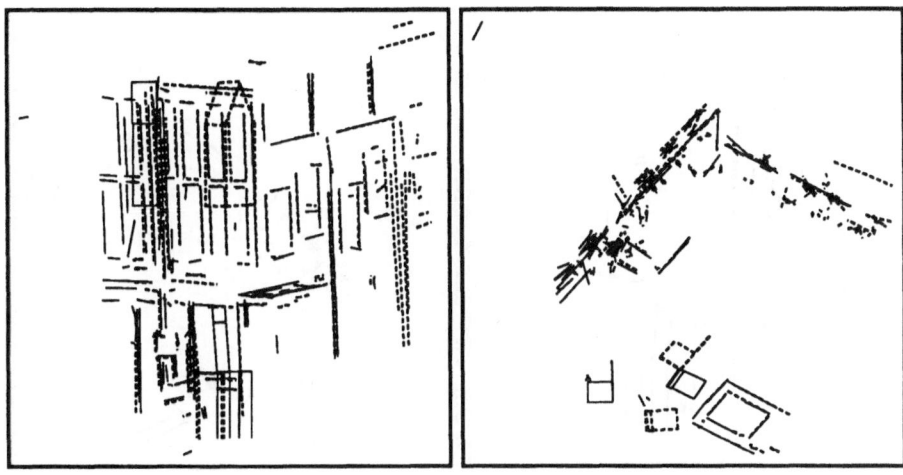

Fig. 9.4. Superposition of the transformed segments of Frame 1 after applying the estimated egomotion and the segments of Frame 2

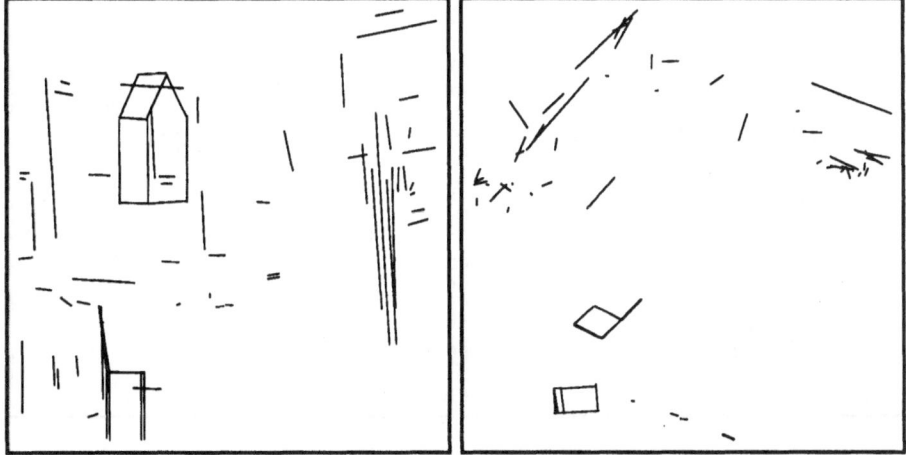

Fig. 9.5. Unmatched segments in Frame 1 after cancellation of the egomotion

- The segments of the moving objects cannot be matched since they undergo different motions from that of the robot.
- Some segments of the static environment are very noisy and the uncertainty measurements given by the stereo program do not correctly describe it. This explains why the matching procedure could not take them into account.
- Some segments of the static environment are visible only in one view, but invisible in another view.

    If we apply the estimate of the displacement of the robot that we have ob-

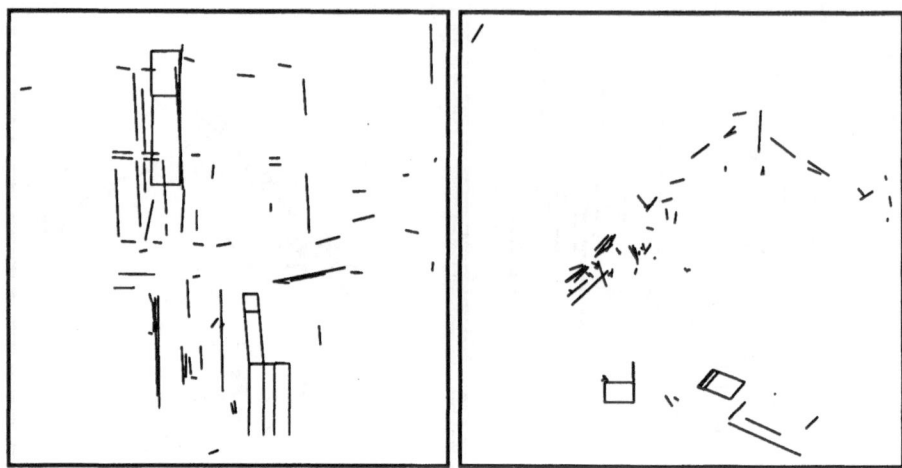

Fig. 9.6. Unmatched segments in Frame 2 after cancellation of the egomotion

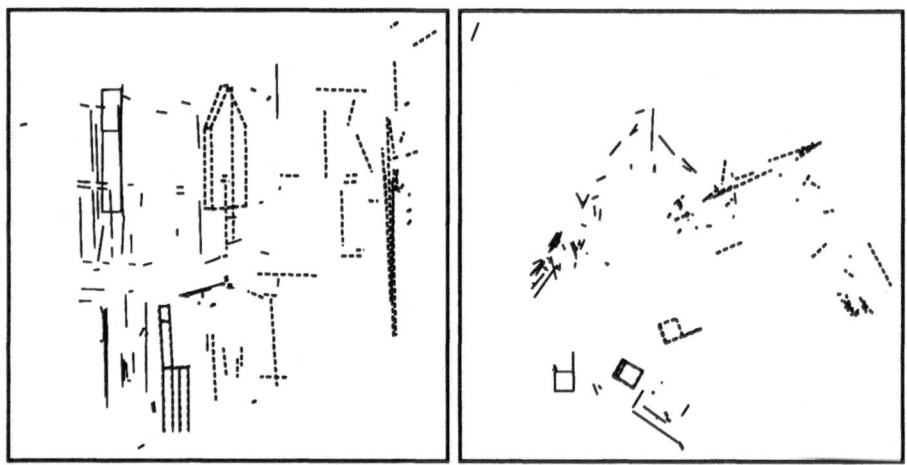

Fig. 9.7. Result of the estimated motion of the model chair

tained to the first frame, and remove the segments matched from both frames, we get two 3D frames which contain the segments of the moving objects (and also some segments of the environment which have not been matched). The influence of the egomotion on the motions of these objects has been cancelled. This is shown in Figs. 9.5 and 9.6. Figure 9.5 displays unmatched segments of the first frame after applying the estimated egomotion. Figure 9.6 displays unmatched segments of the second frame.

At this point, we apply again the procedure of hypothesis generation and verification. All unmatched segments are used in the generation phase. Five

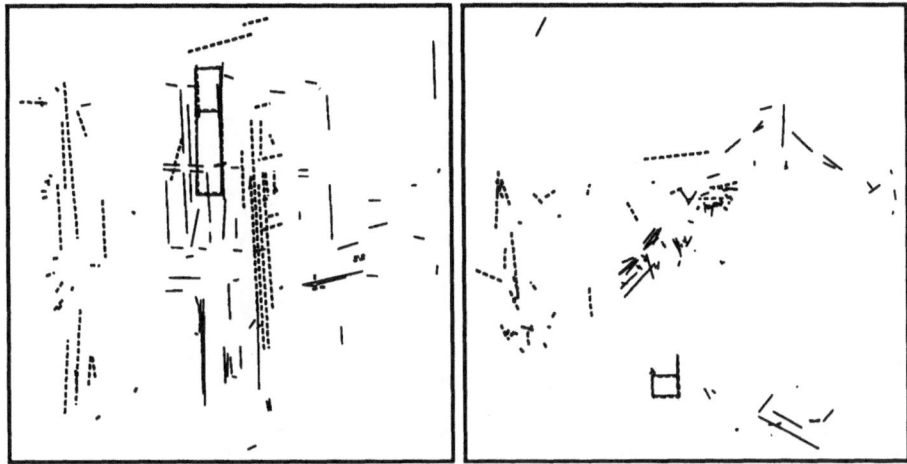

Fig. 9.8. Result of the estimated motion of the house

hypotheses are generated. Two give the motion of the chair. The chair has 12 segments, and 11 segments are matched (one segment has been missed, because it is too noisy). Two others give the motion of the house. The house has 10 segments, and 11 segments are matched (the extra match is due to the occasionally coincidence of two segments from background). Figure 9.7 shows the results for the chair, and Figure 9.8 shows them for the house. Clearly, the two motions have been well recovered.

9.3.2 Real Scene with a Real Moving Object

This example shows an experiment with a real scene containing a real moving object. Figure 9.9 shows two successive images observed by the first camera of the trinocular stereo. Notice that the moving object in the scene is the box in the foreground. It moves from right to left and towards the stereo rig which moves away from the table. Figures 9.10 and 9.11 display the vertical and horizontal projections (front and top views) of the reconstructed 3D scenes in the two positions. Figure 9.12 shows the overlay of the two 3D frames, from which one can observe the difference between them. The triangle in these figures represents the optical centers of the three cameras. There are 168 segments in Frame 1 and 172 segments in Frame 2.

The second approach described in Sect. 9.2 is used to determine the motions. Only half of the longest segments in Frame 1 are used in the hypothesis generation process. Sixteen hypotheses are generated and all of them are evaluated. Eight hypotheses give correctly the estimation of the egomotion. Figure 9.13 shows the result of the estimated egomotion. We remark that the two frames are well superimposed, except for the box. From the top view in Fig. 9.13, we easily observe the motion of the box as indicated by arrows. In fact, five among the 16

Fig. 9.9. Images taken by the first camera of the trinocular stereo: left one is in the first position and the right one is in the second position

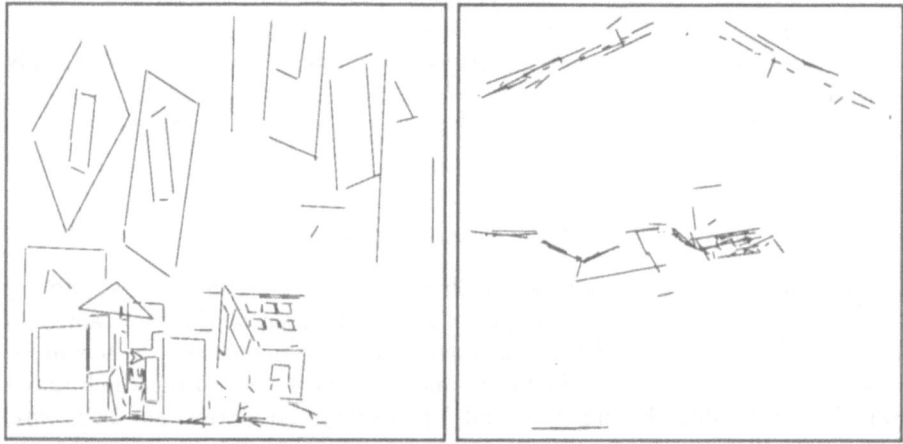

Fig. 9.10. The front and top views of the reconstructed 3D frame in the first position

hypotheses give the estimation of its motion, but we do not intend to recover its motion at this stage, because it may be deteriorated by some occasionally bad alignment of segments belonging to the static environment.

There remain some unmatched segments for the same reasons described in the above example. We apply the estimated egomotion to the first frame, and then remove the segments matched by the egomotion. The result is shown in Fig. 9.14. We also remove the matched segments in the second frame, and the result is shown in Fig. 9.15. We thus obtain two 3D frames containing the segments of

Fig. 9.11. The front and top views of the reconstructed 3D frame in the second position

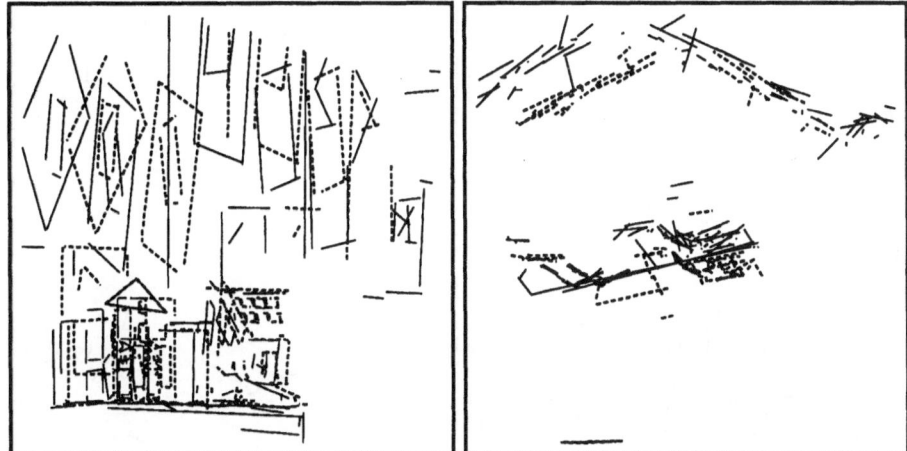

Fig. 9.12. Superposition of the two original frames

the moving object (and several unmatched segments of the environment). The effect of the egomotion has been eliminated. The superposition of these two frames is displayed in Figure 9.16. This figure clearly shows the movement of the box, as indicated by arrows.

We then apply the same process to these two sets of segments. All segments are used in the hypothesis generation phase. Six hypotheses are generated in this phase and they are all evaluated. All yeild the correct motion of the box. The result is shown in Fig. 9.17. We remark that the segments of the box from the two frames coincide very well.

Fig. 9.13. Superposition of the transformed segments of Frame 1 after applying the estimated egomotion and those of Frame 2

Fig. 9.14. Unmatched segments in Frame 1 after egomotion compensation

Now we compare the motion estimations of the box recovered at different stages. Table 9.1 displays different estimated motions. The third row of the table is the observed motion of the box computed at the same time as the egomotion. If we use (9.2), the estimated egomotion (the first row of the table) and the estimated object motion (the second row of the table) to compute the object motion which should be observed by the robot, we get $\mathbf{r}'_O = [5.88\text{e--}03, 3.26\text{e--}02, 8.57\text{e--}03]^T$ and $\mathbf{t}'_O = [-139.90, -3.67, 109.50]^T$. This should be compared with

Fig. 9.15. Unmatched segments in Frame 2 by the egomotion

Fig. 9.16. Superposition of the remaining segments of both frames after those matched through the egomotion-estimation procedure have been eliminated

the third row of the table. The difference between them is relatively big. To know which one is better, we superimpose the two original frames by applying each estimation to the first frame. We observe that the estimation computed above is better than that in the third row of the table. The reason is that when box motion and egomotion are computed simultaneously, several false matches of segments from the environment affect the final estimation of the box motion.

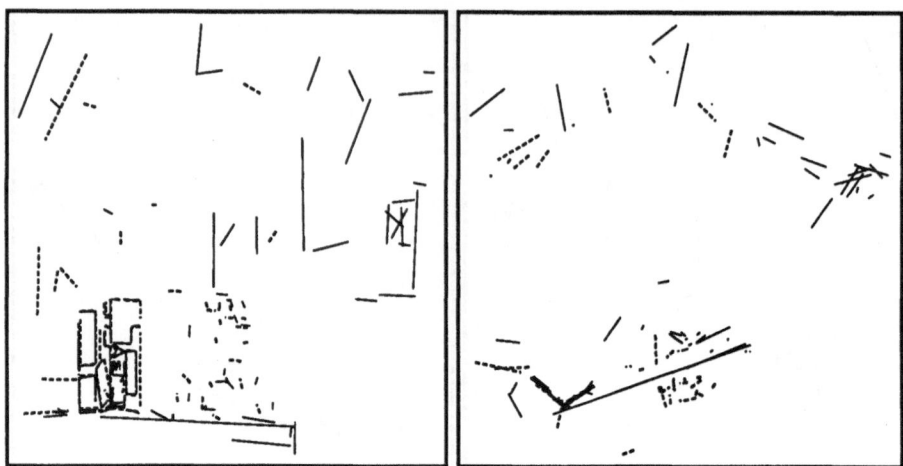

Fig. 9.17. Result of the estimated motion of the box

Table 9.1. Result of the estimated motions

Motion	Rotation (r)	Translation (t)
egomotion	$[\ 1.67e{-}04, -1.22e{-}01, -8.32e{-}05]^T$	$[\ 169.83,\ \ \ 1.73, 132.85]^T$
real object motion	$[\ 5.18e{-}03,\ \ 1.54e{-}01,\ \ 9.01e{-}03]^T$	$[-328.12,\ \ -6.39,\ \ \ \ 4.34]^T$
observed object motion	$[-1.78e{-}02,\ \ 8.98e{-}02,\ \ 5.05e{-}02]^T$	$[-160.74, -42.39, 144.60]^T$

9.4 Summary

In this chapter, we have extended the motion algorithm developed in Chap. 7 to deal with the multiple-object-motions problem. The influence of the egomotion on the object motions observed by the robot has been described in detail. Two approaches have been proposed to recover the correct object motions. The first consists in first recovering all possible motions between frames, including object motions influenced by the egomotion, and then computing the real object motions by using (9.2). The second consists in first finding the egomotion, and then recovering object motions from the remaining data after the egomotion cancellation. The second approach is especially useful when the static environment contains many more segments than the moving objects. Two experimental results have been provided. The object motions as well as the egomotion have been well recovered.

10. Object Recognition and Localization

M odel-based object recognition has been a very active research field in Computer Vision during the past ten years. Several excellent survey papers have been published, for example, [10.1, 2]. *Grimson* has published a book [10.3] describing an extended series of experiments, carried out by himself and his colleagues at MIT, on the role of geometry in solving the object recognition and localization problem. Their approach is mainly based on the search in the interpretation tree, as briefly presented in Sect. 6.1. One chapter of the book by *Faugeras* [10.4] is also devoted to this problem based on the work performed at INRIA. We do not intend in this chapter to deal in detail with this immense subject. We want simply to adapt our motion-determination algorithm developed in Chap. 7 to solve some primitive object recognition problems.

10.1 Model-Based Object Recognition

The object recognition problem can be defined as follows:

Definition 10.1. *We are given a data base of object models and a view of the real world. For each object in the model data base, the* **object recognition and localization** *problem consists in answering the following two questions:*

- *Is the object present in the observed scene ?*
- *If present, what are the 3D pose parameters (translation and rotation parameters) with respect to the sensor coordinate system ?*

If possible, the system should learn unknown objects from the observed data. ◇

From the above definition, a model-based object recognition system can be decomposed into a training phase and a classification phase, according to [10.2], as illustrated in Fig. 10.1. The objective of the training phase is to build, automatically if possible, the object models. The classification phase is to recognize objects stored in the model base from the observed scene.

The three major components of the system are *feature extraction, object modeling*, and *matching*. Due to the large amount of information in the input data

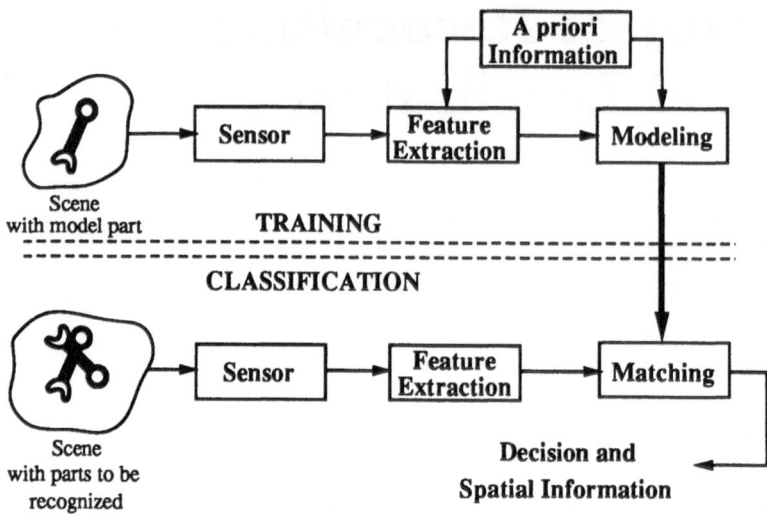

Fig. 10.1. Components of a model-based recognition system (adapted from Chin and Dyer [10.2] ©1986, Association for Computing Machinery, Inc. By permission.)

from the sensor, it is reasonable to work at a level intermediate between the whole object and the pixel values. We use *features*, for example, edge, corner, boundary, line, plane. These features and their spatial relations constitute object descriptions. The description at this level is sometimes referred as the *symbolic scene description*. Because of the unavoidable uncertainty in sensing and feature extraction, it is preferable to include some *a priori* information. In [10.5], for example, knowing that objects are polyhedra, we can impose constraints such as coplanarity on the sensor data to obtain a precise description of the object. The recognition process is then to find a set of features extracted from the observed scene which can be matched with features of one or more model objects.

Most of the current methods for model-based 3D object recognition can be divided into the following paradigms:

1. **Tree search,** for example, the constrained interpretation tree proposed by *Grimson* and *Lozano-Pérez* [10.3, 6]. An interpretation tree is constructed as described in Sect. 6.1. Instead of exploring all interpretations of the observed data, only feasible ones are retained by determining consistent pairings of data features to model features. By consistency is meant the pairing of data and model features satisfying the rigidity constraints as described in Sect. 6.5. Interpretations that are locally inconsistent are rejected, which rapidly reduces the candidate pairings. For each feasible interpretation of the observed data, a rigid transformation is computed to check the global consistency.

2. **Alignment technique.** Representative work includes that of *Ayache* and *Faugeras* [10.7] (the HYPER system), *Faugeras* and *Hébert* [10.8], *Huttenlocher*

and *Ullman* [10.9], and *Lowe* [10.10, 11] (the SCERPO system). The *alignment method* consists of two stages:

(a) compute the possible *alignments*—transformations from model to data coordinate frames—using a minimum number of corresponding model and image features. For example, for m 3D model points and n (2D or 3D) data points, there are $\binom{m}{3}\binom{n}{3}3!$ possible different alignments, resulting from pairing each triple of model points with each triple of data points. Perceptual grouping [10.10] or rigidity constraints [10.7, 8] can be used to reduce the number of hypothetical alignments.

(b) verify each of these hypothetical alignments by transforming the model to the data coordinate frame and comparing it with the data. The verification process consists in examining whether model features transformed by the hypothetical alignment coincide with some data features and how well. It is of critical importance because it must filter out incorrect alignments without missing correct ones.

The RANSAC algorithm proposed by *Fischler* and *Bolles* [10.12] can be considered as a variation of this method. It chooses random samples of matched data features and model features, which are used to compute a hypothetical transformation. The hypothetical transformation is immediately evaluated on the rest of features. If it provides a satisfactory consensus, then the algorithm terminates; otherwise sampling is continued until a satisfactory transformation is found or the algorithm fails to detect the model object.

3. **Focus feature extension** or **Local-Feature-Focus (LFF) method**. Representative work includes that of *Bolles* and *Cain* [10.13] (recognition of 2D objects from an image), and *Bolles* and *Horaud* [10.14, 15] (recognition of 3D objects from range data). The basic principle of the LFF method is to find one salient feature such as hole, corner, or large segment, referred to as the *focus feature*. The selection of a focus feature is based on a function of several factors including the uniqueness of the feature, its expected contribution, the cost of detecting it, and the likelihood of detection. Each focus feature is used to predict a few nearby features to look for. After finding some nearby features, the program constructs a graph of pairwise consistent model and image features, and then uses a graph-matching technique to identify the largest cluster of image features matching a cluster of model features. The largest clique above some minimum size is taken as the best match. Although the maximum-clique algorithm has exponential time-complexity in the number of features, the graph is relatively small since the list of possible model features has been reduced to those near the focus feature, and therefore, the graph can be analyzed efficiently.

4. **Geometric hashing**. Representative work includes that of *Lamdan* and *Wolfson* [10.16]. The idea of *geometric hashing* is to precompile the model information into a hash-table using appropriate representation. Consider, for example, recognition of 2D objects represented by points which have under-

gone a *similarity transformation* (rotation, translation, and a scale). Geometric hashing makes use of the fact that registering a pair of points on the model with a pair of points in the scene uniquely defines the similarity transformation, and the fact that points preserve their coordinates under any similarity transformation if they are represented in the local coordinate frame attached to the same two points. It is divided into an off-line and an on-line procedures:

(a) In the preprocessing step, each ordered pair of points of a model object is used to define an orthogonal coordinate frame, and the pair is referred as a *basis-pair*. All other points of the model object are represented by coordinates with respect to this coordinate frame. Each such coordinate (after a proper quantization) is used as an entry to a hash-table which stores all the pairs (*model, basis-pair*) for every coordinate.

(b) In the recognition step, an arbitrarily ordered point pair in the scene is chosen and the coordinates of all the other scene points are computed in this basis. For each such coordinate, check the appropriate entrance in the hash-table and vote for the pairs (*model, basis-pair*) appearing there. If a certain pair (*model, basis-pair*) scores a large number of votes, this pair is considered to correspond to the one chosen in the scene. The uniquely defined similarity transformation between the coordinate frames corresponding to these basis-pairs can be computed, and is eventually recomputed using additional point matches.

5. **Generalized Hough transform** or **pose clustering**. The Hough transform has been introduced to detect simple geometric primitives such as lines in an image through accumulation of evidence in their parameter space. It has been generalized for object recognition and localization by considering the transformation space [10.17,18]. Such method can be divided into two steps:

(a) The first step is to generate transformation/pose candidates and hash them in the transformation space. In order to compute a unique transformation/pose, a minimum number, say k, of features in a model should be matched with the same number of features in the data. For example, for recognition of 2D objects represented by points from 2D data represented also by points, two matches are required, i.e., $k = 2$. For each pairing of k model and data features, try to compute a transformation/pose. If it exists, hash it in the transformation space (after a proper quantization), which can be represented by an accumulator array. Each cell of the array records the accumulation of evidence for a given transformation/pose (strictly speaking, a set of transformations/poses within the transformation quantization interval).

(b) The second step is to cluster in the transformation space, for example by scanning the accumulator array, to get the best transformation(s)/pose(s) and its (their) supporting evidence(s). In [10.18], *Stockman* proposes a clustering algorithm, based on distances between transformation/pose candidates, to detect the transformation(s)/pose(s) with strong support. This approach avoids the problem due to quantization of the transformation space.

The reader may find some similarity between this method and the last one. The main difference is in the hash space used. The current method uses the transformation/pose space while the last one uses the geometric space (relative coordinates).

6. **CAD-based search**. Representative work includes that of *Hansen* and *Henderson* [10.19], and *Ikeuchi* and *Kanade* [10.20, 21]. The methods in this category exploit geometric knowledge in the CAD models to automatically generate recognition strategies. An object in the model base is usually represented by an *aspect graph*. The *aspect* is the topological appearance of the object from a particular viewpoint. Slight changes in the viewpoint change the size of features (edges and faces), but do not cause them to appear or disappear. When a slight change in viewpoint causes a feature to appear or disappear, an *event* takes place. An aspect graph is formed by representing aspects as nodes and events between aspects as paths between corresponding nodes. Through the analysis of the robustness, completeness, uniqueness, and detection cost of the features extracted from the aspects, a *strategy tree* is constructed for each object. It describes in a systematic manner the search process used for recognition and localization of a particular object in the given scene. During the recognition phase, strategy trees guide the search and thereby increase the speed for recognition.

7. **Evidence-based reasoning**. *Jain* and *Hoffman* [10.22] developed an algorithm based on evidence-based reasoning for recognizing 3D objects from range images. The idea of the method is: rather than use all of the information provided by a representation, it is possible to use only "remarkable" information or evidence, which strongly cues certain objects. A collection of evidences can be used to determine the likely contents of an observed scene. The recognition system described in [10.22] is implemented as a production-rule system. The premise part of a rule consists of bounds of characteristics parameters of a surface patch such as perimeter, area and surface sense (planar, convex, or concave), and relationship between adjacent and jump neighbors such as normal angle and distance between two patches. The action part of a rule consists of evidence weights for each object in the base if the condition is satisfied. After activating all rules, evidence-based reasoning is applied to decide whether an object is observed in the scene or not. The essentiality and also the difficulty of this approach is to capture the properties of an object as general as possible, on one hand, in order not to miss an object due to noise or distortion introduced during scene acquisition, and on the other hand, as discriminating as possible in order not to confuse different objects. The approach has been mainly tested with scenes containing a single object.

One common feature of most of the object recognition methods is a procedure of preprocessing model information in order to simplify the recognition procedure.

The preprocessing procedure is performed off-line, and is usually an expensive one in order to maximizing use of model information. Such matching methods are not appropriate to deal with the registration problem in analyzing image sequences.

10.2   Adapting the Motion-Determination Algorithm

From the brief description of the object recognition problem in the preceding section, we can remark that there exist many similarities between the object recognition and the motion determination problems. We want to solve the following specific problem: given a 3D object model described by 3D line segments and a stereo view of a scene, determine whether the object is present and where it is.

The algorithm described in Chap. 7, which is one of the alignment methods, is directly applicable to the object recognition problem after a minor modification for some specific constraints. In the hypothesis generation phase, the rigidity constraints remain the same, but the additional constraints described in Sect. 7.2.3 do not hold anymore. The constraint on the congruency in segment orientations, for example, is not reasonable, since an object in the model base is described in its specific coordinate system, which may be in any direction with respect to that of the sensor. Another constraint derived from the fact that the robot moves horizontally does not apply here, either. Instead, we can impose a stricter constraint on the lengths of segments, that is, the length of a segment in sensor data should not be longer than that of a segment of models if they can be matched. This is because a segment in the model gives the exactly (with a small error tolerance) length, and the stereo observes usually one part. The algorithm for hypothesis verification does not need any change.

10.3   Experimental Result

In this section, we provide an example of the application of the motion-determination algorithm to the model-based object recognition.

The object we want to recognize consists of two quasi-perpendicular planes on which there are some patterns. The model has been built by simply measuring the object with a ruler. We assume the two planes are completely perpendicular. The standard deviation of the uncertainty due to manual measuring is assumed to be 3 millimeters homogeneous in all directions. The measurements are expressed in the object-centered coordinate system. We have measured 35 line segments to represent the boundaries of the object and the region markings. Figure 10.2 displays the different views of the model we have built. The upper left hand one is the top view (projection on the plane $y = 0$), the lower left hand one is the front view (projection on the plane $z = 0$), the lower right one is the side view

Model.xyzv   Top View (xz)   Model.xyzv   Perspective View

Model.xyzv   Front View (xy)   Model.xyzv   Side View (zy)

Fig. 10.2. Different views of the model object

(projection on the plane $x = 0$) and the upper right is a perspective view (about 40 degrees from the plane $y = 0$) to display the model better.

The observed scene we used is the same as the first frame in the first example in Chap. 7 (see Fig. 7.4). The three images taken are shown in Fig. 11.7. The object on the table in the foreground is the one we want to recognize. Figure 10.3 shows the superposition of the model and the observed scene to see the difference between them. The left is the front view and the right is the top view. The scale in each direction is not uniform in order to display the details to maximum. We then apply our adapted algorithm to the model and the observed scene to see whether the model object is present or not and where it is in the scene if present. The algorithm has generated 5 hypotheses and all of them are evaluated.

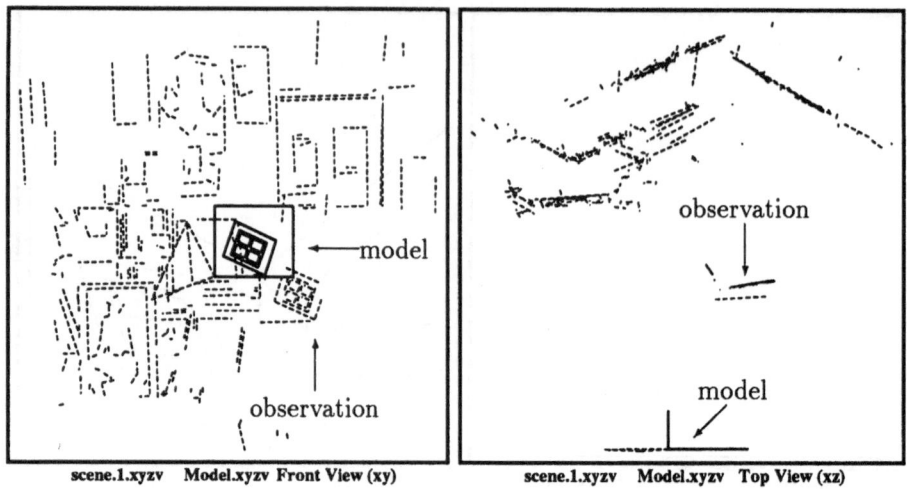

scene.1.xyzv   Model.xyzv   Front View (xy)   scene.1.xyzv   Model.xyzv   Top View (xz)

Fig. 10.3. Superposition of the model and the observed scene: segments of the model are represented in solid lines and those of the observed scene in dashed lines

Three hypotheses find the object and give correctly the transformation between the model and the observed scene. 27 segments have been matched. The estimated transformation is $[-2.08e\text{-}02, -3.09e\text{-}01, -2.77e\text{-}02, 3.412e+02, 2.594e+02, 2.044e+03]^T$. This corresponds to a rotation of 17.8 degrees and a translation of 2088 millimeters (i.e., about two meters). The whole process takes about 35 seconds on a SUN 3/60 workstation.

To see how good the estimated transformation is, we apply it to the segments of the model, transform them to the stereo coordinate system, and compare them with the observed scene. Figure 10.4 displays the superposition of the transformed model object and the observed scene. Figure 10.5 shows the superposition of the matched segments of the model object and those of the observed scene. From these two figures, it is clear that the recovered transformation is quite good, especially in the lateral and vertical directions. In the range direction, the superposition is a little worse. This is reasonable since the stereo reconstructs segments with more uncertainty in the depth. Another observation is that the left plane is not as well superposed as the right one. There are two reasons for this:

- The right plane has more segments than the left one, thus it contributes more to the estimation of the transformation than the left one.
- The model we built is not very precise. Perhaps the angle between the two planes was not exactly 90 degrees.

The other two hypotheses do not give the correct transformation between the model and the observed object. In fact, due to the symmetry of the patterns (the

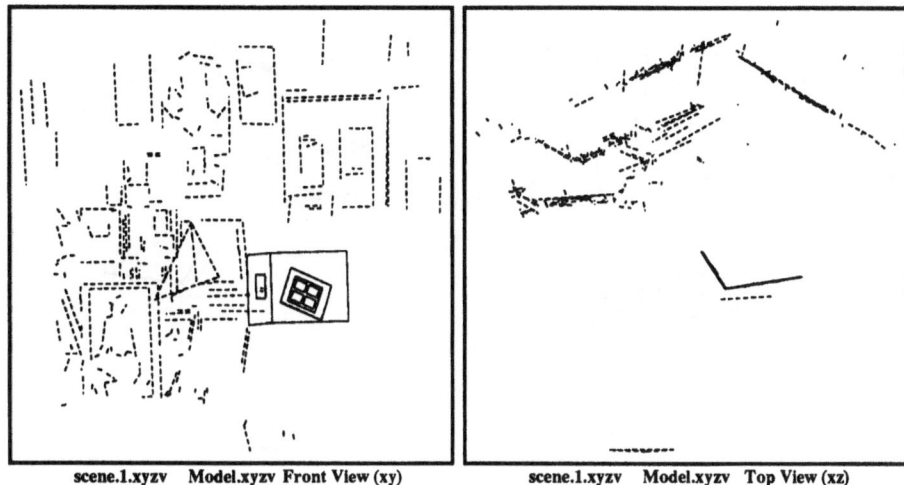

scene.1.xyzv   Model.xyzv Front View (xy)   scene.1.xyzv   Model.xyzv Top View (xz)

Fig. 10.4. Superposition of the transformed model after applying the estimated transformation and the observed scene: segments from the model are represented in solid lines and those from the observed scene in dashed lines

Fig. 10.5. Superposition of the matched segments from the model (in solid lines) after applying the estimated transformation and those from the observed scene (in dashed lines)

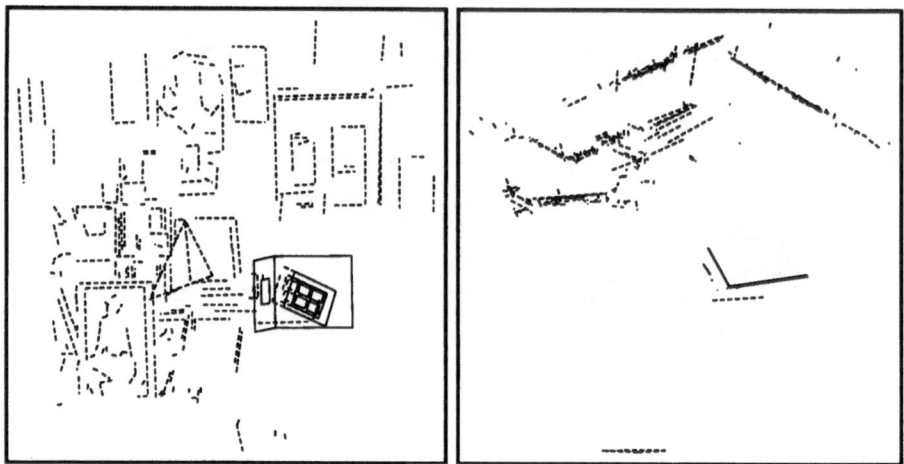

Fig. 10.6. Superposition of the segments from the model (in solid lines) after applying the estimated transformation given by a *wrong* hypothesis and the segments from observed scene (in dashed lines)

rectangles) in the right plane, the two hypotheses match two wrong rectangles. Figure 10.6 shows the result given by a wrong hypothesis, which finally matches 16 segments between the model object and the observed scene. Clearly the two rectangles are well superimposed, but the global matching is not good. From this example, we see that our algorithm may generate many wrong hypotheses if a scene contains many symmetric objects, although we can recover correctly the transformation/motion after evaluating all hypotheses.

10.4 Summary

In this chapter, after a brief description of the model-based object recognition problem, and a brief review of current approaches to this problem reported in the literature, we have adapted our motion-determination algorithm described in Chap. 7 to solve the object recognition problem. We showed that our motion-determination algorithm is directly applicable after a minor modification. To demonstrate this, we have built manually a model of an object. The adapted algorithm is then applied to the model and the real data obtained from stereo. The algorithm correctly recognized the object and gave a precise estimation of the transformation between the model and the observed data.

11. Calibrating a Mobile Robot and Visual Navigation

The calibration problem is very important in robotics for many purposes. There are a variety of calibration problems for stereo [11.1, 2] (see also Sect. 3.2) and for eye-hand systems [11.3]. An intelligent robot system [11.4] can be decomposed into three parts: perception, intelligent decision and control, and action (see Fig. 11.1). A mobile robot must include sensing (vision, sonar) and locomotion [11.5], but we must know the relationship between the coordinate systems corresponding to each perceptual mode in order to combine the different sources. Since the coordinate systems of the perception and action components are different, they should first be calibrated. In this chapter, we address the calibration problem between the stereo and odometric systems of the INRIA mobile robot. We call it the calibration problem for navigation systems or for "eye-wheel" systems. It is essential for many applications, including visual navigation.

In visual navigation applications, the robot uses some visual (monocular or trinocular) sensors to recognize or explore the surrounding environment and then decides where to go and what to do by some task-dependent analysis on the perceived data. The decision is made in the coordinate system attached to the sensor (the stereo system in our case) and should be transformed to the coordinate system attached to the action component (the odometric system in our case). The relation between them is required to be calibrated.

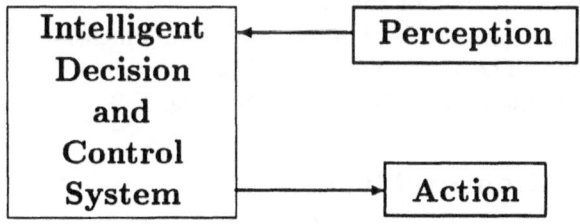

Fig. 11.1. A decomposition of an intelligent robot system

This chapter is organized as follows. Section 11.1 gives a brief presentation of the INRIA mobile robot. Section 11.2 addresses the calibration problem. Section 11.3 shows how to plan the trajectory for some simple navigation problems. As an application of the proposed calibration technique, we describe in Sect. 11.4 a complete visually guided navigation demonstration. Finally, as another application, we describe how to integrate the motion information from the odometric system to speed up the motion-determination process.

11.1 The INRIA Mobile Robot

The INRIA mobile robot (see Fig. 11.2) is, for the moment, dedicated to navigating in an indoor environment and to making visual maps of this environment. The robot is an autonomous vehicle with two driving wheels and two passive rollers. Each of the two driving wheels is independently controlled by a motor, so that the robot can translate and rotate. The following is an example of a fundamental internal command of the robot:

$$\text{MOVE} \quad \text{P} \quad \text{RC=} \ D_r, D_l \quad \text{P=} \ v$$

where D_r, D_l and v are all integers. D_r (D_l) is the run of the right (left) wheel (in millimeters); positive runs correspond to backward motion of the wheel, negative to forward motion. v controls the execution time of the command by controlling the number of cycles taken to execute this command. The following are the four easy-to-use commands:

MOTN F D=d	move forward d mm
MOTN B D=d	move backward d mm
MOTN LE A=a	rotate to the left of a degrees
MOTN RI A=a	rotate to the right of a degrees

Three CCD cameras are mounted on the robot. The trinocular vision system builds a local 3D visual map of the environment. A brief description can be found in Chap. 3. The reader is referred to [11.6, 7] for more details. The robot is also equipped with ultrasonic sensors, but in this chapter, the vision system is the only information source for navigation.

The odometric coordinate system, which is two-dimensional, is attached to the driving wheels. The x-axis goes through the centers of the wheels and the origin is in the middle between the wheels. See Fig. 11.4.

11.2 Calibration Problem

The motion of a mobile robot is equivalent to the motion of the frame of reference of the stereo system (see Fig. 11.3). This reference frame is rigidly attached to the mobile robot and its numerical parameters are determined in the camera calibration phase [11.1].

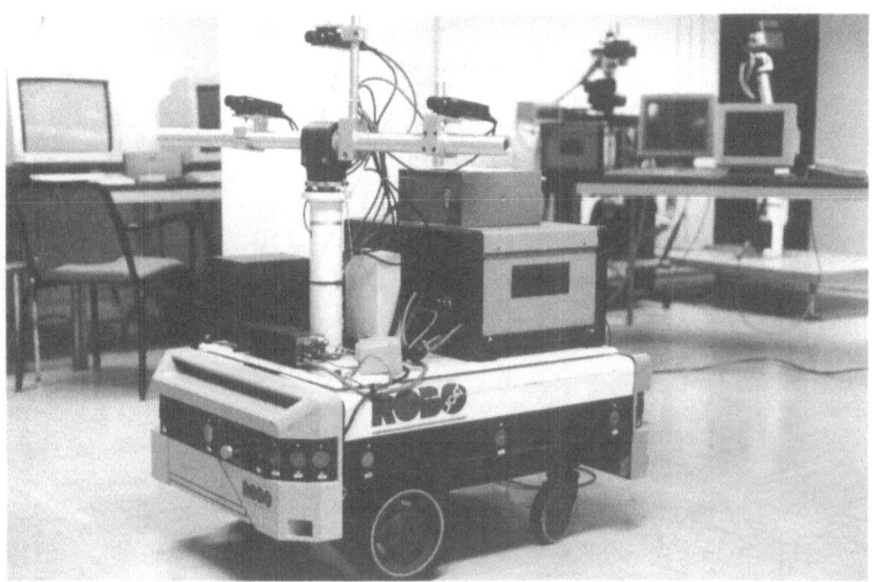

Fig. 11.2. The INRIA mobile robot

The problem of calibration in an eye-wheel system is to determine the transformation between the stereo and the odometric coordinate systems. Solving this problem is important since we want to navigate the robot using visual information from stereo. Figure 11.4 shows the relation between the stereo and odometric coordinate systems. We know (OXY) with respect to the world coordinate system by stereo calibration. In the following, we present a technique to calibrate the eye-wheel system by using motions estimated from stereo as described in Chap. 7. In other words, we should recover $\overrightarrow{OC}$ and $\overrightarrow{Cy}$ as shown in Fig. 11.4, where $\overrightarrow{OC}$ is the center of rotation and $\overrightarrow{Cy}$ is the translation direction of the robot in the stereo coordinate system.

It is well known that there exist an infinite number of decompositions of a 3D motion of a rigid object into rotation and translation. In our motion from stereo algorithm, the displacement of the robot or other moving objects between two successive frames is decomposed into a rotation about an axis through the origin, followed by a translation (see Chap. 4). That is, for all points in correspondence (P_i^1 in the first frame and P_i^2 in the second one, $i = 1, \ldots, n$), we have the following relation:

$$P_i^2 = \mathbf{R}P_i^1 + \mathbf{t} , \qquad (11.1)$$

where $\mathbf{R}$ is the rotation matrix and $\mathbf{t}$ is the translation vector. The following question should be answered:

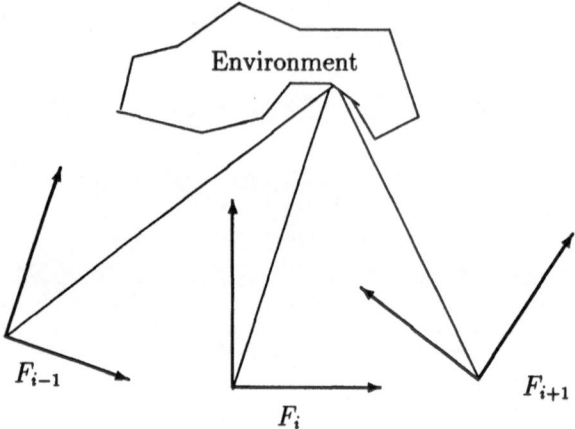

Fig. 11.3. Robot motion and reference in movement

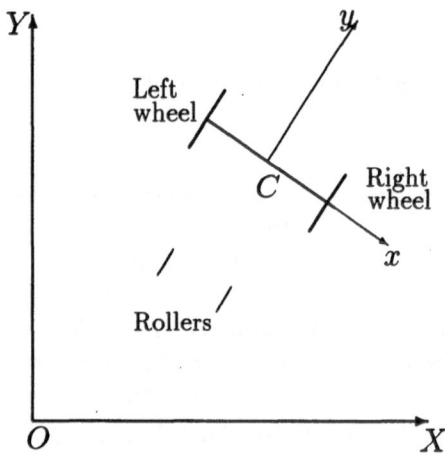

Fig. 11.4. Relation between the stereo and odometric coordinate systems: (x, y) is the robot frame and (X, Y) is the stereo frame

In order to achieve a given motion $\mathbf{R}$ and $\mathbf{t}$ in the stereo coordinate system, what is the required motion in the odometric coordinate system ?

If we have no knowledge about the kinematics of the robot, there are several answers to the question. *Weng* et al. [11.8] have proposed an algorithm to re-

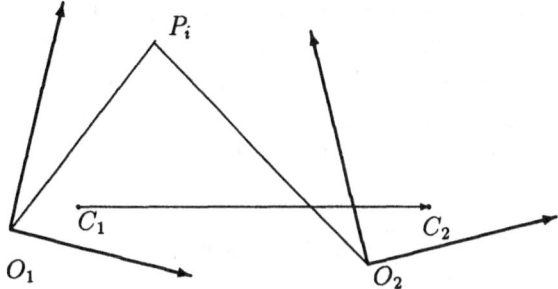

Fig. 11.5. Recovering the rotation axis and the translation vector

cover the center of rotation when the object of interest undergoes a precession movement (see Chap. 14). This is not the case for our robot. It usually moves on a plane, that is, the rotation axis theoretically does not change direction during a short term motion. Thus the rotation center is unrecoverable by their technique.

To simplify the problem, we suppose that the robot first rotates around an axis, and then translates. If the axis goes through the origin, the displacement of the robot can be characterized by $\mathbf{R}^T$, $-\mathbf{t}$ (in the second frame), where $\mathbf{R}^T$ denotes the transpose of $\mathbf{R}$. Suppose the axis passes through a point $\mathbf{C}$ (the center of rotation in a local stereo system, see Fig. 11.5). Let $\mathbf{R}_r, \mathbf{t}_r$ be the rotation and translation of the robot and $\mathbf{I}$ be the 3×3 identity matrix. Then we have the following equations:

$$\mathbf{R}_r = \mathbf{R}^T , \tag{11.2}$$

$$\mathbf{t}_r = \mathbf{C}_2 - \mathbf{C}_1' = \mathbf{C}_2 - (\mathbf{R}\mathbf{C}_1 + \mathbf{t}) = (\mathbf{I} - \mathbf{R})\mathbf{C} - \mathbf{t} , \tag{11.3}$$

where $\mathbf{C}_1$ is the coordinate vector of $\mathbf{C}$ in the first frame, $\mathbf{C}_2$ that in second frame, and $\mathbf{C}_1'$ is the corresponding coordinate vector of $\mathbf{C}_1$ in the second frame (i.e., $\mathbf{C}_1' = \mathbf{R}\mathbf{C}_1 + \mathbf{t}$).

To meet the requirements of navigation, we must recover the rotation center $\mathbf{C}$ and the translation direction of the mobile robot. The key idea of our technique is to use some special robot motions. If the robot makes a pure rotation ($\mathbf{t}_r = 0$), from (11.3) we have

$$(\mathbf{I} - \mathbf{R})\mathbf{C} = \mathbf{t} , \tag{11.4}$$

where $\mathbf{R}, \mathbf{t}$ are the displacement between the two 3D views, estimated by the motion-determination algorithm which we have described in Chap. 7. But from (11.4), we cannot completely solve for $\mathbf{C}$, because $Rank(\mathbf{I} - \mathbf{R}) \leq 2$. Indeed, if the robot undergoes a pure translation, then $\mathbf{R} = \mathbf{I}$, i.e., $Rank(\mathbf{I} - \mathbf{R}) = 0$. If the motion includes rotation, there exists a vector $\mathbf{r} \neq 0$ (the direction vector of the rotation axis) such that $\mathbf{R}\mathbf{r} = \mathbf{r}$, i.e., $Rank(\mathbf{I} - \mathbf{R}) \leq 2$. This is in agreement with our earlier remark. Indeed, Equation (11.4) defines the rotation axis if $\mathbf{R} \neq \mathbf{I}$. In practice, it is sufficient to recover a point on the rotation axis. The intersection point between the line given by (11.4) and a plane parallel to the ground plane (we choose $y = 0$ for simplicity) can be taken as point **C**. The direction of the rotation axis $\hat{\mathbf{r}} = \mathbf{r}/\|\mathbf{r}\|$ between the two views is also available.

The translation direction can be easily recovered. Let the robot undergo a pure translation, i.e., $\mathbf{R} = \mathbf{I}$. From (11.3), we have $\mathbf{t}_r = -\mathbf{t}$, and the direction of translation is

$$\hat{\mathbf{t}} = \frac{\mathbf{t}_r}{\|\mathbf{t}_r\|} = -\frac{\mathbf{t}}{\|\mathbf{t}\|} \ .$$

Up to now, we have obtained one estimate of the real calibration parameters $(\hat{\mathbf{r}}, \mathbf{C}, \hat{\mathbf{t}})$. Since our motion-determination algorithm provides also the uncertainty measure (the covariance matrix) of the motion estimate $(\mathbf{r}, \mathbf{t})$ (see Chap. 7), we can easily compute the covariance matrix of the calibration parameters $(\hat{\mathbf{r}}, \mathbf{C}, \hat{\mathbf{t}})$. In order to reduce the uncertainty of these parameters due to the vision system and especially the mechanical system, we iterate this procedure by letting the robot perform several pure rotations and pure translations to compute several estimates of $(\hat{\mathbf{r}}, \mathbf{C}, \hat{\mathbf{t}})$, and carry out a weighted least-squares adjustment to obtain a more precise estimate. The weight for each estimate is the inverse of its covariance matrix. Now we have $\hat{\mathbf{r}}$, the direction of the rotation axis, **C**, a point on the rotation axis and $\hat{\mathbf{t}}$, the translation direction, which will be used to solve navigation problems.

11.3 Navigation Problem

Using the motion-determination algorithm and the results of Sect. 11.2, we can recover $\hat{\mathbf{r}}, \mathbf{C}$ and $\hat{\mathbf{t}}$ of the mobile robot. Now, the navigation problem can be solved.

By using a stereovision system as a perception tool to interact with the environment, the robot can determine the next position of interest. More precisely, if the robot must go to another position and orientation (from position A to B in Fig. 11.6) described by **R** and **t** in the stereo coordinate system, this can be achieved in three steps:

- rotate first to the required direction of translation ($\overrightarrow{CC'} \triangleq \mathbf{d}$ in Fig. 11.6),
- translate by **d**, and
- rotate again to the required orientation of the mobile robot.

Here we suppose that no obstacles exist between A and B. The trajectory generated by the above three steps is certainly very primitive, but it is sufficient to demonstrate the potential application. For more sophisticated approaches to trajectory planning, the reader is referred to, for example, [11.9, 10]. The translation **d** can be computed as follows:

$$\mathbf{d} = \mathbf{RC} + \mathbf{t} - \mathbf{C} = (\mathbf{R} - \mathbf{I})\mathbf{C} + \mathbf{t} \ .$$

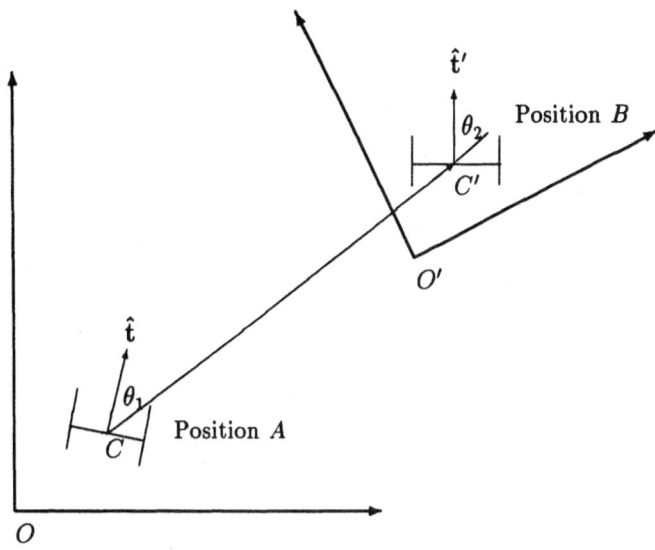

Fig. 11.6. Visual navigation problem

So the first rotation angle θ_1 is given up to its sign by

$$\theta_1 = \cos^{-1}(\hat{t}^T \frac{d}{\|d\|}) \; .$$

The rotation angle θ_1 is positive if $\hat{t}$, d and r satisfy the right rule, i.e., r has the same direction as $\hat{t}$ cross d and the reverse if they do not.

Theoretically, the rotation axis r defined by R is parallel to $\hat{r}$, because the robot moves in the ground plane. In practice, there is a slight difference. We can project r on $\hat{r}$ to compute the total rotation angle θ, i.e., $\theta = r^T \hat{r}$. Then the second rotation angle is:

$$\theta_2 = \theta - \theta_1 \; .$$

Suppose there is no obstacle between between A and B, three commands (two for rotation and one for translation, as described in Sect. 11.1) are sufficient for the robot to accomplish its displacement of R and t from position A to position B.

11.4 Experimental Results

In this section, a complete visual navigation loop is presented. Such a loop brings together the techniques developed earlier and provides an interesting test for the vision algorithms.

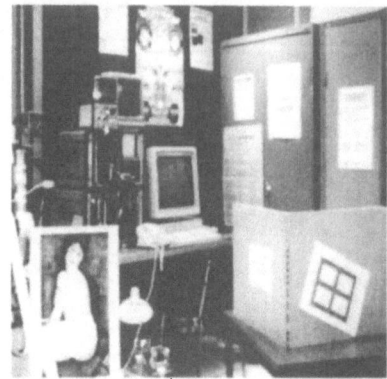

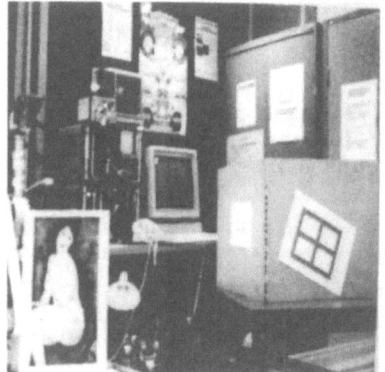 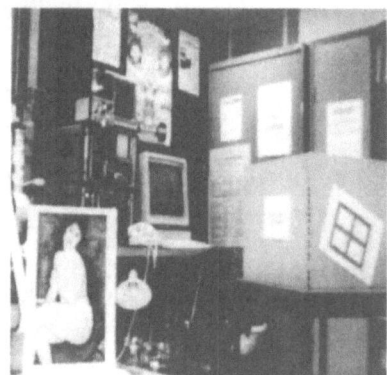

Fig. 11.7. Three images taken by the robot in position A

Table 11.1. The calibration result of the INRIA mobile robot

	x component	y component	z component
$\hat{\mathbf{r}}$	1.612e−02	9.999e−01	−1.994e−03
$\mathbf{C}$	−1.124e+02	−4.897e−04	−1.159e+02
$\hat{\mathbf{t}}$	1.178e−01	5.371e−03	9.930e−01

The calibration is done by three pure rotations and only one translation. The rotation angles and the translation distance are of little importance, since we use the $\mathbf{R}$ and $\mathbf{t}$ given by the motion-determination algorithm. We carry out the weighted least-squares adjustment only on the parameters $\hat{\mathbf{r}}$ and $\mathbf{C}$ as described in Sect. 11.2. The calibration results are displayed in Table 11.1.

The time needed for the calibration depends essentially on the time spent on the motion-estimation process described in Chap. 7. The run time of the motion

Fig. 11.8. Three images taken by the robot in position B

estimation algorithm depends on the number of segments in each stereo frame and on the number of hypotheses generated. The bigger the number of segments and the number of hypotheses, the more time is needed in the motion-estimation process. Usually, more hypotheses are generated when the motion between two frames is small than when the motion is large. The time to verify one hypothesis increases with the number of segments. The reader is referred to Sect. 7.5 for the processing time required to register two stereo frames.

The calibration results are used in the visual navigation demonstration. The robot is first in position A. It takes three images (see Fig. 11.7) and the reconstructed 3D frame in that position is portrayed in Fig. 7.4. Then we move the robot to another position B with about 10 degrees of rotation and 750 millimeters of translation, where the robot again takes three images (see Fig. 11.8). The corresponding 3D frame is reconstructed and is displayed in Fig. 7.5. The objective of this demonstration is to recover the initial position A from position B and

Front View (xy)   Top View (xz)

Fig. 11.9. Front and top views of the superposition of the two 3D frames after applying the computed motion

Table 11.2. The computed motion parameters between the positions A and B

	x component	y component	z component
r	−5.427e−03	−1.625e−01	1.268e−03
t	3.057e+02	−2.519e+00	6.887e+02

to return to position A. Note that there is a large displacement between these two positions, which can be noticed by simply superimposing the two frames (see Fig. 7.6).

The motion-estimation algorithm is applied to these two 3D frames and the displacement is recovered as shown in Table 11.2. From **r** we know that there is a rotation of −9.32 degrees and that the direction of the rotation axis is $[-3.338\text{e}{-}02, -9.994\text{e}{-}01, 7.798\text{e}{-}03]^T$, almost vertical. We apply this final estimated displacement to the first frame and display it in the reference attached to the second frame. Figure 11.9 shows the result: the segments from the first frame are in dotted lines, while those from the second frame are in solid lines. The displacement between the two frames can be noticed by the shift of the triangle. One can observe the very good accuracy of this estimate.

From these results, three robot commands are generated (as described in Sect. 11.3) and are executed. The total rotation is −9.32 degrees and the translation is 747 millimeters. The robot is now in position A', not far from A. If there are no errors in motion estimation, calibration and mechanical system, A' should be exactly the same as A. To measure the error of this demonstration, the robot

Fig. 11.10. Three images taken by the robot in position A'

Table 11.3. The error of the complete navigation loop

	x component	y component	z component
r	1.481e−04	−1.060e−02	2.762e−06
t	2.487e+01	−2.209e−01	−5.883e+00

takes again three images (see Fig. 11.10). They should be compared with those in Fig. 11.7. If there is no error in the vision and mechanical systems, the two triplets should be identical. Visually, the difference between them is very small. The corresponding 3D frame is reconstructed in this position (see Fig. 11.11) and the motion-estimation algorithm is applied again to the 3D frames of A and A'. The motion between A and A' (i.e., the errors in motion of the whole demonstration) we obtain are shown in Table 11.3. That is, we have an error in

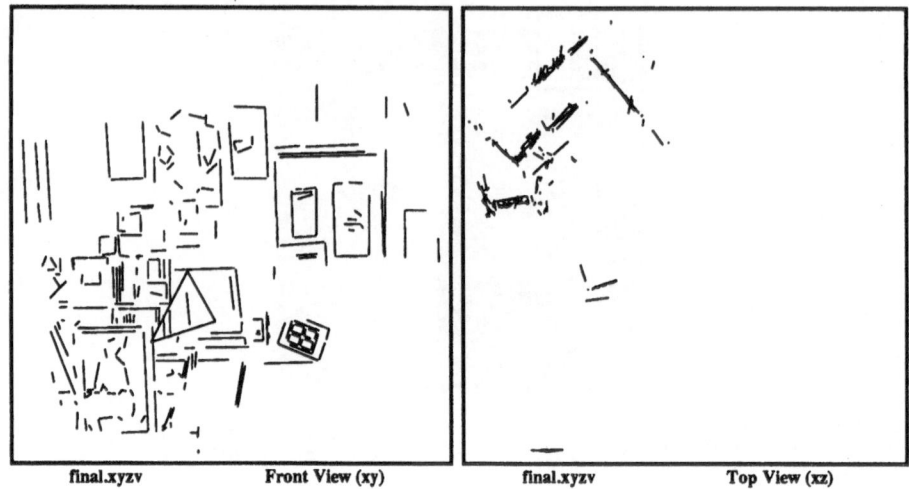

final.xyzv   Front View (xy)   final.xyzv   Top View (xz)

Fig. 11.11. Front and top views of the 3D frame in position A' (uniform scale)

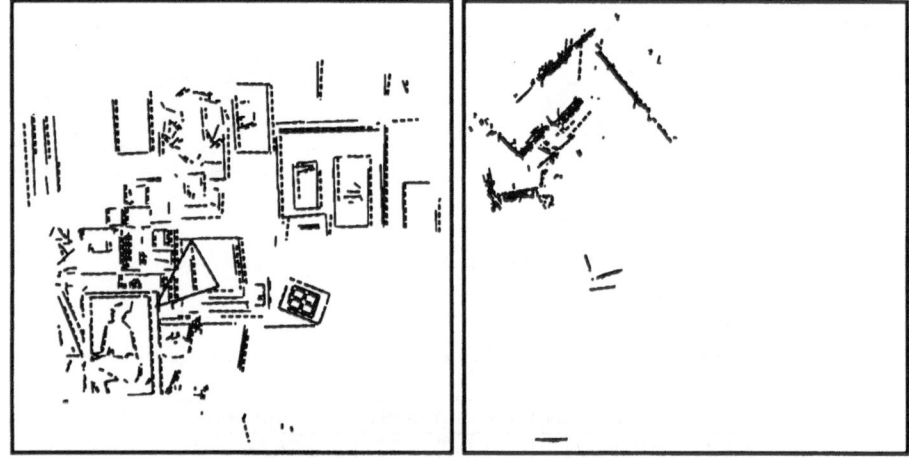

Fig. 11.12. Comparison between the 3D frame reconstructed in position A and that in position A' (uniform scale)

rotation equal to 0.6 degrees and an error in translation equal to 27 millimeters, both less than 10%. Figure 11.12 shows the difference between the two frames. Segments from the last frame are drawn in solid lines, while those from the first frame are drawn in dashed lines. Note that this error is the cumulation of errors in all different phases of our navigation loop, in particular the errors due to the mechanical system of the robot.

Table 11.4. Calibration results of another set of experiments

	x component	**y** component	**z** component
$\hat{r}$	−4.80397e−03	9.99948e−01	−8.94829e−03
C	−7.36543e+02	−9.38340e−04	−1.12195e+03
$\hat{t}$	6.54501e−01	5.02050e−02	7.54393e−01

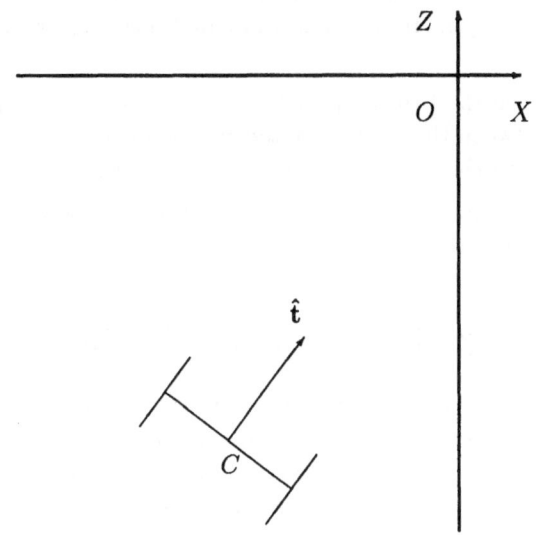

Fig. 11.13. The relation between the stereo and odometric coordinate systems

The robot calibration (i.e., the relation between the stereo and odometric systems) changes with the stereo calibration. Each time the stereo calibration is redone, we must rerun the eye-wheel calibration process to get the updated relation. For example, in another navigation demonstration, we obtained the calibration results as shown in Table 11.4. The results are obtained after 3 pure rotations and 3 pure translations. Figure 11.13 depicts the relation between the stereo odometric coordinate systems (only the projection on the plane $y = 0$ is shown). The results are completely different from those shown in Table 11.1 except for the rotation axis. The rotation axis is always almost perpendicular to the plane $y = 0$ which represents the ground plane. The dramatic difference between the two calibration results is due to the change of the grid (pattern-board) for the stereo calibration. In our stereo calibration technique [11.1, 11], the stereo coordinate system is attached to the calibration grid.

In order to know the robustness of our calibration process, we have done another set of displacements (3 pure rotations and 3 pure translations). The stereo calibration remains the same, but the positions, the angles of rotations and the distance of translations in this robot calibration demonstration are different

Table 11.5. Calibration results in the same demonstration as in Table 11.4

	x component	y component	z component
$\hat{r}$	5.20855e−04	9.99989e−01	4.64361e−03
C	−7.46351e+02	2.84762e−03	−1.12478e+03
$\hat{t}$	6.52346e−01	3.98895e−02	7.56871e−01

from those in the previous demonstration. The calibration results are shown in
Table 11.5. Comparing the results with those in Table 11.4, we find the following
differences:

- The angle between the two computed rotation axes is 0.84 degrees,
- The distance between the two computed rotation centers is 9.8 millimeters,
- The angle between the two computed translation directions is 0.62 degrees.

From the above presentation, we conclude that the proposed calibration process
for the mobile robot is robust.

11.5 Integrating Motion Information from Odometry

We have presented in the previous sections how to use the motion information
from stereo to guide the navigation of our mobile robot. In this section, we show
how we can integrate the motion information from the odometric system of the
robot to speed up the motion-determination process.

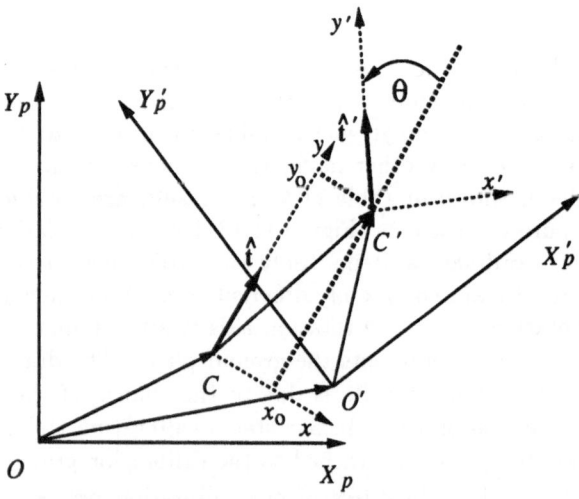

Fig. 11.14. Computing an initial motion estimate from odometry

The problem addressed here is dual to that of Sect. 11.3. In Fig. 11.14, we portray the different coordinate systems necessary for ulterior computation. Only the projection on the ground plane is shown, which we assume is perpendicular to the rotation axis $\hat{\mathbf{r}}$. Cxy is the odometric coordinate system and OX_pY_p is that of the stereo system (the subscript p denotes the *projection*). The information about the robot displacement given by the odometric system is expressed in Cxy reference, that is, (x_o, y_o, θ), where (x_o, y_o) are the coordinates of the new position C' and θ is the rotation angle of the frame (see Fig. 11.14). Note that θ is a signed value according to the direction of the rotation: θ is positive if the rotation axis has a positive Z_p component, and negative otherwise.

Let us first compute the robot displacement in the stereo frame, which is denoted by $(\mathbf{r}_r, \mathbf{t}_r)$ for the rotation and translation. After verifying the sign consistence, the rotation $\mathbf{r}_r$ is simply given by

$$\mathbf{r}_r = \theta\hat{\mathbf{r}} \ . \tag{11.5}$$

The translation $\mathbf{t}_r$ is a little bit more complicated. We find that

$$\mathbf{t}_r = \overrightarrow{OO'} = \overrightarrow{OC} + \overrightarrow{CC'} - \overrightarrow{O'C'} \ , \tag{11.6}$$

where all vectors must be evaluated in the stereo coordinate system. $\overrightarrow{OC}$ is known equal to $\mathbf{C}$ from the calibration. The vector $\overrightarrow{O'C'}$ is equal to $\mathbf{C}$ in the new stereo frame, so $\overrightarrow{O'C'}$ in the original stereo frame is given by

$$\overrightarrow{O'C'} = \mathbf{R}_r\mathbf{C} \ , \tag{11.7}$$

where $\mathbf{R}_r$ is the rotation matrix defined by $\mathbf{r}_r$. It remains to compute $\overrightarrow{CC'}$. Because the y-axis of the odometric coordinate system is equal to $\hat{\mathbf{t}}$ and the x-axis is equal to $\hat{\mathbf{t}} \wedge \hat{\mathbf{r}}$ (where $\wedge$ denotes the cross product), $\overrightarrow{CC'}$ is then given by

$$\overrightarrow{CC'} = x_o(\hat{\mathbf{t}} \wedge \hat{\mathbf{r}}) + y_o\hat{\mathbf{t}} \ . \tag{11.8}$$

Therefore, the translation of the robot is computed by

$$\mathbf{t}_r = (\mathbf{I} - \mathbf{R}_r)\mathbf{C} + x_o(\hat{\mathbf{t}} \wedge \hat{\mathbf{r}}) + y_o\hat{\mathbf{t}} \ . \tag{11.9}$$

Seeing that the observed scene motion is the inverse of the robot displacement, the motion between the first and second stereo frames is given by

$$\mathbf{r} = -\mathbf{r}_r \ , \quad \text{and} \quad \mathbf{t} = -\mathbf{R}_r^T\mathbf{t}_r = -\mathbf{R}\mathbf{t}_r \ , \tag{11.10}$$

where $\mathbf{R}$ is the rotation matrix defined by $\mathbf{r}$.

Now we provide an experimental example. The two stereo frames observed by the robot are superimposed in Fig. 11.15 to display the difference. The odometry gives the motion parameters as $[-289.0, 0.0, 0.0873]^T$, or, 289 millimeters

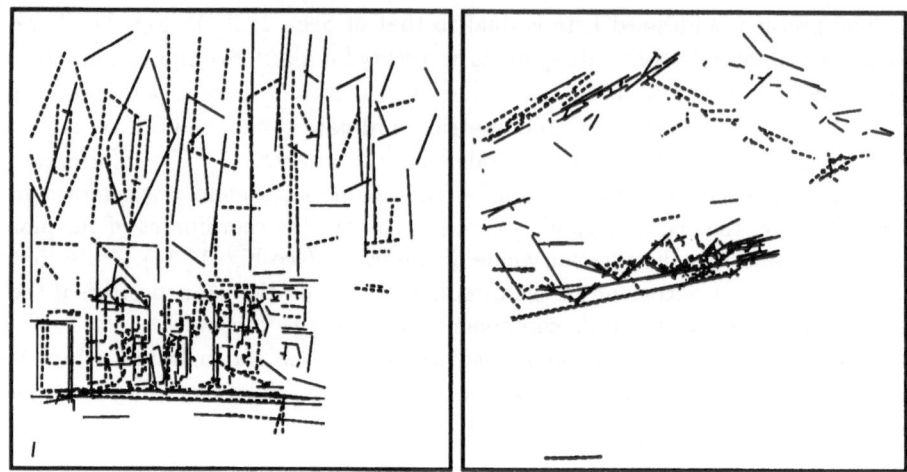

Fig. 11.15. Superposition of two original stereo frames

Fig. 11.16. Result after applying the motion estimate from odometry

of translation and 5 degrees of rotation. The calibration results are given in Table 11.6.

Based on the formulas described above, we compute an initial motion estimate as $[-1.397e-03, \ 8.725e-02, \ 1.776e-04, \ 3.228e+02, \ 1.280e+01, \ 1.018e+02]^T$. The result is shown in Fig. 11.16, where we have applied the initial estimate to the first frame and superimposed it on the second. We observe that the odometry gives approximately the real displacement of the robot. Since we have

Fig. 11.17. Result after applying the final motion estimate

Table 11.6. Another calibration result of the INRIA mobile robot

	x component	y component	z component
$\hat{r}$	−1.601e−02	9.999e−01	2.036e−03
C	−6.901e+02	5.278e−05	−1.022e+03
$\hat{t}$	7.649e−01	3.807e−02	6.430e−01

already an initial estimate, the generation phase of the motion-determination algorithm described in Chap. 7 is not needed anymore (but we should remember that we need it in the eye-wheel calibration phase). We use directly the verification algorithm (see Algorithm 7.3) to find the matches between the two frames and to update the estimate. The final estimate given by the algorithm is $[-1.306\mathrm{e}{-03}, 8.948\mathrm{e}{-02}, -1.363\mathrm{e}{-03}, 3.383\mathrm{e}{+02}, 3.741\mathrm{e}{+00}, 1.060\mathrm{e}{+02}]^T$. The result is shown in Fig. 11.17, where we have applied the final estimate to the first frame and superimposed it onto the second. We observe that the motion between the two frames is very well recovered. The execution of the above process on a SUN 3/60 workstation takes about 10 seconds. If we apply the whole motion algorithm on these two frames, we get almost the same result, but the execution time is about 130 seconds. This shows the advantage of using information from the odometry system.

11.6 Summary

In this chapter, we have addressed the calibration problem between the stereo co-ordinate system and the odometric system of the INRIA mobile robot, which we call the eye-wheel calibration problem. The calibration is essential for many applications, including visually guided navigation. The proposed method recovers the direction of the rotation axis, a point on the rotation axis, and the direction of translation of the robot in the stereo coordinate system, using some simple maneuvers (pure rotations and pure translations). The motion-estimation algorithm registers two successive stereo frames and determines the displacement between them. The displacement estimate is then used in the calibration algorithm. To reduce the uncertainty in the calibration parameters due to the vision algorithm and especially the mechanical system, a weighted least-squares adjustment is carried out. A complete visual navigation loop including the calibration technique has been presented. The result is shown to be satisfactory and promising. Though this is still simple, it is the first step towards solving more complicated navigation problems.

We have also demonstrated that we could considerably speed up the motion-determination process by integrating the odometric information from the robot. This is possible only when the relation between the stereo and odometric coordinate systems is calibrated. We have derived several formulas to compute an initial estimate of motion from the odometry system. Since an initial estimate is available, the generation phase is skipped over, and we use directly the verification algorithm to find matches between two stereo frames and to update the motion estimate. An experimental example has been provided to show the improvement.

12. Fusing Multiple 3D Frames

In this chapter, we present a system, called the *world-model builder system*, which builds automatically and incrementally a model of an unknown environment. Since most parts of the system have been described previously, we concentrate, after a brief description of the system, on how to fuse multiple 3D frames obtained by a stereo system.

The world model is for the moment based on 3D line segments, reconstructed by the trinocular stereovision system described in Chap. 3. The world-model builder system uses the watch-and-explore strategy: based on the information from stereo, it decides where is the interesting space to explore, plans the trajectories, navigates, and updates the global model of the environment with the currently observed stereo frame. This world model may be used later for navigation or recognition. Besides stereovision, this system involves several important issues in Robotics and Vision:

1. Matching consecutive 3D frames and estimating motion from 3D data,
2. Finding areas of interest for further exploration,
3. Planning trajectories and controlling the robot,
4. Fusing 3D data obtained at different instants,
5. Interpreting and possibly editing 3D data.

The matching and motion-estimation problems have already been addressed in the previous chapters (Chaps. 5 to 7). The solution to the second problem is usually goal-dependent. How to displace the robot with minimum energy (or other criteria) while avoiding stationary or moving obstacles is a challenging subject of research. Some work related to this issue includes [12.1–3]. A simple example of trajectory planning for navigation has been described in Chap. 11. Data fusion, i.e., how to model sensor noise and how to integrate optimally data from multiple sensors or from the same sensor at different instants, has attracted many researchers [12.4–8]. Finally, 3D data should be organized and interpreted (possibly edited) by using some *a priori* geometric constraints in order to obtain a consistent and higher level representation [12.9, 10].

The problem of integrating a sequence of *monocular* views to build a global map was addressed by *Jezouin* and *Ayache* [12.11], where the motion is assumed given with very good accuracy by inertia sensors. A sequence of images of an outdoor scene generated by a realistic image synthesis system was used, and they were able to reconstruct primitives (points and line segments) of buildings at a distance of several thousands of meters with an accuracy of a couple of meters. As pointed by the authors, the accuracy of 3D reconstruction depends heavily on camera motion.

The problem of integrating sonar and stereo range data was addressed by *Matthies* and *Elfes* [12.12]. They used a cellular representation called the *Occupancy Grid* to describe the vicinity of a robot. Range information from sonar and one-dimensional stereo is combined into such a 2D map. Each cell in the map contains a probabilistic estimate of whether it is empty or occupied by an object in the environment. These estimates are obtained from sensor models that describe the uncertainty in the range data. A Bayesian estimate scheme is applied to update the current map using successive range readings from each sensor. The occupancy grid representation is suitable for robot navigation application. It is, however, very limited in its descriptive ability if we want to interpret and understand the 3D environment of a robot.

A similar system to ours, called the 3D Mosaic scene understanding system, was designed by *Herman* and *Kanade* at CMU [12.13, 14]. A comment to this system can be found in the summary section.

In the remainder of this chapter, we first present our system. After a brief explanation of each module of the system, our emphasis is on the fusion of line segments measured at different instants. Two experimental results are provided. The first one is the fusion of two stereo views, which will give us a good feeling of the performance of the fusion process. The second shows the result of fusing 35 different stereo views of a room.

12.1 System Description

Our system is being developed on the INRIA mobile robot. The *perception* tools include 3 cameras and 24 ultrasound cells. The three cameras form a trinocular stereovision system providing a local 3D map of the environment. Ultrasound is, for the moment, used only to avoid obstacles. The *action* is to move the robot to the desired position. More details about the robot can be found in Sect. 11.1. The tasks of the *intelligent decision and control system* include finding points of interest and planning trajectories. Since the coordinate systems of the perception and action components are different, these coordinate systems should be first calibrated (see Chap. 11).

Figure 12.1 shows the architecture of our world-model builder. As indicated by the name, the goal of our system is to incrementally build a model of its environment. We, for the moment, restrict the domain to the indoor environment. The system is simply composed of several individual modules. First, the *Stereovision Module* (**SM**) [12.15, 16] builds a local visual map which is a set of 3D segments. The *Analysis and Decision Module* (**ADM**) then analyzes the local map. By combining the information from the *Data Fusion Module* (**DFM**) and *Interpretation Module* (**IM**), the **ADM** decides where the robot should go in

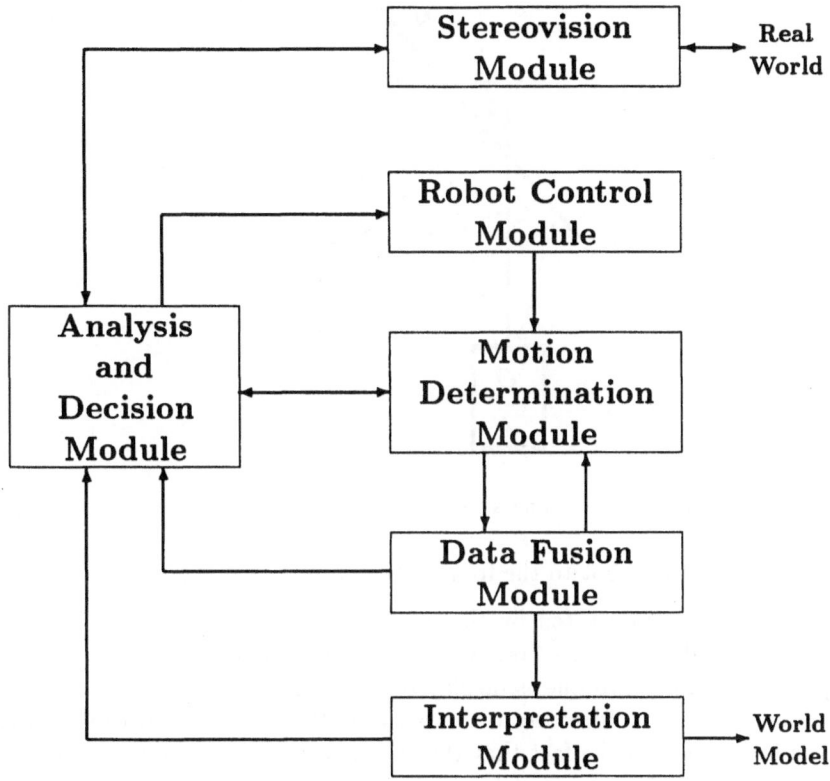

Fig. 12.1. System architecture

order to explore further the environment and to obtain more precise data. This module is being under development [12.17]. For the moment, it is the user who accomplishes its task with the aid of a graphic interface. The graphic interface displays the projection of the 3D segments on the ground plane, the borders of objects recovered by making the Delaunay triangulation on the projected data, and also the position of the robot. The borders are updated when new observations become available. It is then sufficient for the user to click the mouse on the next preferred position. The following three strategies will be implemented in the system:

— **If** segments of an object are not accurate enough,
   and the space between the object and the robot is free,
   then move the robot ahead.
— **If** there is room for the robot to rotate θ degrees to the right (or left),
   then rotate the robot to the right (or left) $\min(\theta, \Theta/4)$ degrees.
   (Θ is the field of view of the stereovision system)
— **If** the angle between the optical direction OZ and the object to observe MN is small enough (less than 45 degrees, for example),

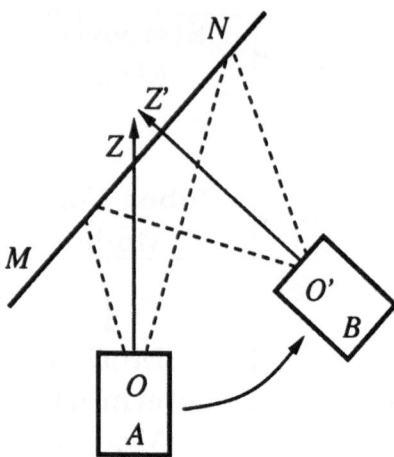

Fig. 12.2. Illustration of a planning strategy: the robot is represented by a rectangle

then move the robot to the front of the object. (see Fig. 12.2)

These strategies are motivated by the need to reduce the uncertainty due to the stereo triangulation. The first two are easy to understand. The situation corresponding to the third one is displayed in Fig. 12.2. The robot is found in the position A and should move to the position B. There are two reasons for this. The first is due to the fact that the cameras are mainly calibrated for the central parts. The reconstructed 3D information is less reliable in the border than in the center. This effect is more severe in position A than in position B. The second reason is that when the angle between OZ and MN is small, a part of the observed object is far from the camera and the reconstructed 3D information is not very useful.

The **ADM** module then sends a command of displacement to the *Robot Control Module* (**RCM**). Since the command of displacement is related to the stereo coordinate system and the **RCM** works in the odometric one, we should first calibrate the two coordinate systems (see Chap. 11). When the **RCM** module receives a command of displacement, it plans the trajectories, as smooth as possible, to reach the goal. The ultrasound sensors are used to avoid local obstacles on the trajectory [12.18]. The real trajectories are displayed by the graphic interface.

The task of the *Motion-Determination Module* (**MDM**) is to match successive 3D frames and to obtain a precise estimate about the robot displacement. In Chap. 7, we have presented a motion-estimation algorithm based on the hypothesize-and-verify paradigm. We can also make use of the odometric information as an initial estimate of the motion to speed up considerably the process, as described in Sect. 11.5. Later, the initial estimate of motion may come directly from the **ADM**.

Motion estimation and segment correspondences are then provided to the *Data Fusion Module*. Section 12.2 describes in detail this module. The fused data are fed back to the *Motion-Determination Module* to track the position of the robot at the next instant. The fused data are also supplied to the *Analysis and Decision Module* and the *Interpretation Module*.

Now we have only a set of 3D line segments. It is not useful enough for navigation or object recognition. 3D line segments should be organized and interpreted in order to obtain a higher level representation. Since the 3D line segments we have are still noisy, we can impose some *a priori* geometric constraints on them to reduce the uncertainty as in [12.7]. All these are the tasks of the *Interpretation Module*. This module is being developed. We have implemented some primitive procedures: parallelism of segments, perpendicularity of segments, coplanarity of segments, etc. Some preliminary results have already been presented in Chap. 8. We note also several researchers working on this subject [12.9, 10].

12.2 Fusing Segments from Multiple Views

The problem addressed in this section is known as the multiple viewpoint problem. The objective is to build a consistent, accurate and complete description (model) of objects or environments by combining the observations taken by the stereo system from multiple stereo views. One stereo view can only provide partial imprecise information about the environment, which is not sufficient for interpretation, recognition or navigation.

Assume the robot navigates in a static environment. Suppose that at time t_i we have built up a model $\mathcal{F}_{i-1}$ from previous views, we now want to integrate it with the new view $\mathcal{S}_i$. Figure 12.3 shows a more detailed description of the relation between the motion-determination and data-integration modules. The stereo frame at the current instant $\mathcal{S}_i$ and that of the preceding instant $\mathcal{S}_{i-1}$ are used at the first stage to generate hypotheses of motion between successive instants. We do not use the fused data $\mathcal{F}_{i-1}$ since the complexity of the hypothesis generation algorithm in the worst case is $O(n^2m^2)$ (n and m are the number of primitives in $\mathcal{F}_{i-1}$ and $\mathcal{S}_i$, respectively) and $\mathcal{F}_{i-1}$ has more primitives than $\mathcal{S}_i$. $\mathcal{S}_i$ and $\mathcal{F}_{i-1}$ (at time t_2, $\mathcal{F}_1$ is the same as $\mathcal{S}_1$) are used at the verification stage which provides as its output an optimal estimate of the egomotion $\mathbf{d}_i$ and also segment

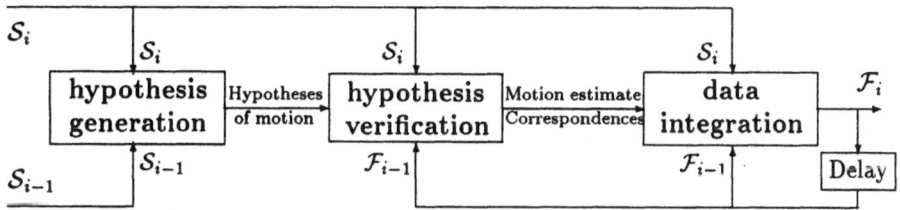

Fig. 12.3. Relation between motion estimation and data integration

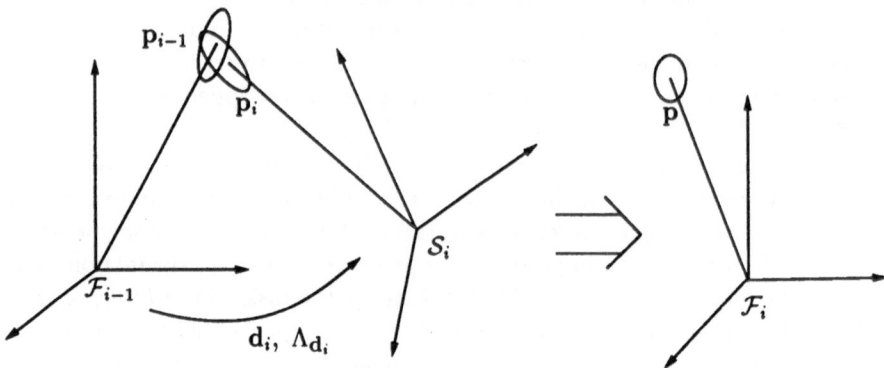

Fig. 12.4. Fusing data from two different views

correspondences between $\mathcal{S}_i$ and $\mathcal{F}_{i-1}$. The integration module then builds a more accurate and complete model $\mathcal{F}_i$ by combining all available information. We choose the coordinate system of the last observed frame $\mathcal{S}_i$ as that of the global model $\mathcal{F}_i$ being updated.

Fused segments have a label which indicates their number of instances up to the current instant. For example, if a segment in $\mathcal{F}_{i-1}$ with a label k is matched to a segment in $\mathcal{S}_i$, then the fused segment in $\mathcal{F}_i$ has a label $k + 1$. Unmatched segments in $\mathcal{S}_i$ are simply copied into $\mathcal{F}_i$ and are labeled with 1. These segments are always retained since they are likely to appear at the next instant. Unmatched segments in $\mathcal{F}_{i-1}$ with a label 1 may or may not be retained depending upon whether the number of primitives in $\mathcal{F}_i$ becomes too big or not. Those segments are less likely to appear at the next instant than those unmatched segments in $\mathcal{S}_i$. In fact, a segment which appeared only once during several preceding instants is very likely not to correspond to any real segment in space. It may have been reconstructed from false matches in stereo or its image contours have been poorly segmented. Other unmatched segments in $\mathcal{F}_{i-1}$ are retained in $\mathcal{F}_i$ after applying the motion estimate $\mathbf{d}_i$ and its covariance matrix $\Lambda_{\mathbf{d}_i}$ to them. See the appendix to Chap. 7 for how to transform a 3D line segment.

There remains the problem of fusing the primitives in correspondence into a new primitive while taking into account $\mathbf{d}_i$ and its covariance matrix $\Lambda_{\mathbf{d}_i}$ (see Fig. 12.4).

12.2.1 Fusing General Primitives

We use a procedure based on the Kalman filter theory to integrate multiple observations. Suppose we have two independent observations $\mathbf{x}_1$ and $\mathbf{x}_2$ with their covariance matrices Λ_1 and Λ_2 of a state vector $\mathbf{x}$. The modified Kalman minimum-variance estimator says that the optimal estimate is given by[1]

$$\hat{\mathbf{x}} = [\Lambda_1^{-1} + \Lambda_2^{-1}]^{-1}[\Lambda_1^{-1}\mathbf{x}_1 + \Lambda_2^{-1}\mathbf{x}_2] \,,$$
$$\Lambda = [\Lambda_1^{-1} + \Lambda_2^{-1}]^{-1} \,. \tag{12.1}$$

That is, the optimal estimate is just a weighted average of the observations and the corresponding information matrix (i.e., the inverse of the covariance matrix) is the sum of the information matrices of the observations. The covariance matrix is always reduced in integrating an observation. Equation (12.1) can be easily proved using the standard Kalman filter (see Sect. 2.2) by setting the transformation matrix H_i and the measurement matrix F_i to identity matrices. Note that we can apply iteratively the above procedure if we have more than two observations. First we use (12.1) to compute a new estimate $\hat{\mathbf{x}}$ and its associated covariance matrix Λ, then for every other observation we compute an up-to-date estimate by integrating the old one with the observation based on (12.1).

One requirement to use the above procedure is that all observations are expressed in the same coordinate system. Consider Fig. 12.4. Let $\mathbf{p}_{i-1}$ and $\mathbf{p}_i$ be primitives in correspondence in $\mathcal{F}_{i-1}$ and $\mathcal{S}_i$ and $\mathbf{p}$ the new primitive to be included in $\mathcal{F}_i$ (we choose the coordinate system of $\mathcal{S}_i$ as that of $\mathcal{F}_i$). We see that $\mathbf{p}_{i-1}$ and $\mathbf{p}_i$ are represented in two different coordinate systems which are related by the motion vector $\mathbf{d}_i$ and its covariance matrix $\Lambda_{\mathbf{d}_i}$. In order to use the above procedure, we should first transform $\mathbf{p}_{i-1}$ in the coordinate system of $\mathcal{F}_i$ based on $\mathbf{d}_i$ and $\Lambda_{\mathbf{d}_i}$, denoted by $\mathbf{p}'_{i-1}$. The appendix to Chap. 7 explains how to transform a 3D line segment.

Another requirement is that the observations are independent. Strictly speaking, $\mathbf{p}'_{i-1}$ and $\mathbf{p}_i$ are not independent, although $\mathbf{p}_{i-1}$ and $\mathbf{p}_i$ can be reasonably assumed independent. This is because $\mathbf{p}'_{i-1}$ is computed from the original primitive $\mathbf{p}_{i-1}$ by applying the motion estimate $\mathbf{d}_i$ and both $\mathbf{p}_{i-1}$ and $\mathbf{p}_i$ contribute to the estimation of $\mathbf{d}_i$. This results in a correlation between $\mathbf{p}'_{i-1}$ and $\mathbf{p}_i$. However this correlation is negligible since the motion vector $\mathbf{d}_i$ is usually estimated from more than 30 correspondences (150 correspondences, sometimes).

12.2.2 Fusing Line Segments

A 3D line segment is represented by $(\psi, \mathbf{m}, l)$ and $(\Lambda_\psi, \Lambda_\mathbf{m})$ (see Sect. 4.2). No modelization of the uncertainty of the segment lengths has been carried out

---

[1]These formulae can be extended for n independent observations $\mathbf{x}_i$ $(i = 1, \ldots, n)$:

$$\hat{\mathbf{x}} = (\Sigma_{i=1}^n \Lambda_i^{-1})^{-1}(\Sigma_{i=1}^n \Lambda_i^{-1}\mathbf{x}_i) \,,$$
$$\Lambda = (\Sigma_{i=1}^n \Lambda_i^{-1})^{-1} \,.$$

When $n = 2$, the following formulae are computationally less expensive

$$\hat{\mathbf{x}} = \Lambda_2(\Lambda_1 + \Lambda_2)^{-1}\mathbf{x}_1 + \Lambda_1(\Lambda_1 + \Lambda_2)^{-1}\mathbf{x}_2 \,,$$
$$\Lambda = \Lambda_1(\Lambda_1 + \Lambda_2)^{-1}\Lambda_2 \,.$$

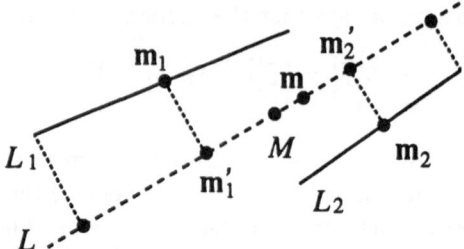

Fig. 12.5. Union of two matched segments

(we do not need it). Let the *transformed* segment from $\mathcal{F}_{i-1}$ be represented by $(\psi_1, \mathbf{m}_1)$ with their covariance matrices $(\Lambda_{\psi_1}, \Lambda_{\mathbf{m}_1})$ and the segment from $\mathcal{S}_i$ be represented by $(\psi_2, \mathbf{m}_2)$ with their covariance matrices $(\Lambda_{\psi_2}, \Lambda_{\mathbf{m}_2})$. Then the parameters of the fused segment are given by

$$
\begin{aligned}
\psi &= [\Lambda_{\psi_1}^{-1} + \Lambda_{\psi_2}^{-1}]^{-1}(\Lambda_{\psi_1}^{-1}\psi_1 + \Lambda_{\psi_2}^{-1}\psi_2) \,, \\
\Lambda_\psi &= [\Lambda_{\psi_1}^{-1} + \Lambda_{\psi_2}^{-1}]^{-1} \,, \\
\mathbf{m} &= [\Lambda_{\mathbf{m}_1}^{-1} + \Lambda_{\mathbf{m}_2}^{-1}]^{-1}(\Lambda_{\mathbf{m}_1}^{-1}\mathbf{m}_1 + \Lambda_{\mathbf{m}_2}^{-1}\mathbf{m}_2) \,, \\
\Lambda_{\mathbf{m}} &= [\Lambda_{\mathbf{m}_1}^{-1} + \Lambda_{\mathbf{m}_2}^{-1}]^{-1} \,.
\end{aligned}
\tag{12.2}
$$

They give the orientation and position of the fused segment.

In the case of the discontinuity of ϕ (see Sect. 4.2.2, page 47), care must be taken before we use the above procedure. The idea is the following. If a segment is represented by $\psi = [\phi, \theta]^T$, it is also represented by $[\phi - 2\pi, \theta]^T$. Therefore, when fusing two segments, we perform the following tests and actions. If $\phi_1 < \pi/2$ and $\phi_2 > 3\pi/2$, then set $\phi_2 = \phi_2 - 2\pi$; else if $\phi_1 > 3\pi/2$ and $\phi_2 < \pi/2$, then set $\phi_1 = \phi_1 - 2\pi$; else do nothing. Notice that adding a constant to a random variable does not affect its covariance matrix. Using ϕ_1 and ϕ_2, we can compute a ϕ' using the above procedure. The ϕ of the fused segment is then equal to ϕ' if $\phi' \geq 0$, or equal to $2\pi + \phi'$ if $\phi' < 0$.

Due to the reasons described in Sect. 4.2, we can conclude that a reconstructed segment is only part of a real segment in space. Two segments are considered to be matched if they have a common part (see Sect. 7.4). Their corresponding segment in space can be expected not to be shorter than either of the two segments. That is, the union of the two segments can be reasonably considered as a better estimate of the corresponding segment in space.

Consider two segments to be fused, L_1 and L_2 in Fig. 12.5. First an infinite line L is computed using (12.2). Then the midpoints of the two segments are projected on L, and we get $\mathbf{m}'_1$ and $\mathbf{m}'_2$. The lengths of the projected segments are also computed, and are denoted by l'_1 and l'_2. Now we want to compute the real midpoint M and the length l of the fused segment L. After studying all cases, we have Algorithm 12.1, in which $\overrightarrow{\mathbf{m}'_1\mathbf{m}'_2} = \mathbf{m}'_2 - \mathbf{m}'_1$.

Algorithm 12.1: Union of Two Segments

- **Input:** $(\boldsymbol{\psi}_1, \mathbf{m}_1, l_1)$ and $(\boldsymbol{\psi}_2, \mathbf{m}_2, l_2)$
- **Output:** $(\boldsymbol{\psi}, \mathbf{m}, l)$
- **Define** local variables: $\mathbf{m}'_1$, $\mathbf{m}'_2$, l'_1, l'_2 and M

- **Compute**, using (12.2), the infinite line L: $(\boldsymbol{\psi}, \mathbf{m})$
- Project $\mathbf{m}_1$ and $\mathbf{m}_2$ on L: $\mathbf{m}'_1$ and $\mathbf{m}'_2$
- Project l_1 and l_2 on L: l'_1 and l'_2
- Compute the real midpoint M and length l of L:

$\rightarrow$**if** $(\|\overrightarrow{\mathbf{m}'_1\mathbf{m}'_2}\| + l'_2/2 < l'_1/2)$
    then $\hookrightarrow M = \mathbf{m}'_1, \quad l = l'_1$

    else if $(\|\overrightarrow{\mathbf{m}'_1\mathbf{m}'_2}\| + l'_1/2 < l'_2/2)$
    then $\hookrightarrow M = \mathbf{m}'_2$
        $\hookrightarrow l = l'_2$

    else $\hookrightarrow M = \mathbf{m}'_1 + \dfrac{\|\overrightarrow{\mathbf{m}'_1\mathbf{m}'_2}\| + l'_2/2 - l'_1/2}{2} \dfrac{\overrightarrow{\mathbf{m}'_1\mathbf{m}'_2}}{\|\overrightarrow{\mathbf{m}'_1\mathbf{m}'_2}\|}$

        $\hookrightarrow l = l'_1 + (\|\overrightarrow{\mathbf{m}'_1\mathbf{m}'_2}\| + l'_2/2 - l'_1/2)$
$\leftarrow$**endif**
- Set $\mathbf{m} = M$

The covariance matrix of M can be extrapolated from that of $\mathbf{m}$, $\Lambda_\mathbf{m}$, and that of the unit direction vector $\mathbf{u}$, $\Lambda_\mathbf{u}$. Indeed, M can always be represented by

$$M = \mathbf{m} + s\mathbf{u} \ , \tag{12.3}$$

where s is an arbitrary scalar. The above equation can be interpreted as the addition of a *biased* noise to $\mathbf{m}$, therefore the covariance matrix of M is given by

$$\Lambda_M = \Lambda_\mathbf{m} + s^2(\Lambda_\mathbf{u} + \mathbf{u}\mathbf{u}^T) \ . \tag{12.4}$$

12.2.3 Example

In this section, we give an example of fusing two segments.

Figure 12.6 shows two segments which are taken from two real stereo frames. Pictures in the first row are the front views and those in the second row are the top views. Figure 12.6a shows the superposition of the two original segments and Figure 12.6b shows their superposition after applying the estimated motion to the first segment (in dashed line). The solid segment in Fig. 12.6c is the fused one which is, for reason of comparison, superimposed with its original segments (in dashed line). The uncertainties of the midpoints are represented by ellipses.

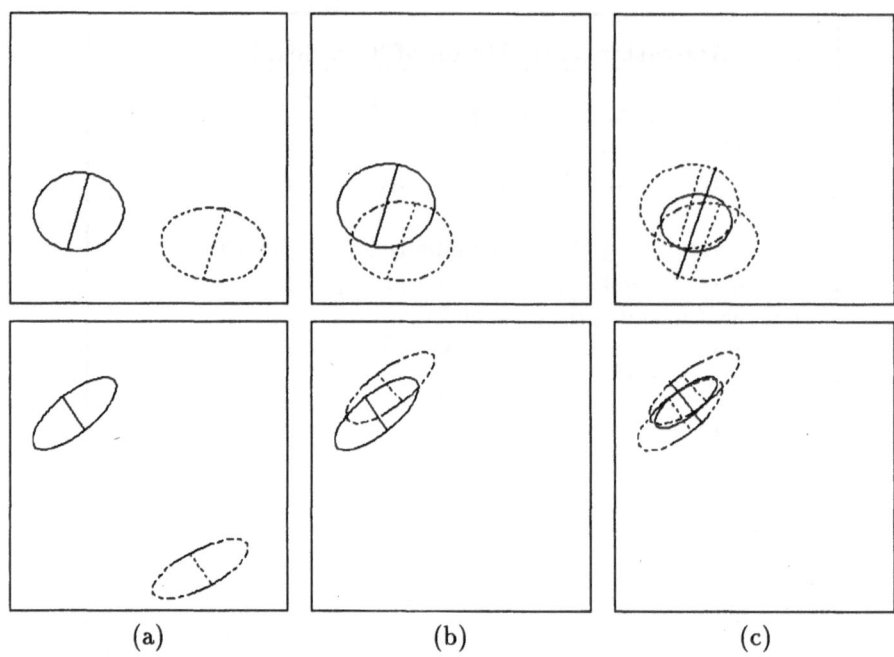

Fig. 12.6. Fusing two segments: evolution of the uncertainty of the midpoint

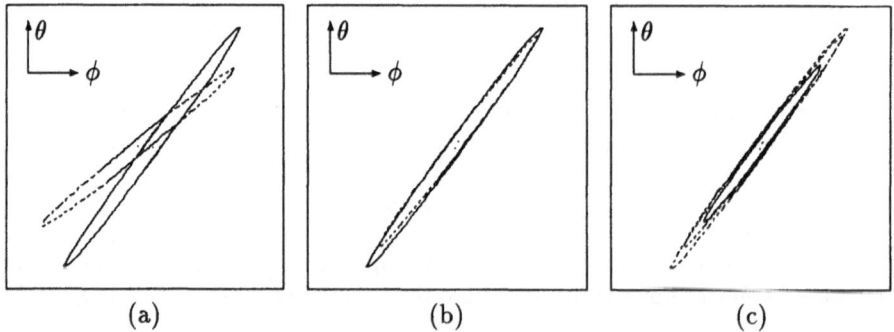

Fig. 12.7. Fusing two segments: evolution of the uncertainty of the orientation

Figure 12.7 shows the evolution of uncertainty in the orientation parameters ψ. Figure 12.7a shows the superposition of the two original ψ's (represented by points) and their uncertainty ellipses, and Figure 12.7b shows their superposition after applying the estimated motion to the first ψ (in dashed line). Figure 12.7c displays the ψ and its uncertainty ellipse of the fused segment (in solid line) with its original observations (in dashed line). We can observe how the uncertainty is reduced.

The reader is referred to [12.19] for a more extended series of experiments with synthetic data which shows how fusion can considerably improve the precision in measurements.

12.2.4 Summary of the Fusion Algorithm

A fused 3D line segment has a label called the *instance* which indicates the number of instances, i.e., the number of original segments which have been fused, until the current instant. Algorithm 12.2 gives a more formally description on how the fusion process works. The fused frame $\mathcal{F}_1$ at t_1 is set to be the observed stereo frame $\mathcal{S}_1$.

Algorithm 12.2: Integration of Multiple Stereo Views

- **Input:** Stereo frames $\mathcal{S}_{i-1}$ and $\mathcal{S}_i$
  Previous fused frame $\mathcal{F}_{i-1}$
- **Output:** Updated fused frame $\mathcal{F}_i$

- /* Algorithm 7.2 */
  Generate motion hypotheses between $\mathcal{S}_{i-1}$ and $\mathcal{S}_i$
- /* Algorithm 7.3 */
  $\rightarrow$ **for** each hypothesis generated, using $\mathcal{F}_{i-1}$ and $\mathcal{S}_i$
  $\hookrightarrow$ Compute the optimal motion estimate $\mathcal{D}$
  $\hookrightarrow$ Find line segment matches $\{\mathcal{M}\}$
  $\leftarrow$ **endfor**
- Choose the best hypothesis which gives $\mathcal{D}_{\text{best}}$ and $\{\mathcal{M}_{\text{best}}\}$
- Get $\mathcal{F}'_{i-1}$ by applying $\mathcal{D}_{\text{best}}$ to $\mathcal{F}_{i-1}$
  /* the last frame has been chosen as the global reference */
- Consider segments which have been matched
  $\rightarrow$ **for** each pairing (S', S) in $\{\mathcal{M}_{\text{best}}\}$
  $\hookrightarrow$ Fuse them as S_{F} /* Algorithm 12.1 */
  $\hookrightarrow$ Retain S_{F} in $\mathcal{F}_i$ and set its *instance* to that of S' plus 1
  $\leftarrow$ **endfor**
- Copy unmatched segments from $\mathcal{F}'_{i-1}$ to $\mathcal{F}_i$
- Copy unmatched segments in $\mathcal{S}_i$ into $\mathcal{F}_i$
  and initialize their *instances* all to 1

12.3 Experimental Results

In this section, we provide two experimental results. The first one is the integration of two stereo views, which will give us a good idea of the performances of the integration process. The second shows the result of integrating 35 stereo views of a room.

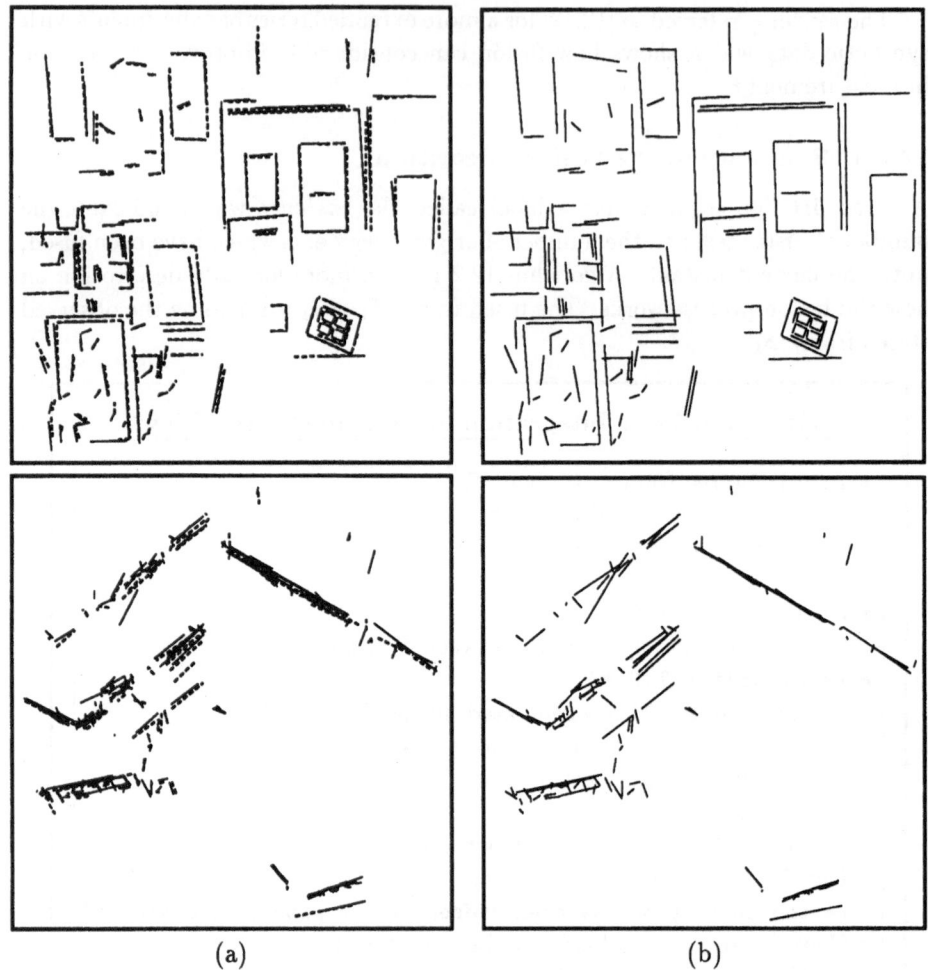

(a)   (b)

Fig. 12.8. Example 1: (a) superposition of the matched segments after applying the estimated motion to the first frame, (b) fused segments (unmatched segments are not displayed)

12.3.1 Example 1: Integration of Two Views

In this experiment, we use the two 3D scenes described in Chap. 7 (they are displayed in Figs. 7.4 and 7.5). We apply our motion-determination algorithm to these two frames. Figure 12.8a shows the superposition of the matched segments after applying the estimated motion to the first frame. One observes a very good accuracy of the motion estimate. We obtain, at the same time, 170 matches between the two frames. Among the recovered matches, 14 are multiple matches. Our algorithm admits multiple matches, that is, a segment in the first frame

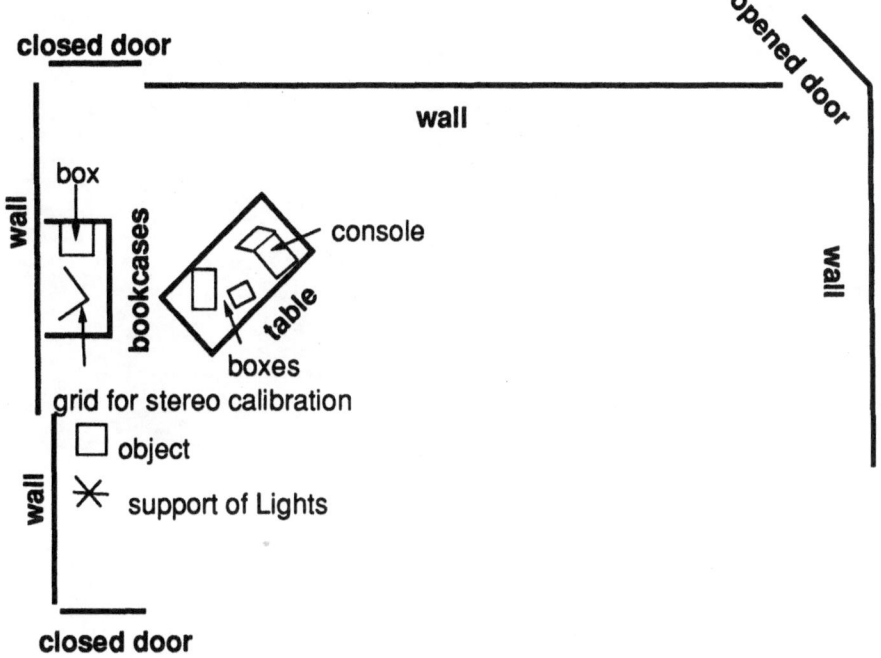

Fig. 12.9. Semantic description of the room

can have two or more correspondences in the second frame, and *vice versa*. Two broken segments and a long segment can be matched by our algorithm. Segments of multiple matches are fused into a single segment, so we get 156 fused segments in total. The fusion result is displayed in Fig. 12.8b. Comparing Fig. 12.8b with Figs. 7.4 and 7.5 (or Fig. 12.8a), we observe the improvement in the accuracy of segment measurements (examining, for example, segments on the right).

12.3.2 Example 2: Integration of a Long Sequence

We now describe the result of the integration of 35 stereo views taken when the robot navigates in a room. Merged segments are represented in the coordinate system related to the last position. As described in Sect. 12.1, the displacement of the robot is manually controlled with the aid of a graphic interface, such that there exists some common part of the view fields in the successive positions (the necessary condition that the motion algorithm succeeds). A labeled map of the room is shown in Fig. 12.9. In Fig. 12.10, we show four sample images taken by the first camera. Figure 12.11 shows the final 3D map obtained by integrating 35 stereo views of the room. One can easily establish the relation between the descriptions in Fig. 12.11 and in Fig. 12.9. In order to give an idea about the displacements of the robot in this experiment, we also show the position of the cameras at each instant in the room in the top view which is represented by a

(a) first sample image   (b) second sample image

(c) third sample image   (d) fourth sample image

Fig. 12.10. Four sample views of the room

line segment in the middle part. Amongst all the displacements effectuated, the biggest rotation is of 17.4 degrees and the largest translation is of 683 millimeters.

From Fig. 12.11, several remarks can be made. Consider the upper left corner of the room in the top view. Large uncertainty can be observed, which is due to the fact that the robot is distant to that corner. Consider now the same corner in the front view. In fact, the front view displays the height information observed by the robot. Since the robot is more distant to the corner than to the other parts, it observes things in the corner higher than in the other parts of the room. The low part of the corner, however, is not seen by the robot, because the table in front of it hides it. The left wall is in fact composed of two noncoplanar

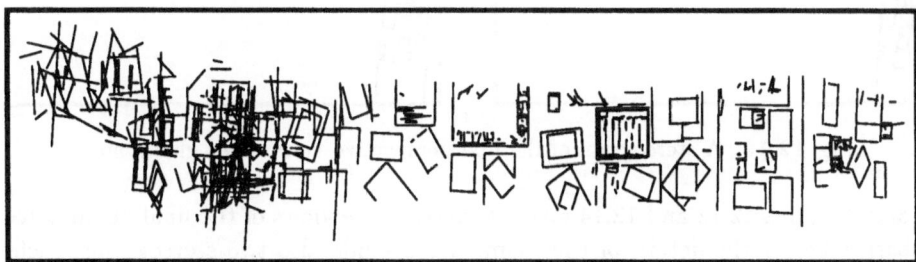

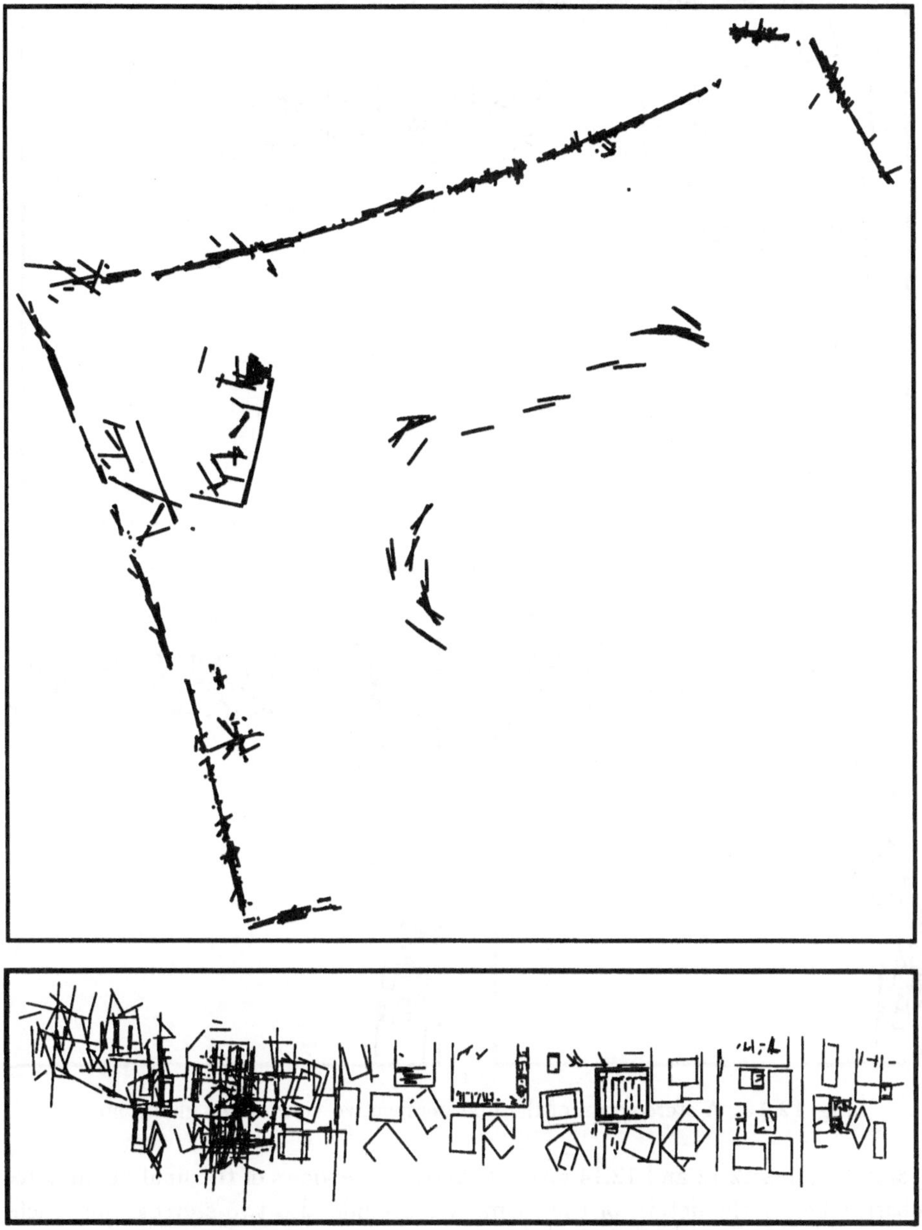

Fig. 12.11. The final 3D map of the room by integrating 35 3D frames obtained from stereo: top and front views (line segments in the middle part of the top view indicate the positions of the cameras)

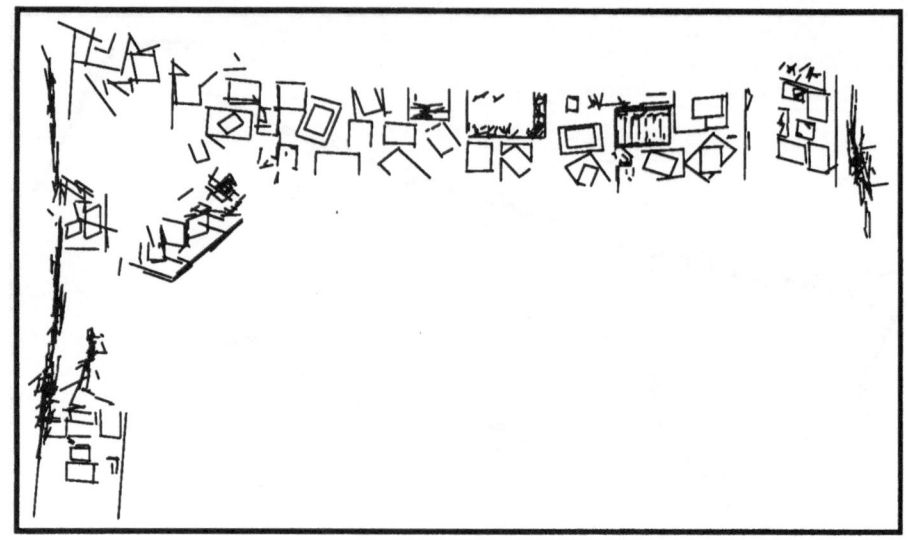

Fig. 12.12. First perspective view of the global map

Fig. 12.13. A stereogram of the first perspective view of the global map

parts. Figures 12.12 and 12.14 give two perspective views of the final 3D map to observe better the details of the room. We provide also two stereograms, each corresponding to a perspective view (Figs. 12.13 and 12.15), which allow the reader to perceive the depth through cross-eye fusion.

There are 3452 segments in total in all stereo frames and we have only 839 segments seen at least twice in the final map (segments which appear only once are not counted). This shows that fusion is also very useful in reducing the memory requirements. Although the result is very encouraging, one can observe some distortion in the global map, very probably due to the stereo calibration or 3D reconstruction.

Fig. 12.14. Second perspective view of the global map

Fig. 12.15. A stereogram of the second perspective view of the global map

12.4 Summary

A system to incrementally build a world model with a mobile robot in an unknown environment has been described. The model is, for the moment, segment-based. A trinocular system is used to build a local map of the environment. A global map is obtained by integrating a sequence of 3D frames taken by a stereo system when the robot navigates in the environment. The emphasis has been on the integration of segments from multiple views. Our approach to the integration of multiple stereo views is very similar to those reported in the literature based on Bayesian estimation [12.5, 7, 20–22], although there is a slight difference in technical details. An important characteristic of our integration strategy is that a segment observed by the stereo system corresponds only to one part of the segment in space if it exists, so the union of different observations gives a better estimate on the segment in space. We have succeeded in integrating 35 stereo frames taken in our robot room. Although the results are very encouraging, several points need to be investigated further:

- The distortion in the global map should be analyzed in detail. It may result from camera calibration, 3D reconstruction or motion estimation.
- Geometric constraints (parallelism of segments, coplanarity of segments, etc.) can be imposed on the segment set to improve the accuracy of measurements.
- The resulting world model needs to be represented in a more symbolic manner, identifying, for example, walls, tables, doors, in it.
- The analysis and decision module should be developed in order for the mobile robot to build the world model *automatically.* This is something we are doing [12.17].

The technique presented in this chapter has been adapted in [12.17] to fuse 2D line segments. The primary goal in [12.17] is obstacle avoidance and trajectory planning for an indoor mobile robot. We first project, on the ground plane, 3D line segments reconstructed by stereo to obtain a two-dimensional map. Those 2D segments are used to construct a tessellation of the ground plane through the Delaunay triangulation. We then determine free space by marking those triangles which are empty and generate collision-free trajectories. As the mobile robot navigates, more 2D segments are available, and a technique similar to that developed in this chapter is used to build incrementally a global 2D map. The process is iterated until the task is accomplished.

Several similar ideas can be found in the 3D Mosaic scene understanding system [12.13]. That system is intended for incrementally generating a 3D model of a complex scene from multiple images. The primitives used are edges and vertices. It differs from the work described in this article in at least the following points. Firstly, no extended experiments have been carried out using this system. Only the merging result of a 3D frame obtained from a pair of stereo aerial images and a manually generated 3D frame has been reported. Secondly, the motion (coordinate transformation) between successive views has been assumed to be known, which makes the matching problem trivial. Finally, uncertainty in the model and in the measurements has not been systematically addressed. One important point of the 3D Mosaic system is that part of the knowledge of planar-faced objects has been explicitly formulated. Such knowledge may constitute a good starting point for us to interpret fused data towards derivation of a symbolic model.

13. Solving the Motion Tracking Problem: A Framework

U p to now, we have been interested in the two-view motion analysis. We have used more feature correspondences between two frames than the minimum in order to obtain robust solutions (see Chap. 5). An alternative method is to use long sequences of frames (usually fewer feature correspondences are required), as in [13.1–3]. Furthermore, one can use both more frames and more feature correspondences. In other word, one can use all the available (both spatial and temporal) information to derive a robust estimation. When more information is used, some minimization algorithms are required: least-squares estimation, singular value decomposition, Kalman filter, or other general minimization procedures such as conjugate gradient descent methods or Gauss-Newton methods. In the above, we assumed feature correspondences over frames were given. However, the correspondence problem is a very difficult one (see Chaps. 6 and 7), and far from being solved. It is at this point, as we shall see later, that using long sequences of frames shows its advantage.

Indeed, there is a significant difference between using short sequences (2 or 3 frames) and using long sequences. We have described the short-sequence analysis problem in detail in the second part of this monograph. The most difficult problem is the correspondences of features. Some *a priori* assumptions such as rigidity and similarity must be exploited. In this part of the monograph, we deal with the long-sequence analysis problem. In this case, a sequence of frames is taken with a very short sampling time, and there is only a small change between adjacent frames. A primitive of an object in one frame must appear in the neighborhood of its occurrence in the succeeding frame. Even if the interval is not very short, we can use the information extracted from previous frames to predict the occurrence in the next instant under some kinematics model. This temporal continuity makes the correspondence problem much easier [13.4].

13.1 Previous Work

We find two groups of work in the literature on using long sequences of frames. The first one is called sometimes the *token tracking* approach, which tracks (2D or 3D) tokens from one frame to another based on some model of the evolution of token parameters. Velocity changes are usually assumed to be smooth. The related work includes [13.5–9]. The problem with this approach is that token tracking is based on a model on the change in the token parameters, and this change does not reflect explicitly the real motion. The relation between the change in the token parameters and 3D motion parameters is not linear. For example, if a token undergoes a constant rotation, we cannot model the change in the token parameters as constant velocity or even constant acceleration. Thus the tracking process will diverge after certain time, or a big noise disturbance term must be added in the model equations. The second approach assumes that (2D or 3D) feature correspondences are already established. Object motion is then computed based on some model of the evolution of 3D motion parameters (kinematic models). Some related work includes [13.1–3, 10, 11]. These two approaches can be combined in order to solve matching and motion simultaneously. *Toscani* et al. [13.12] address the problem of solving automatically 3D motion and structure from motion with the aid of a token tracker. In this monograph, we propose a new approach to analyze long sequences of stereo frames, which *unifies* the above two approaches to some extent: tracking features from frame to frame and estimating motion parameters. The idea is to embed directly a model of 3D motion kinematics into the token tracking process, instead of using a model on the changes in the token parameters.

We are not the first ones to investigate this problem by using a sequence of stereo frames. *Young* and *Chellappa* [13.3] describe the computer simulation of a system that uses a number of noisy 3D points assumed to belong to the same rigid object to estimate its motion. In their work, the problem of obtaining the matches from frame to frame and the problem of multiple objects are not addressed.

13.2   Position of the Problem and Primary Ideas

We address the motion tracking problem that arises in the context of a mobile vehicle navigating in an unknown environment where other mobile bodies such as human beings or robots may also be moving. A stereo rig mounted on the mobile vehicle provides a sequence of 3D maps of the environment. As described earlier, the 3D primitives (tokens) we use at the moment are line segments. Although the framework to solve the motion tracking problem developed here arises in this specific context and that in the details of the current implementation the line segment assumption plays an important role, we believe it should be applicable in other contexts:

- Other 3D primitives, for example, points, or combination of points and lines,
- Motion tracking in a sequence of monocular images.

The situation is illustrated in Fig. 13.1. The static environment is represented by cross-hatched regions. Only one moving object is drawn, which is represented

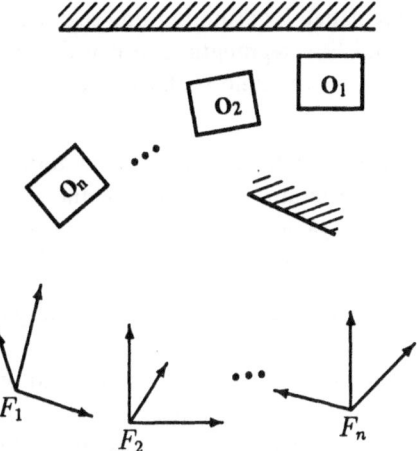

Fig. 13.1. Illustration of the motion tracking problem

by a square. The mobile robot is represented by a frame of reference, since the motion of a mobile robot is equivalent to the motion of the frame of reference of the stereo system (Chap. 11). The reference frame is rigidly attached to the mobile robot and its numerical parameters are determined in the camera calibration phase [13.13]. In the figure, the object undergoes a motion from right to left, and the robot moves from left to right. We want to solve the following problems:

> 1. Find the positions of static and moving objects in each stereo frame,
> 2. Determine the motion of the robot as well as those of the moving objects with respect to the static environment.

Those problems can be solved at the level of objects: **object tracking**, or at the level of features which constitute objects: **token tracking**. In the object tracking approach, the scene must first be segmented into objects, which in general requires high-level knowledge about the characteristics of objects such as rigidity and geometry (planar world). This approach is in general difficult. In some special cases, such as Radar imagery and in the experiment reported in [13.8] using tennis balls, objects can be easily detected and can be replaced by points (usually their centers of gravity). In the token tracking approach, no such knowledge is required and the tracking process can be carried out in parallel for each token. We track 3D line segments instead of 3D objects for this reason and also for the followings:

1. Objects can be later identified by grouping line segments with similar motion,
2. After tracking individual line segments, one can detect multiple moving objects, articulated objects or even deformable objects based on common motion characteristics.

Because of this, the hypothesis which assumes that objects are moving rigidly can be somewhat relaxed in the analysis of long sequences.

We also tackle the following problems that arise when we deal with (long) sequences of images:

- **Occlusion:** A moving object may be partially or totally occluded by the background or by other objects.
- **Disappearance:** A moving object in the current field of view may move partially or totally out of it in the next frames.
- **Appearance:** A previously unseen object may partially or totally come into view.

Clearly, occlusion is related to disappearance and appearance, since when we talk about the occlusion of an object, we mean that some of its features disappear for a moment and may eventually reappear in the future. Those three events are due to regular transformations of the scene. We must add to them a fourth one which is due to the failure of the algorithms that produce the description:

- **Absence:** When features which should be present are not due to the failure of the feature extraction (or reconstruction) process.

These remarks bring forward an interesting aspect of the problem, namely that there are always two kinds of tokens: those which have been seen for a sufficiently long time so that the system has been able to build a good model of their kinematics, and those which have just entered the field of view and for which no kinematics information is available. The first kind of tokens is "easily" dealt with since it is likely that the prediction stage will help to cut down heavily the number of tentative candidates to a match in the next frame. For the second kind a computational explosion is likely to happen: in order to find the right match, we may have to explore a large number of possibilities and if we make the wrong choice we will lose track of the token. Therefore our system can be seen as operating in two modes, the first one called the *continuous* mode and the second called the *bootstrapping* mode.

The continuous mode applies to tokens for which the system has built up a kinematic model with low uncertainty. The model at time t is used to predict the position and orientation of the token in the scene at time $t + \Delta t$. Since the uncertainty of the model is small, the search for corresponding tokens can be restricted to a small zone around the predicted token.

The bootstrapping mode assumes no knowledge of the kinematics of the token, i.e., assumes that it is not moving, with a large uncertainty. Its position and orientation at time $t + \Delta t$ are predicted to be the same as those at time t but, since the uncertainty of the model is large, the search for corresponding tokens is

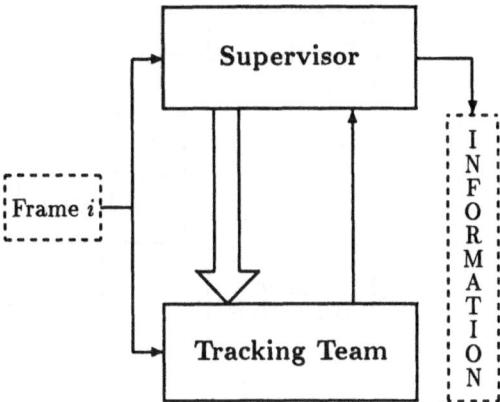

Fig. 13.2. An architecture for solving the motion tracking problem

conducted in a larger zone than in the previous mode leading to the possibility of many candidates.

One interesting feature of both modes is that they use the idea of *least-commitment* and, instead of forcing a decision, may make multiple correspondence choices and use the time continuity to throw away later the ones which are not confirmed by the measurements (see the following section and Sect. 15.2).

13.3   Solving the Motion Tracking Problem: A Framework

To clarify the presentation, we call the 3D line segments being tracked the *tokens* and the currently observed 3D line segments the *scene tokens*.

13.3.1   Outline of the Framework

The framework in which we propose a solution to the motion tracking problem consists of two levels (see Fig. 13.2). In the figure, single arrows represent data flow, and the double arrow represents the flow of control.

The low level is called the *tracking team*. A token being tracked can be considered as one of the team members. As we discussed earlier, the token tracking process uses a model of the 3D kinematics (see Chap. 14), instead of a model of the evolution of the token parameters. A token is then characterized by its *position* and *orientation* in the current frame, its *kinematic parameters*, and a positive number called *support*. The support indicates the degree of support for the existence of the token, which will be described in Sect. 15.2.

When a new frame is acquired, each token being tracked searches the whole frame for a correspondence. The search space can be considerably reduced by using a kinematic model and information from previous frames: before the new

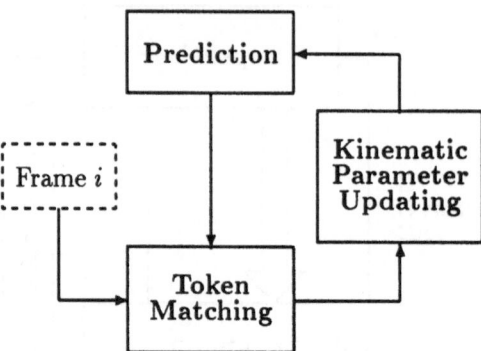

Fig. 13.3. Block diagram of the token tracking algorithm

frame is available, one can predict the occurrence of each token in the new frame based on the kinematic model. When the new frame is obtained, one only needs to look for scene tokens in the neighborhood of the predicted position.

When a match is found, the parameters of the token kinematic model are updated, the position parameters are replaced by those of its match[1], and the support parameter is also updated. The prediction and update of the position and kinematic parameters are done by using an extended Kalman filter (see Sects. 2.2 and 14.4). The matching process is based on the Mahalanobis distance (see Sect. 15.1). Due to occlusion or absence of some scene tokens in the current frame, a token may not find any match in the neighborhood of its predicted position in the current frame. Of course, this phenomenon may also occur due to the disappearance of the token. To handle the occlusion and absence of scene tokens, it is necessary to hypothesize the existence of the token and continue to change its kinematic parameters according to the kinematic model and update its support (see Sect. 15.2).

Figure 13.3 shows the block diagram of the algorithm, which is performed in a "matching-update-prediction" loop. The details will be described in the following chapters. As we can observe, the above process can be performed independently for each token to be tracked, and this allows a completely parallel implementation.

The high level is called the *supervisor*. It has three main functions:

• Grouping tokens with similar kinematic parameters as a single object. If there exist multiple moving objects, they can be segmented on the grounds that they undergo different motions. We describe later the details about how to group tokens (Sect. 15.4).

---

[1] One could also update the position parameters by modifying a little the state vector in the formulation of Sect. 14.4. The new state vector would have five dimensions more. The computation would be more expensive, as the complexity of the EKF is $O(n^3)$, where n is the dimension of the state vector.

- Monitoring the tracking team by detecting the following events:

  1. **Appearances**: When a new token appears, i.e., when a scene token in the current frame cannot be matched with any token being tracked, then the supervisor activates an additional token in the tracking team. This new token starts the same process as the others.

  2. **False matches**: When a token loses its support of existence (see Sect. 15.2), the supervisor then deactivates this token. Usually such tokens have been activated due to false matches in the previous frames. In a parallel implementation, the processor occupied by this token would be freed, and could be used by some new token.

  3. **Disappearances**: When a token moves out of the field of view, the supervisor deactivates this token. We can easily determine whether a token being tracked is out of the field of view by projecting it onto one of the camera planes. Just as in the previous case, in a parallel implementation, the processor occupied by this token would be freed, and could be reutilized.

  4. **Multiple matches**: A token being tracked may find multiple matches in the current frame with a criterion defined *a priori* (the Mahalanobis distance, for example), especially when there are several scene tokens which are near to each other. A common way to solve this problem is to choose the scene token which is the nearest to the predicted position (*best-first search*), as in [13.6, 14]. The problem with such approach is that the nearest scene token is not always the correct match. Therefore, the kinematic parameters will be sometimes updated using incorrect matches while believing that they are correct. This may lead to unpredictable results. A more robust approach is to keep tracking the token using several nearest scene tokens in the current frame; thus a token can be split. This approach can be called *beam search*. In Radar target tracking literature [13.15], such approach is referred as *track-split approach*. In our implementation, we choose the two nearest scene tokens to the predicted position in the sense of the Mahalanobis distance (see Sect. 15.1), if their distances are both less than some threshold. The token updates its kinematic parameters by incorporating the nearest scene token. If the second nearest scene token exists, then the token reports it to the supervisor. The supervisor activates an additional token by integrating the original token and the matched one. The beam search strategy is utilized in other research fields, such as in the HARPY speech understanding system [13.16]. This strategy has been found to be efficient as well as robust.

  5. **About changes**: A potential capacity of the supervisor to monitor the tracking team is to detect abrupt changes in the motion of a token due, for example, to collision, and to reinitialize its kinematic parameters. For the moment, we handle, only in a *passive* manner, the abrupt changes of motion (see Sect. 13.5).

- Providing information to other components of the global system. For example, in an active tracking application, one may need to control the motion of the robot or adjust the camera parameters to adapt the changing situation based on the information provided by the motion tracking algorithm. The information may include the kinematics of the robot (egomotion) and the kinematics and relative positions of the moving objects.

13.3.2 A Pedagogical Example

Figure 13.4 shows an example of how the tracking team works. At t_2, token 1 is split into two (token 1 and token 1') due to ambiguous matches. At t_3, token 1' cannot find a correspondence in the current frame, and it makes an hypothetical extension to cope with the occlusion problem. But because too many such hypothetical extensions are made consecutively, it loses its support of existence at t_5 and is then deactivated (see Sect. 15.2). At t_5, token 1 cannot find its correspondence in the current frame, and it makes an hypothetical extension. It finds its correspondence at t_6. Thus the occlusion problem is handled gracefully.

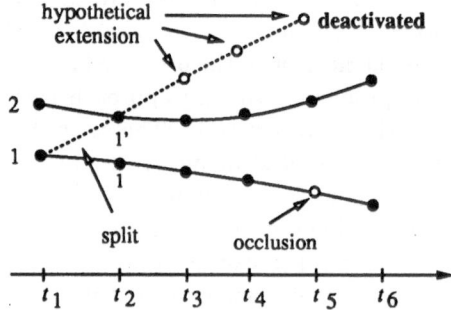

Fig. 13.4. An example of motion tracking

13.4 Splitting or Merging

In the above discussion about multiple matches, we presented the concept of *splitting* a token. The main problem with the splitting approach is that the number of tokens may grow exponentially with time. We should rethink before doing splitting. In fact, two cases of multiple matches should be distinguished. Figure 13.5 shows such cases. The token L is represented by a thick line segment. It has two matches S_1 and S_2 (represented by thin line segments). In the first case, the two matches S_1 and S_2 are not collinear. The splitting technique can be applied to handle the problem. That is, the token L is duplicated, one pursuing the tracking by incorporating S_1, and another pursuing the tracking by incorporating S_2.

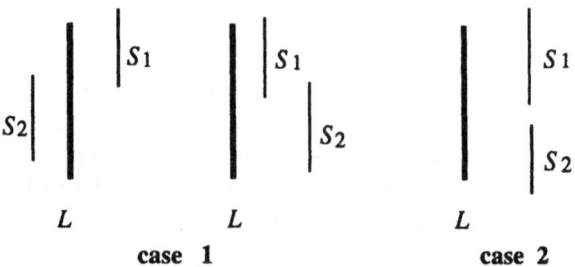

Fig. 13.5. Two different cases of multiple matches

In the second case, the two matches S_1 and S_2 are collinear. Of course, the splitting technique is still applicable, but we do not want to apply it (see below). A reasonable interpretation is that S_1 and S_2 are both parts of a single longer segment in space. The segment is broken into two because of the edge detection process, the line segment fitting process or other reasons. Note that not *any* two collinear segments can be interpreted this way. We use the fact that there exists a token L observed previously which links S_1 and S_2 together because it has been matched to both of them. Based on the above consideration, we first *merge* S_1 and S_2 into a single segment S. The token L continues to track by matching S without splitting. The *merging* concept is very important. It can avoid the abnormal growth of the number of tokens due to splitting. For example, in an extreme case as shown in Fig. 13.6, we have 2 tokens at t_2, 4 tokens at t_4, 8 tokens at t_6, ...

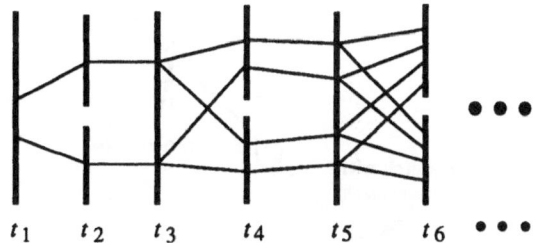

Fig. 13.6. An example to show the importance of merging

Now we return to the technical details of merging. First, how to detect the collinearity. Using the line segment representation described in Sect. 4.2.2, two segments S_1 and S_2 are collinear if and only if the following relations hold[2]

$$\psi_1 = \psi_2 \, , \tag{13.1}$$

---

[2]Note that if

$$\psi_1 + \psi_2 = \begin{bmatrix} 2\pi \\ \pi \end{bmatrix} \, ,$$

i.e., S_1 and S_2 have opposite orientations, they are also collinear if $(\mathbf{m}_1 - \mathbf{m}_2) \wedge \mathbf{u}_1 = \mathbf{0}$. But since our segments are oriented, two segments with opposite orientations cannot come from a single segment in space.

$$(\mathbf{m_1} - \mathbf{m_2}) \wedge \mathbf{u_1} = \mathbf{0} \;, \tag{13.2}$$

where $\mathbf{u_1} = \mathbf{h}(\boldsymbol{\psi_1})$ defined by (4.20). The discontinuity of ϕ is treated in the same fashion as in Sect. 12.2.2. In practice, the equality can seldom or never hold, so we use the Mahalanobis distance instead (see Sect. 2.2.5). It remains now how to merge S_1 and S_2 into a single segment S. The technique used here is exactly the same as described in Sect. 12.2.2.

13.5 Handling Abrupt Changes of Motion

In the above framework, we assume that objects move under some kinematic models. Smooth evolution in kinematic parameters is allowed. In reality, one may encounter the case where objects change their motion abruptly due to, for example, collision. Another example is vehicle guidance: before a turn, a vehicle moves straightly and it must move almost circularly during the turn. For such cases, some module should be developed to predict the fast change in motion parameters.

For the moment, our algorithm passively handles the abrupt changes of motion of a token. If a token changes abruptly its motion, its corresponding feature in the current frame usually cannot find matches to the tokens being tracked and the supervisor initializes it as a new token. This token then starts the tracking process.

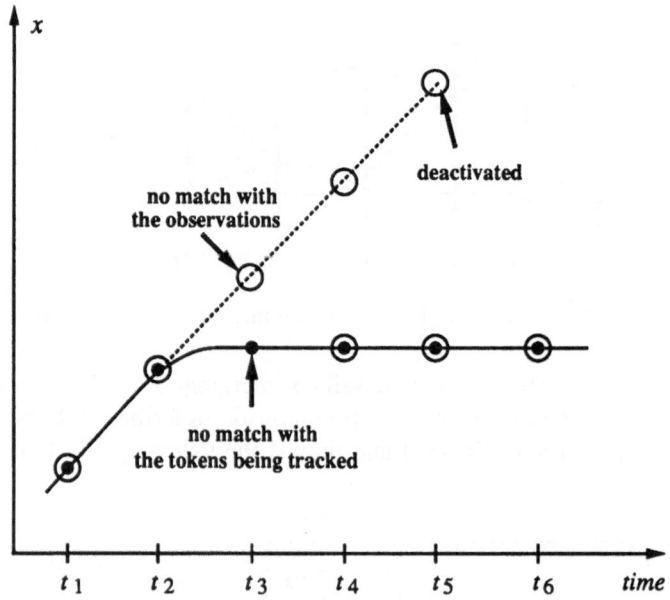

Fig. 13.7. Passive handling of abrupt changes of motion

Figure 13.7 displays an example of a unidimensional tracking to show how to handle passively abrupt changes of motion in our framework of tracking. In the figure, dots indicate the observed positions and small circles indicate the predicted positions of the token in tracking. The real trajectory of the token is shown by a solid curve. Until t_2, the token moves uniformly. Between t_2 and t_3, some external force makes the token remain stationary. At t_2, the token being tracked predicts its position for t_3 based on the information gathered previously, thus in general it cannot find any match with the observations. That token will continue its tracking by hypothetical extension (in dashed lines) until it loses the support of existence (say at t_5). On the other hand, the observed feature cannot find any matches at t_3 with the tokens being tracked. It is then initialized as a new token which is put in the tracking team. The abrupt change of motion is thus handled, but more tokens exist temporarily during several frames after the change.

13.6 Discussion

Several ideas developed in the above are not completely new, and can be found in the literature [13.6, 9, 17]. Consider, for example, the concept of "support". In [13.9], *Hwang* postulates that the correspondence process in human vision is local and opportunistic. The correspondence of two image features in two consecutive frames should be determined only by the contextual information collected during some short time interval in the past. And the correspondence algorithm should allow multiple solutions to compete with each other. As more frames are observed, the correspondence that best fits the observed data should eventually win. Due to occlusion or absence of features or false match in the past, a trajectory may not find any match in the current frame by extension. In his algorithm, he uses a concept called *age* to indicate the number of times that a trajectory (similar to our concept of token) does not find continuously a match in the observed frame. Any trajectory whose age is greater than MAXAGE (a fixed integer) is removed for further consideration. Our concept of support differs significantly from the age concept in that it takes into account not only the number of times a token has not been present, but also the number of times it has been present in the past and how well the measurements agreed with the prediction. This will be summarized in the log-likelihood function defined in Sect. 15.2.

In [13.4], *Jenkin* and *Tsotsos* propose an approach different from ours to handle the multiple matching problem. They call it a wait-and-see approach: multiple matches are first hypothesized to be correct and are later disambiguated based on a temporal smoothness constraint by considering all possible *temporal* combinations. Their approach requires to store data observed during the last instants (at least three frames) in memory. As we shall see, only the log-likelihood function is updated and retained in our approach.

Our approach to tracking is similar to several methods which exist in the literature [13.2, 3, 11, 18] in the sense that they are also based on the estimation theory and that the state parameters are estimated by filtering over time. The main difference is that the other approaches assume that there exist only one moving object in the scene or that all objects are known *a priori*. Such an assumption is not used in our approach, in which objects are segmented after tracking, making it more flexible.

13.7 Summary

In this chapter, we have reviewed several approaches reported in the literature to deal with the motion analysis using long sequences. Some important problems of interest which arise in this context have been highlighted, such as occlusion, disappearance, appearance and absence of features. We have then proposed a framework to solve the motion tracking problem. It consists of two levels. The low level tracks tokens from frame to frame and estimates their kinematic parameters. The processing at the low level is completely parallel for each token. The high level groups tokens into objects based on kinematic parameters, controls the processing at the low level to cope with problems such as occlusion, disappearance, appearance and absence of features, and provides information to other components of the system. The advantages of our approach over the previous ones have been discussed. In the following chapters, we develop in detail our algorithm within this framework.

14. Modeling and Estimating Motion Kinematics

A common approach to modeling the motion kinematics is to divide the motion into two parts: a rotation around a point (called the *center of rotation*) and a translation of the center of rotation. The rotation is often assumed to be constant angular velocity or constant precession[1]. The trajectory of the rotation center is assumed to be well approximated by the first k terms of a polynomial ($k \geq 0$). See [14.1–5] for such a modeling. The problem with such a modeling is that rotation center is not always uniquely recoverable, as will be described in Sect. 14.2 (see page 222). We shall show that, in the case of constant angular velocity, that modeling is a special case of the one described in this chapter. In [14.6], the *fixed axis assumption* is used to recover the 3D structure of moving rigid and jointed objects from several single camera views. The fixed axis assumption is stated as follows: every rigid object movement consists of a translation plus a rotation about an axis that is fixed in direction for short periods of time. In this chapter, we describe the well-known model of classical rigid-bodies kinematics and then derive the closed-form solutions for some special motions. We then use those forms to solve the motion tracking problem by the EKF.

14.1 The Classical Kinematic Model

Given a Cartesian system of reference $Oxyz$ for rigid bodies in which a rigid body is in motion, we are interested in the kinematics of the rigid body, i.e., the nature and characteristics of its motion. For the moment, we are not interested in how such motion is generated.

---

[1]The constant precession describes the following special motion of an object possessing an axis of symmetry. The object rotates about its symmetric axis L with constant angular velocity and at the same time the symmetric axis L rotates about another axis L_p with constant angular velocity. The composition of the two motions is called *constant precession*, which is shown in Fig. 14.1. The movement of the Moon can be approximately considered as a constant precession, as it rotates about the Earth and the Earth rotates about the Sun. The intersection of L and L_p is the rotation center C. Maybe due to the good definition of the rotation center, many researchers use this modelization in their simulations. In applications of a mobile robot, we seldomly encounter such situations.

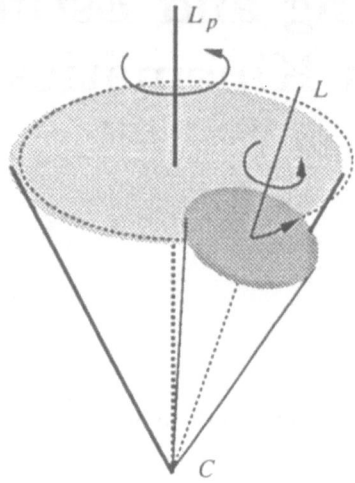

Fig. 14.1. Illustration of a constant precession motion.

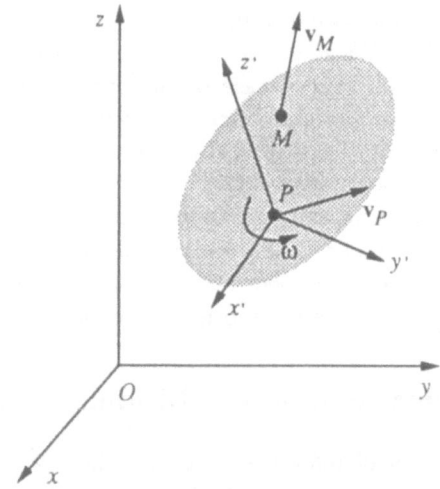

Fig. 14.2. Illustration of the classic kinematics

Choose a point on the rigid body, denoted by P. Define a Cartesian system of reference $Px'y'z'$ in the rigid body by taking P as its origin. The motion of the rigid body is completely specified by the change of the origin P and the orientation of $Px'y'z'$ relative to $Oxyz$ as time progresses (see Fig. 14.2).

If we fix the point P in $Oxyz$, or for an observer fixed at the origin of the system of reference $Px'y'z'$, the motion of the rigid body is, based on *Euler's theorem*, a rotation about some axis [14.7]. This axis is called the *axis of ro-*

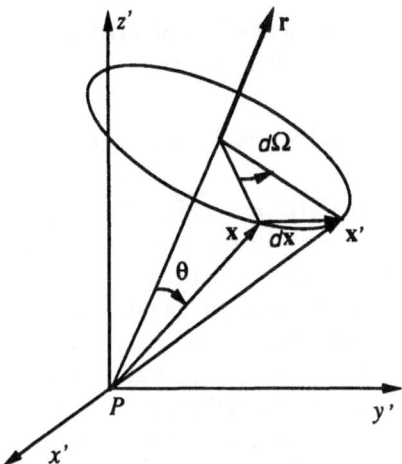

Fig. 14.3. Change in a vector by a rotation

tation. Assume at time t the rotation axis is $\mathbf{r}$ (see Fig. 14.3) (known as the *instantaneous* axis of rotation). Consider the change in a vector $\mathbf{x}$ upon rotating it counterclockwise through an infinitesimal angle $d\Omega$ about $\mathbf{r}$ during a differential time dt. The magnitude of $d\mathbf{x}$ (the change in vector $\mathbf{x}$), to first order in $d\Omega$, is

$$\|d\mathbf{x}\| = \|\mathbf{x}\| \sin \theta \, d\Omega \ .$$

Further, $d\mathbf{x}$ is perpendicular to $\mathbf{r}$ and $\mathbf{x}$ and they construct a right-hand system, thus we have

$$d\mathbf{x} = d\Omega \, \mathbf{r} \wedge \mathbf{x} \ ,$$

where $\wedge$ denotes the cross product of two vectors. Dividing the above equation by the differential time dt, we obtain the velocity of the vector $\mathbf{x}$:

$$\frac{d\mathbf{x}}{dt} = \boldsymbol{\omega} \wedge \mathbf{x} \ . \tag{14.1}$$

Here $\boldsymbol{\omega}$ is called the *instantaneous angular velocity* of the rigid body, defined as

$$\boldsymbol{\omega} = \frac{d\Omega}{dt} \mathbf{r} \ . \tag{14.2}$$

Now free the point P, and without loss of generality, let the axes of $Px'y'z'$ have the same orientations as those of $Oxyz$ at time t. The observer is fixed in $Oxyz$. Consider an arbitrary point M of the rigid body, and we have

$$\overrightarrow{OM} = \overrightarrow{OP} + \overrightarrow{PM} \ . \tag{14.3}$$

Differentiate the above equation with respect to time, we get

$$\frac{d}{dt}\overrightarrow{OM} = \frac{d}{dt}\overrightarrow{OP} + \frac{d}{dt}\overrightarrow{PM} \ . \tag{14.4}$$

However, $\frac{d}{dt}\overrightarrow{PM}$ is the rate of change in $\overrightarrow{PM}$ arising solely from rotation and is exactly what is given by (14.1), thus we have

$$\mathbf{v}_M(t) = \mathbf{v}_P(t) + \boldsymbol{\omega}(t) \wedge \overrightarrow{PM} \,, \tag{14.5}$$

where $\mathbf{v}_M(t) = \frac{d}{dt}\overrightarrow{OM}$ and $\mathbf{v}_P(t) = \frac{d}{dt}\overrightarrow{OP}$. That is, the velocity $\mathbf{v}_M$ of point M is the sum of the velocity $\mathbf{v}_P$ of the point P and the effect of the rotation around the point P. Note that in (14.5), we have added the time argument to emphasize that the velocities and the angular velocity are all *instantaneous*. Equation (14.5) describes the kinematics of the motion of a rigid body, which allows us to obtain the velocity of any point of a rigid body when $\boldsymbol{\omega}(t)$ and the velocity of an arbitrary point of the rigid body are known.

The above equation is true for any point P. For simplicity, we choose the origin as the point P, i.e., $P = O$, and we have the kinematic model as following:

$$\mathbf{v}_M(t) = \mathbf{v}(t) + \boldsymbol{\omega}(t) \wedge \overrightarrow{OM} \,. \tag{14.6}$$

The kinematics of any point M is completely characterized by $\mathbf{v}(t)$, the velocity of the point of the rigid body coinciding with the origin of the reference system, and $\boldsymbol{\omega}(t)$, the angular velocity of the point M around the origin. The pair $(\boldsymbol{\omega}(t), \mathbf{v}(t))$ is called the *kinematic skew* of the rigid body.

Let us replace $\overrightarrow{OM}$ in (14.6) by $\mathbf{p}(t)$, and $\mathbf{v}_M(t)$ by $\dot{\mathbf{p}}(t)$, where $\dot{\mathbf{p}}(t)$ denotes the time derivative of $\mathbf{p}(t)$ with respect to t, i.e., $\frac{d\mathbf{p}(t)}{dt}$. For the sake of clarity, we write the time argument as a subscript. For instance, $\mathbf{p}(t)$ is written as $\mathbf{p}_t$. Recall that "$\tilde{\mathbf{v}}$" represents the antisymmetric matrix associated with $\mathbf{v}$, see (4.2). Equation (14.6) can therefore be rewritten as a first order differential equation in $\mathbf{p}_t$:

$$\dot{\mathbf{p}}_t \;=\; \widetilde{\omega}_t \mathbf{p}_t + \mathbf{v}_t \,. \tag{14.7}$$

It is very difficult to get the solution of (14.7) for a general motion. Instead, we will derive in the following sections the closed form of the kinematic models for some special motions.

Recall the definition of the exponential of a matrix e^M, see (4.3). If M is a constant matrix, it can be easily shown that

$$\frac{d}{dt}e^{Mt} = M e^{Mt} = e^{Mt} M \,, \tag{14.8}$$

since we have

$$\frac{d}{dt}(Mt)^n = \frac{d}{dt}(M^n t^n) = nM^n t^{n-1} = nM(Mt)^{n-1} = n(Mt)^{n-1}M \,.$$

Note that in general, if M is not a constant matrix, Equation (14.8) does not hold. The Rodrigues' formula (see Theorem 4.2) is useful in the following derivations.

14.2   Closed-Form Solutions for Some Special Motions

In this section, we shall derive closed-form solutions for motion with constant angular velocity and polynomial translational velocity. In other words we show how to integrate (14.7). We have not yet found closed-form solutions for more complex motions and we think they probably do not exist, although the sufficient condition for the existence of unique solutions to (14.7) is not very difficult to fulfill (see Sect. 14.2.4).

14.2.1   Motion with Constant Angular and Translational Velocities

In the case of constant angular and translational velocities, we have a simple form of the solution. Let $\omega_t = \omega$ and $v_t = v$.

Theorem 14.1.   *The trajectory of a point* p_t *given by (14.7) is given, in the case of constant angular velocity* ω *and constant translational velocity* v, *by the following formula:*

$$p_t = W p_0 + V v , \qquad (14.9)$$

where

$$W \;=\; I_3 + \frac{\sin(\theta \Delta t)}{\theta}\tilde{\omega} + \frac{1 - \cos(\theta \Delta t)}{\theta^2}\tilde{\omega}^2 , \qquad (14.10)$$

$$V \;=\; I_3 \Delta t + \frac{1 - \cos(\theta \Delta t)}{\theta^2}\tilde{\omega} + \frac{\theta \Delta t - \sin(\theta \Delta t)}{\theta^3}\tilde{\omega}^2 , \qquad (14.11)$$

and $\theta = \|\omega\|$, $\Delta t = t - t_0$, p_0 *is the position at time* t_0 *and* I_3 *is the* 3×3 *identity matrix.*   ∎

Proof.   Let

$$y_t = e^{-\tilde{\omega}(t-t_0)} p_t .$$

This yields $y_{t_0} = p_0$ and based on (14.8), we have

$$\dot{y}_t = e^{-\tilde{\omega}(t-t_0)}\dot{p}_t - \tilde{\omega} e^{-\tilde{\omega}(t-t_0)} p_t .$$

We can easily show using Theorem 4.2 that

$$\tilde{\omega} e^{-\tilde{\omega}(t-t_0)} = e^{-\tilde{\omega}(t-t_0)}\tilde{\omega} ,$$

therefore we have

$$\dot{y}_t = e^{-\tilde{\omega}(t-t_0)}(\dot{p}_t - \tilde{\omega} p_t) = e^{-\tilde{\omega}(t-t_0)} v .$$

Integrating the above equation, we get

$$y_t = p_0 + \int_{t_0}^{T} e^{-\tilde{\omega}(s-t_0)} v \, ds .$$

222   14. Modeling and Estimating Motion Kinematics

From the definition of $\mathbf{y}_t$, we thus have

$$\mathbf{p}_t = e^{\tilde{\omega}(t-t_0)}\mathbf{y}_t = e^{\tilde{\omega}(t-t_0)}\mathbf{p}_0 + \int_{t_0}^T e^{\tilde{\omega}(t-s)}\mathbf{v}ds .$$

If we develop $e^{\tilde{\omega}(t-t_0)}$ using Theorem 4.2, we find

$$e^{\tilde{\omega}(t-t_0)} = W .$$

It remains to show that

$$\int_{t_0}^T e^{\tilde{\omega}(t-s)}\mathbf{v}ds = V\mathbf{v} .$$

If we note the term on the left hand of the above equation by $\mathbf{b}_t$, then

$$\mathbf{b}_t = \int_{t_0}^T e^{\tilde{\omega}(t-s)}\mathbf{v}ds$$

(using Theorem 4.2)

$$= \int_{t_0}^T \left[\mathbf{I}_3 + \frac{\sin[\theta(t-s)]}{\theta}\tilde{\omega} + \frac{1-\cos[\theta(t-s)]}{\theta^2}\tilde{\omega}^2 \right] \mathbf{v}ds$$

$$= \left[\mathbf{I}_3 s + \frac{\cos[\theta(t-s)]}{\theta^2}\tilde{\omega} + \frac{\theta s + \sin[\theta(t-s)]}{\theta^3}\tilde{\omega}^2 \right]\Bigg|_{s=t_0}^{s=t} \mathbf{v}$$

$$= \left[\mathbf{I}_3(t-t_0) + \frac{1-\cos[\theta(t-t_0)]}{\theta^2}\tilde{\omega} + \frac{\theta(t-t_0)-\sin[\theta(t-t_0)]}{\theta^3}\tilde{\omega}^2 \right]\mathbf{v} .$$

This is exactly $V\mathbf{v}$.   □

From Theorem 14.1, we can observe that when $\omega = 0$ (i.e., pure translation), then

$$\mathbf{p}_t = \mathbf{p}_0 + \mathbf{v}(t-t_0) . \tag{14.12}$$

This is the well-known equation for a point moving on a straight line with constant velocity.

Figure 14.4 gives a 3D plot of the trajectories of a 3D moving point undergoing two different motions starting from the origin of the coordinate system: (a) ω: $[0.05, 0, -0.02]^T$, $\mathbf{v}$: $[100, -300, 200]^T$, viewpoint of the plot: $[1, 10, 2]^T$, observation time: 40 seconds; (b) ω: $[0, 0.1, 0]^T$, $\mathbf{v}$: $[10, 20, 10]^T$, viewpoint of the plot: $[2, 7, -1]^T$, observation time: 60 seconds.

In the introduction of the present chapter, we said that the modelization of motion by two parts (rotation about a point called rotation center and translation of that center) cannot meet our requirement, because the rotation center cannot be determined uniquely in most of our applications. For example, objects usually move in a plane. This implies that the rotation axis does not change, so any point on that axis can be considered as its rotation center. Furthermore, if an object rotates about some (fixed) point $\mathbf{p}$ with constant angular velocity ω and the instantaneous velocity of $\mathbf{p}$, denoted by $\mathbf{v_p}$, is constant, we can always represent

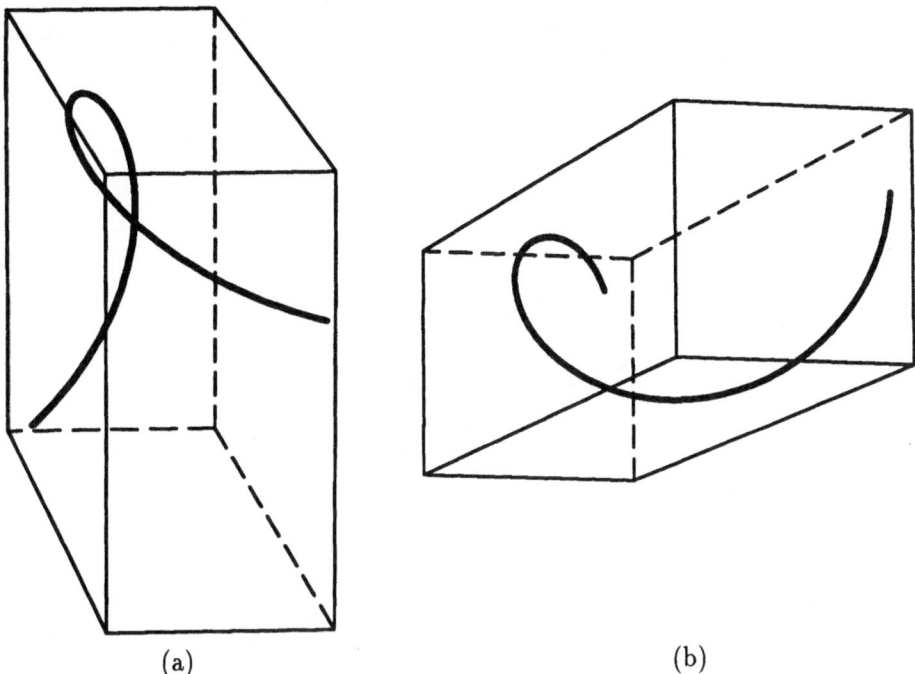

(a)   (b)

Fig. 14.4. 3D plot of the trajectories of a 3D moving point starting from the origin of the coordinate system

that motion using an arbitrary point $\mathbf{q}$ as the rotation center. The angular velocity about $\mathbf{q}$ remains the same, i.e., $\boldsymbol{\omega}$. The velocity of $\mathbf{q}$ is still constant, but is given by

$$\mathbf{v_q} = \mathbf{v_p} - \boldsymbol{\omega} \wedge (\mathbf{p} - \mathbf{q}) \ .$$

This can be easily justified using (14.5). In fact, for any point M on the object, its velocity is given by (using that of $\mathbf{q}$)

$$\begin{aligned} \mathbf{v}_M &= \mathbf{v_q} + \boldsymbol{\omega} \wedge (M - \mathbf{q}) \\ &= \mathbf{v_p} - \boldsymbol{\omega} \wedge (\mathbf{p} - \mathbf{q}) + \boldsymbol{\omega} \wedge (M - \mathbf{q}) \\ &= \mathbf{v_p} + \boldsymbol{\omega} \wedge (M - \mathbf{p}) \ , \end{aligned}$$

which describes exactly the same motion as if the point $\mathbf{p}$ is used. Thus we can choose the origin as the rotation center without loss of the *property* of motions. However, the ambiguity of the rotation center in using the two-parts motion modelization is avoided if our motion modelization is used. Take an example: an object rotates about a point $C = [0, 3, 0]^T$ with constant angular velocity $\boldsymbol{\omega} = [0, 0, 0.1]^T$ and *zero* translational velocity. The trajectory of an arbitrary point on the object, say $[0, 5, 0]^T$, is a circle about C with radius equal to 2 units. Interestingly and as expected, the same trajectory is obtained based on Theorem 14.1 with $\boldsymbol{\omega} = [0, 0, 0.1]^T$ and $\mathbf{v} = -\boldsymbol{\omega} \wedge C = [0.3, 0, 0]^T$, which is shown in Fig. 14.5.

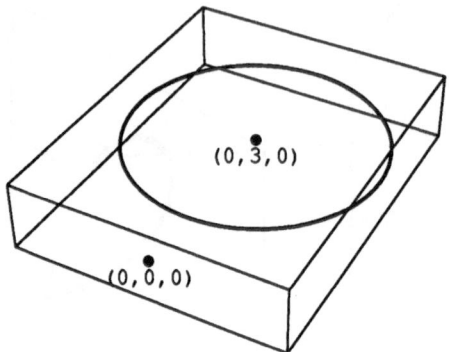

$(0, 3, 0)$

$(0, 0, 0)$

Fig. 14.5. Ambiguity in rotation center

14.2.2 Motion with Constant Angular Velocity and Constant Translational Acceleration

When angular velocity and translational acceleration are constant, we have the following equations:

$$
\begin{aligned}
\boldsymbol{\omega}_t &= \boldsymbol{\omega} \,, \\
\mathbf{v}_t &= \mathbf{v} + \mathbf{a}(t - t_0) \,,
\end{aligned}
\tag{14.13}
$$

where $\boldsymbol{\omega}$ denotes the constant angular velocity, $\mathbf{v}$ denotes the translational velocity at $t = t_0$, and $\mathbf{a}$ denotes the constant translational acceleration. The trajectory of a point in this case is defined by the following theorem:

Theorem 14.2. *The trajectory of a point $\mathbf{p}_t$ given by (14.7) is given, in the case of constant angular velocity $\boldsymbol{\omega}$ and constant translational acceleration $\mathbf{a}$, by the following formula:*

$$
\mathbf{p}_t = W\mathbf{p}_0 + V\mathbf{v} + A\mathbf{a} \,,
\tag{14.14}
$$

where W is the same as in (14.10), V is the same as in (14.11), and

$$
A = \frac{\Delta t^2}{2}\mathbf{I}_3 + \frac{\theta\Delta t - \sin(\theta\Delta t)}{\theta^3}\widetilde{\omega} + \frac{(\theta\Delta t)^2 - 2(1 - \cos(\theta\Delta t))}{2\theta^4}\widetilde{\omega}^2 \,,
\tag{14.15}
$$

and $\theta = \|\boldsymbol{\omega}\|$, $\Delta t = t - t_0$ and $\mathbf{I}_3$ is the 3×3 identity matrix. ∎

Proof. After manipulations similar to those in Theorem 14.1, we get the solution for the current case, which is

$$
\mathbf{p}_t = e^{\widetilde{\omega}(t-t_0)}\mathbf{p}_0 + \int_{t_0}^{T} e^{\widetilde{\omega}(t-s)}[\mathbf{v} + \mathbf{a}(s - t_0)]ds \,.
$$

We observe that the only difference between the solution of the current case and that of the previous case is the following term:

$$\mathbf{d}_t = \int_{t_0}^T e^{\widetilde{\omega}(t-s)}\mathbf{a}(s - t_0)ds .$$

We need only to show that $\mathbf{d}_t = A\mathbf{a}$. Indeed, using Theorem 4.2, we have

$$\mathbf{d}_t = \int_{t_0}^T e^{\widetilde{\omega}(t-s)}\mathbf{a}(s - t_0)ds$$

$$= \int_{t_0}^T \left[\mathbf{I}_3 + \frac{\sin[\theta(t-s)]}{\theta}\widetilde{\omega} + \frac{1 - \cos[\theta(t-s)]}{\theta^2}\widetilde{\omega}^2\right]\mathbf{a}(s - t_0)ds$$

$$= \left[\frac{(t-t_0)^2}{2}\mathbf{I}_3 + \frac{(t-t_0) + \frac{1}{2}\widetilde{\omega}(t-t_0)^2}{\theta^2}\widetilde{\omega} - \frac{\sin[\theta(t-t_0)]}{\theta^3}\widetilde{\omega}\right.$$

$$\left. - \frac{1 - \cos[\theta(t-t_0)]}{\theta^4}\widetilde{\omega}^2\right]\mathbf{a} .$$

After a little algebra, we get $A\mathbf{a}$. □

From Theorem 14.2, we observe that when $\omega = 0$ (i.e., pure translation), then

$$\mathbf{p}_t = \mathbf{p}_0 + \mathbf{v}(t - t_0) + \mathbf{a}\frac{(t-t_0)^2}{2} . \tag{14.16}$$

This is the well-known equation for a point moving on a straight line with constant acceleration.

Now we show that the following motion is a special case of Theorem 14.2: rotating with a constant angular velocity (say ω) about a point (say $\mathbf{c}$) which moves with a constant velocity (say $\mathbf{v_c}$). This is the common motion used in the two-parts motion modelization [14.1–5]. Let the point $\mathbf{c}$ be $\mathbf{c}_0$ at time t_0, then we have $\mathbf{c}(t) = \mathbf{c}_0 + \mathbf{v_c}(t - t_0)$. Using (14.5), we can compute the velocity at the origin O of the system of reference as follows:

$$\mathbf{v}(t) = \mathbf{v_c} + \omega \wedge (O - \mathbf{c}) = \mathbf{v_c} - \omega \wedge [\mathbf{c}_0 + \mathbf{v_c}(t - t_0)]$$
$$= \mathbf{v_c} - \omega \wedge \mathbf{c}_0 - \omega \wedge [\mathbf{v_c}(t - t_0)] .$$

We are then able to describe the above motion by Theorem 14.2 with the same angular velocity ω, but

$$\mathbf{v} = \mathbf{v_c} - \omega \wedge \mathbf{c}_0 , \tag{14.17}$$

$$\mathbf{a} = -\omega \wedge \mathbf{v_c} . \tag{14.18}$$

One should note that there exists an ambiguity in choosing the rotation center $\mathbf{c}$ in the two-parts motion modelization. In fact, any point on the rotation axis can be chosen without affecting the motion model. More concretely, if ω, $\mathbf{c}$ and $\mathbf{v_c}$ describe a motion, then ω, $\mathbf{c} + \lambda\omega$ and $\mathbf{v_c}$ describe the same motion, where λ

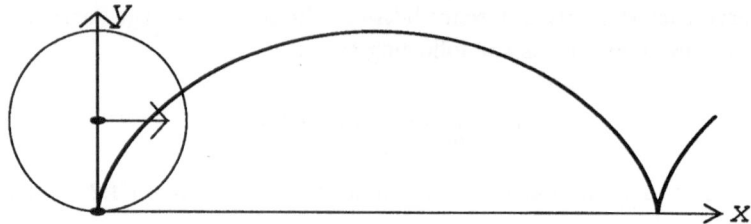

Fig. 14.6. A cycloid can be generated by a motion with constant angular velocity and translational acceleration

is an arbitrary scalar. Interestingly enough, there does not exist any ambiguity in parameters using our modelization. Indeed, from (14.17), we have

$$\mathbf{v} = \mathbf{v_c} - \boldsymbol{\omega} \wedge (\mathbf{c_0} + \lambda \boldsymbol{\omega}) = \mathbf{v_c} - \boldsymbol{\omega} \wedge \mathbf{c_0} \ .$$

Taking a concrete example. The movement of a circle (in xy plane) rolling along a straight line (say OX) with a constant velocity (say v) is an example of the above motion. It can be described as a rotation with constant angular velocity about the axis, passing through the center of the circle, parallel to OZ and a translation of the center of the circle with constant velocity v along OX. Let the radius of the circle be r, then the angular velocity $\omega = v/r$. The trajectory of a point on the circumference of the circle is called a *cycloid*, which can be described as follows

$$\begin{cases} x &= r(\omega t - \sin(\omega t)) \\ y &= r(1 - \cos(\omega t)) \\ z &= 0 \ . \end{cases}$$

Figure 14.6 displays the cycloid generated by a point $[0, 0, 0]^T$ at time t_0 with $v = 0.4$. The center of the circle at time t_0 is $\mathbf{c_0} = [0, 5, 0]^T$. Since $\mathbf{v_c} = [0.4, 0, 0]^T$ and $\boldsymbol{\omega} = [0, 0, -0.08]^T$, we get $\mathbf{v} = [0, 0, 0]^T$ and $\mathbf{a} = [0, 0.032, 0]^T$ from (14.17) and (14.18). Using Theorem 14.2 we get exactly the same curve as shown in Fig. 14.6.

One simple example to show that our modelization is more general than the two-parts motion modelization is a pure translation with constant acceleration. In a word, if the same number of parameters is used, our modelization is able to describe more general motion than the two-parts one, and the ambiguity in parameters is avoided.

14.2.3   Motion with Constant Angular Velocity and General Translational Velocity

Now we show the following important theorem.

Theorem 14.3.   *The trajectory of a point $\mathbf{p}_t$ given by (14.7) can be described in closed form if the angular velocity is constant and if the translational velocity is a polynomial of degree n ($n \geq 0$).*   ∎

Proof. Let the motion be described by $\boldsymbol{\omega}_t = \boldsymbol{\omega}$ and

$$\mathbf{v}_t = \mathbf{v}_0 + \mathbf{v}_1(t - t_0) + \mathbf{v}_2\frac{(t - t_0)^2}{2!} + \cdots + \mathbf{v}_n\frac{(t - t_0)^n}{n!} \, , \tag{14.19}$$

where $\mathbf{v}_0$ is the velocity at time t_0. Define a new variable $\mathbf{y}_t$ so that

$$\mathbf{y}_t = e^{-\widetilde{\omega}(t-t_0)}\mathbf{p}_t \, .$$

This yields $\mathbf{y}_{t_0} = \mathbf{p}_0$, the position of the point at t_0. Based on (14.8), we have

$$\dot{\mathbf{y}}_t = e^{-\widetilde{\omega}(t-t_0)}(\dot{\mathbf{p}}_t - \widetilde{\omega}\mathbf{p}_t) = e^{-\widetilde{\omega}(t-t_0)}\mathbf{v}_t \, .$$

Integrating the above equation, we have

$$\mathbf{y}_t = \mathbf{p}_0 + \int_{t_0}^{T} e^{-\widetilde{\omega}(s-t_0)}\mathbf{v}_s ds \, .$$

From the definition of $\mathbf{y}_t$, we thus have

$$\mathbf{p}_t = e^{\widetilde{\omega}(t-t_0)}\mathbf{y}_t = e^{\widetilde{\omega}(t-t_0)}\mathbf{p}_0 + \int_{t_0}^{T} e^{\widetilde{\omega}(t-s)}\mathbf{v}_s ds \, . \tag{14.20}$$

Using Theorem 4.2, we find that

$$e^{\widetilde{\omega}(t-t_0)} = W \, ,$$

where W is given by (14.10). Denoting the last term of (14.20) by $\mathbf{t}_t$ and using Theorem 4.2, we have

$$\mathbf{t}_t = \int_{t_0}^{T} \left[\mathbf{I}_3 + \frac{\sin[\theta(t - s)]}{\theta}\widetilde{\omega} + \frac{1 - \cos[\theta(t - s)]}{\theta^2}\widetilde{\omega}^2\right]$$
$$\left[\mathbf{v}_0 + \mathbf{v}_1(s - t_0) + \mathbf{v}_2\frac{(s - t_0)^2}{2!} + \cdots + \mathbf{v}_n\frac{(s - t_0)^n}{n!}\right] ds \, .$$

Let us define

$$L_k = \int_{t_0}^{t}(s - t_0)^k \sin[\theta(t - s)] ds$$

and

$$J_k = \int_{t_0}^{t}(s - t_0)^k \cos[\theta(t - s)] ds \, .$$

We can express $\mathbf{t}_t$ in terms of L_k and J_k ($k = 0, \ldots, n$). If we obtain closed-form expressions for L_k and J_k, this will be also true for $\mathbf{t}_t$. Indeed,

$$L_k = \frac{1}{\theta}\int_{t_0}^{t}(s - t_0)^k d(\cos[\theta(t - s)])$$
$$= \frac{1}{\theta}(s - t_0)^k \cos[\theta(t - s)]\Big|_{s=t_0}^{s=t} - \frac{k}{\theta}\int_{t_0}^{t}(s - t_0)^{k-1}\cos[\theta(t - s)] ds$$

$$\qquad = \frac{1}{\theta}(t - t_0)^k - \frac{k}{\theta}J_{k-1} , \tag{14.21}$$

$$J_k \;=\; -\frac{1}{\theta}\int_{t_0}^t (s - t_0)^k d(\sin[\theta(t - s)])$$

$$\qquad = -\frac{1}{\theta}(s - t_0)^k \sin[\theta(t - s)]\Big|_{s=t_0}^{s=t} + \frac{k}{\theta}\int_{t_0}^t (s - t_0)^{k-1}\sin[\theta(t - s)]ds$$

$$\qquad = \frac{k}{\theta}L_{k-1} , \tag{14.22}$$

$$L_0 \;=\; \int_{t_0}^t \sin[\theta(t - s)]ds = \frac{1}{\theta}\{1 - \cos[\theta(t - t_0)]\} , \tag{14.23}$$

$$J_0 \;=\; \int_{t_0}^t \cos[\theta(t - s)]ds = \frac{1}{\theta}\sin[\theta(t - t_0)] . \tag{14.24}$$

By iteration, we can obtain closed forms of L_k and J_k for all $k > 0$. Indeed, if we use the following notations:

$$\mathbf{x}_k = \begin{bmatrix} L_k \\ J_k \end{bmatrix} , \quad A = \begin{bmatrix} 0 & -1 \\ 1 & 0 \end{bmatrix} , \quad \mathbf{b}_k = \begin{bmatrix} (t - t_0)^k/\theta \\ 0 \end{bmatrix} ,$$

$$\text{and} \quad \mathbf{d} = \begin{bmatrix} 1 - \cos[\theta(t - t_0)] \\ \sin[\theta(t - t_0)] \end{bmatrix} ,$$

we have

$$\mathbf{x}_k \;=\; \frac{k}{\theta}A\mathbf{x}_{k-1} + \mathbf{b}_k \quad \text{for } k > 0 , \tag{14.25}$$

$$\mathbf{x}_0 \;=\; \frac{1}{\theta}\mathbf{d} . \tag{14.26}$$

After some algebra, we obtain

$$\mathbf{x}_k \;=\; \frac{k!}{\theta^{k+1}}A_k\mathbf{d} + B_k \quad \text{for } k > 0 , \tag{14.27}$$

where

$$A_k \;=\; (-1)^{\lfloor (k+1)/2 \rfloor}\begin{bmatrix} (k+1)\bmod 2 & k\bmod 2 \\ -k\bmod 2 & (k+1)\bmod 2 \end{bmatrix} ,$$

$$B_k \;=\; \begin{bmatrix} \displaystyle\sum_{i=0}^{\lfloor k/2 \rfloor} \frac{(-1)^i}{\theta^{2i+1}}\frac{d^{2i}}{dt^{2i}}(t - t_0)^k \\[4mm] \displaystyle\sum_{i=0}^{\lfloor (k-1)/2 \rfloor} \frac{(-1)^i}{\theta^{2(i+1)}}\frac{d^{2i+1}}{dt^{2i+1}}(t - t_0)^k \end{bmatrix} .$$

Here $\lfloor j \rfloor$ denotes the largest integer not exceeding j, "mod" denotes the modulo function, and $\frac{d^i}{dt^i}(\cdot)$ denotes the i-th derivative with respect to t. This yields a closed-form expression for $\mathbf{t}_t$ and therefore, also for $\mathbf{p}_t$. $\qquad\qquad \Box$

Theorem 14.1 is a special case of Theorem 14.3 for $\mathbf{v}_t = \mathbf{v}$ (i.e., $n = 0$). The reader can easily verify it.

Theorem 14.2 describes the special case $\mathbf{v}_t = \mathbf{v} + \mathbf{a}(t - t_0)$ (i.e., $n = 1$). From (14.21) to (14.24), we have

$$L_1 = \frac{1}{\theta^2} \{\theta(t - t_0) - \sin[\theta(t - t_0)]\} , \tag{14.28}$$

$$J_1 = \frac{1}{\theta^2} \{1 - \cos[\theta(t - t_0)]\} . \tag{14.29}$$

After some algebra, we get Theorem 14.2.

14.2.4 Discussions

Up to now, we have shown that a closed-form expression exists to describe the trajectory of a rigid object undergoing a motion with constant angular velocity and polynomial translational velocity.

If the angular velocity is also a polynomial of degree, say, m ($m \geq 0$), that is

$$\omega(t) = \sum_{i=0}^{m} \omega_i \frac{(t - t_0)^i}{i!} , \tag{14.30}$$

where ω_i ($i = 0, \ldots, m$) are constant vectors and ω_0 is the angular velocity at t_0, then Equation (14.7) is rewritten as

$$\dot{\mathbf{p}}_t = B_t \mathbf{p}_t + \mathbf{v}_t , \tag{14.31}$$

where

$$B_t = \sum_{i=0}^{m} \tilde{\omega}_i \frac{(t - t_0)^i}{i!} . \tag{14.32}$$

The solution to (14.31) depends on the solution to its homogeneous equation

$$\dot{\mathbf{x}}_t = B_t \mathbf{x}_t . \tag{14.33}$$

A *sufficient* condition for the existence of a unique solution to (14.33) is to require that all elements of B_t be continuous [14.8]. That condition is satisfied in our case as all elements of B_t given by (14.32) are polynomial in time. However, closed-form solutions to (14.33) are in general not possible except maybe in some special cases (which are currently under investigation). The special case where $m = 0$ has been addressed in the foregoing sections.

The solution U_t to the following matrix equation is called the *fundamental solution matrix*:

$$\dot{U}_t = B_t U_t \tag{14.34}$$

with the initial condition $U_0 = \mathbf{1}$, the identity matrix. We can obtain an approximation for U_t by using a sequence of approximations for the solutions to (14.34). As the zeroth approximation, let $U_t^0 = I$ (the superscript indicates the number of approximations). The first approximation is obtained by solving $\dot{U}_t^1 = B_t U_t^0$, which is given by

$$U_t^1 = U_0 + \int_{t_0}^{t} \dot{U}_{\tau_0}^1 d\tau_0 = \mathbf{I} + \int_{t_0}^{t} B_{\tau_0} d\tau_0 \; .$$

Let $\dot{U}_t^2 = B_t U_t^1$. Then

$$U_t^2 = U_0 + \int_{t_0}^{t} \dot{U}_{\tau_1}^2 d\tau_1 = \mathbf{I} + \int_{t_0}^{t} B_{\tau_0} d\tau_0 + \int_{t_0}^{t} B_{\tau_0} \int_{t_0}^{\tau_0} B_{\tau_1} d\tau_1 d\tau_0 \; .$$

Continuing this procedure with $\dot{U}_t^{k+1} = B_t U_t^k$ leads to

$$U_t = \mathbf{I} + \int_{t_0}^{t} B_{\tau_0} d\tau_0 + \int_{t_0}^{t} B_{\tau_0} \int_{t_0}^{\tau_0} B_{\tau_1} d\tau_1 d\tau_0$$
$$+ \int_{t_0}^{t} B_{\tau_0} \int_{t_0}^{\tau_0} B_{\tau_1} \int_{t_0}^{\tau_1} B_{\tau_2} d\tau_2 d\tau_1 d\tau_0 + \cdots \; .$$

Truncating the series after a finite number of terms yields an approximation for U_t.

Assuming that the fundamental solution matrix is available, the solution to (14.33) with an arbitrary initial condition $\mathbf{x}_0$ is

$$\mathbf{x}_t = U_t \mathbf{x}_0 \; . \tag{14.35}$$

This is easily verified. Checking the initial condition,

$$\mathbf{x}_{t_0} = U_0 \mathbf{x}_0 = \mathbf{I} \mathbf{x}_0 = \mathbf{x}_0 \; .$$

Checking whether the solution satisfies the differential equation (14.33),

$$\dot{\mathbf{x}}_t = \dot{U}_t \mathbf{x}_0 = B_t U_t \mathbf{x}_0 = B_t \mathbf{x}_t \; .$$

Since both the initial condition and the differential equation are satisfied, Equation (14.35) is the unique solution to (14.33).

The solution to (14.31) is given by

$$\mathbf{p}_t = U_t \mathbf{p}_0 + \int_{t_0}^{t} U_t U_\tau^{-1} \mathbf{v}_\tau d\tau \; . \tag{14.36}$$

This can be easily verified by checking that the initial condition and the differential equation are satisfied.

In [14.9], we proposed to use time polynomials to describe the angular and translational velocities of object motions (see (14.30) and (14.19)). An extended Kalman filter involving a *continuous-time* system with *discrete-time* measurements [14.10] was applied to estimate the kinematic parameters. A numerical integration procedure (we use the Runge-Kutta method) is required. One advantage of this approach is that we are able to describe complex motions even if the angular velocity is not constant. However, from the several experiments we have carried out, better results have been obtained using the closed-form solution than using the numerical integration if, of course, the angular velocity is constant.

14.3   Relation with Two-View Motion Analysis

In two-view motion analysis, we can decompose uniquely any rigid motion by a rotation around the origin of the reference followed by a translation (see Sect. 4.1). Let $\mathbf{p}$ be the coordinates of a point in the first frame and $\mathbf{p}'$ in the second frame, then we have the following equation:

$$\mathbf{p}' = \mathbf{R}\mathbf{p} + \mathbf{t} , \tag{14.37}$$

where $\mathbf{R}$ is the rotation matrix and $\mathbf{t}$ is the translation vector.

If the first frame is taken at time t_0 and the second frame at time t, and the object is undergoing a motion with constant angular velocity and translational acceleration, then we have the following formulae to relate the two-view motion to the kinematic model described in Theorem 14.2:

$$\begin{aligned} \mathbf{R} &= W , \\ \mathbf{t} &= V\mathbf{v} + A\mathbf{a} . \end{aligned} \tag{14.38}$$

The reader can easily obtain similar relations for other special motions. We can show that the matrix W has all properties of a rotation matrix described in Theorem 4.1.

14.4   Formulation for the EKF Approach

In this section, we formulate the motion tracking problem in such a way that we can apply the extended Kalman filter formulated in Sect. 2.2.2. The token is assumed, as an example, to undergo a motion with constant angular velocity and constant translational acceleration (see Sect. 14.2.2). We are given a sequence of stereo frames taken at t_0, t_1, $\cdots$, t_{i-1}, t_i, $\cdots$. Without loss of generality, the interval between t_{i-1} and t_i is constant and is denoted by Δt. Consider a token being tracked which is matched with a segment observed at t_i. The formulation in this section allows the token to update its kinematic parameters by incorporating its match.

14.4.1   State Transition Equation

Let the angular velocity at time t_i be $\boldsymbol{\omega}_i$, the translational velocity $\mathbf{v}_i$ and the translational acceleration $\mathbf{a}_i$. Define the state vector as

$$\mathbf{s}_i = [\boldsymbol{\omega}_i^T \ \mathbf{v}_i^T \ \mathbf{a}_i^T]^T . \tag{14.39}$$

We have the state transition equation as follows:

$$\mathbf{s}_i = H\mathbf{s}_{i-1} + \mathbf{n}_{i-1} , \tag{14.40}$$

where

$$H = \begin{bmatrix} \mathbf{I}_3 & \mathbf{0} & \mathbf{0} \\ \mathbf{0} & \mathbf{I}_3 & \mathbf{I}_3\Delta t \\ \mathbf{0} & \mathbf{0} & \mathbf{I}_3 \end{bmatrix} .$$

We then replace $\frac{\partial h_i}{\partial s_i}$ in Algorithm 2.2 by H, since the transition function is linear. The n_{i-1} in (14.40) is the random disturbance, with

$$E[\mathbf{n}_{i-1}] = 0 \quad \text{and} \quad \Lambda_{\mathbf{n}_{i-1}} = Q_{i-1} .$$

This random noise $\mathbf{n}_i$ is used first to model noise due to, for example, vibration of objects during motion. Furthermore, our constant acceleration kinematic model is also in general only an approximation. By adding $\mathbf{n}_{i-1}$ in the dynamic model, we can partially take into account the approximation error.

14.4.2 Measurement Equations

As described in Sect. 4.2, a segment S is represented by ψ and $\mathbf{m}$ and their covariance matrix. Suppose the token being considered has been matched with S_1 at t_{i-1} and is now matched with S_2 at t_i. We define the measurement vector as

$$\mathbf{x} = [\psi_1^T \ \mathbf{m}_1^T \ \psi_2^T \ \mathbf{m}_2^T]^T . \tag{14.41}$$

From Theorem 14.2, we have the following equation for the midpoints

$$\mathbf{m}_2 = W\mathbf{m}_1 + V\mathbf{v} + A\mathbf{a} . \tag{14.42}$$

Let $\mathbf{u}_1$ be the unit direction vector of segment S_1 and $\mathbf{u}_2$ that of segment S_2. We have the following relation for the orientation

$$\mathbf{u}_2 = W\mathbf{u}_1 . \tag{14.43}$$

The above equation can be easily justified. Indeed, if we denote the endpoints of a segment by $\mathbf{b}$ and $\mathbf{e}$, then we have $\mathbf{b}_2 = W\mathbf{b}_1 + V\mathbf{v} + A\mathbf{a}$ and $\mathbf{e}_2 = W\mathbf{e}_1 + V\mathbf{v} + A\mathbf{a}$. Thus $\mathbf{e}_2 - \mathbf{b}_2 = W(\mathbf{e}_1 - \mathbf{b}_1)$. It can be shown that the determinant of W is $+1$, then $\|\mathbf{e}_2 - \mathbf{b}_2\| = \|\mathbf{e}_1 - \mathbf{b}_1\|$. Since $\mathbf{u}_1 = (\mathbf{e}_1 - \mathbf{b}_1)/\|\mathbf{e}_1 - \mathbf{b}_1\|$ and $\mathbf{u}_2 = (\mathbf{e}_2 - \mathbf{b}_2)/\|\mathbf{e}_2 - \mathbf{b}_2\|$, we get immediately (14.43).

If we define two functions $\mathbf{g}$ and $\mathbf{h}$ to relate ψ and $\mathbf{u}$ together (see (4.20) and (4.21)) so that

$$\psi = \mathbf{g}(\mathbf{u}) \quad \text{and} \quad \mathbf{u} = \mathbf{h}(\psi) , \tag{14.44}$$

then we have the following measurement equation

$$\mathbf{f}(\mathbf{x}, \mathbf{s}) = \left[\begin{array}{c} \mathbf{g}(W \, \mathbf{h}(\psi_1)) - \psi_2 \\ W\mathbf{m}_1 + V\mathbf{v} + A\mathbf{a} - \mathbf{m}_2 \end{array} \right] = \mathbf{0} . \tag{14.45}$$

This is a 5-dimensional vector equation. In the following, the first two elements in $\mathbf{f}(\mathbf{x}, \mathbf{s})$ are denoted by $\mathbf{f}_1$ and the last three elements by $\mathbf{f}_2$.

The relation between $\mathbf{s}$ and $\mathbf{x}$ described by (14.45) is not linear. In order to apply the EKF algorithm, it is necessary to compute the derivatives of $\mathbf{f}(\mathbf{x}, \mathbf{s})$

with respect to **s** and **x**. It is easy to show that

$$\frac{\partial \mathbf{f}}{\partial \mathbf{s}} = \begin{bmatrix} \frac{\partial \mathbf{f}_1}{\partial \omega} & \mathbf{0} & \mathbf{0} \\ \frac{\partial \mathbf{f}_2}{\partial \omega} & V & A \end{bmatrix}, \tag{14.46}$$

$$\frac{\partial \mathbf{f}}{\partial \mathbf{x}} = \begin{bmatrix} \frac{\partial \mathbf{f}_1}{\partial \psi_1} & \mathbf{0} & -\mathbf{I}_2 & \mathbf{0} \\ \mathbf{0} & W & \mathbf{0} & -\mathbf{I}_3 \end{bmatrix}, \tag{14.47}$$

where

$$\frac{\partial \mathbf{f}_1}{\partial \omega} = \frac{\partial \mathbf{g}}{\partial \mathbf{u}_1'} \frac{\partial (W\mathbf{u}_1)}{\partial \omega},$$

$$\frac{\partial \mathbf{f}_2}{\partial \omega} = \frac{\partial (W\mathbf{m}_1)}{\partial \omega} + \frac{\partial (V\mathbf{v})}{\partial \omega} + \frac{\partial (A\mathbf{a})}{\partial \omega},$$

$$\frac{\partial \mathbf{f}_1}{\partial \psi_1} = \frac{\partial \mathbf{g}}{\partial \mathbf{u}_1'} W \frac{\partial \mathbf{h}}{\partial \psi_1},$$

with $\mathbf{u}_1 = \mathbf{h}(\psi_1)$ and $\mathbf{u}_1' = W\mathbf{u}_1$. $\frac{\partial \mathbf{g}}{\partial \mathbf{u}_1'}$ is given by (4.24). $\frac{\partial \mathbf{h}}{\partial \psi_1}$ is given by (4.29). Since $\frac{\partial (W\mathbf{u}_1)}{\partial \omega}$ is similar to $\frac{\partial (W\mathbf{m}_1)}{\partial \omega}$, we then only need to compute $\frac{\partial (W\mathbf{u}_1)}{\partial \omega}$, $\frac{\partial (V\mathbf{v})}{\partial \omega}$ and $\frac{\partial (A\mathbf{a})}{\partial \omega}$.

After some simple computations, we get:

$$\begin{aligned}
\frac{\partial (W\mathbf{u})}{\partial \omega} = {}& -\frac{\sin(\theta \Delta t)}{\theta} \tilde{\mathbf{u}} + \frac{\theta \Delta t \cos(\theta \Delta t) - \sin(\theta \Delta t)}{\theta^3} (\tilde{\omega}\mathbf{u})\omega^T \\
& + \frac{\theta \Delta t \sin(\theta \Delta t) - 2(1 - \cos(\theta \Delta t))}{\theta^4} (\tilde{\omega}(\tilde{\omega}\mathbf{u}))\omega^T \\
& + \frac{1 - \cos(\theta \Delta t)}{\theta^2} \left[-\widetilde{\tilde{\omega}\mathbf{u}} + (\omega \cdot \mathbf{u})\mathbf{I}_3 - \mathbf{u}\omega^T \right],
\end{aligned} \tag{14.48}$$

$$\begin{aligned}
\frac{\partial (V\mathbf{v})}{\partial \omega} = {}& -\frac{1 - \cos(\theta \Delta t)}{\theta^2} \tilde{\mathbf{v}} + \frac{\theta \Delta t \sin(\theta \Delta t) - 2(1 - \cos(\theta \Delta t))}{\theta^4} (\tilde{\omega}\mathbf{v})\omega^T \\
& + \frac{3\sin(\theta \Delta t) - \theta \Delta t(2 + \cos(\theta \Delta t))}{\theta^5} (\tilde{\omega}(\tilde{\omega}\mathbf{v}))\omega^T \\
& + \frac{\theta \Delta t - \sin(\theta \Delta t)}{\theta^3} \left[-\widetilde{\tilde{\omega}\mathbf{v}} + (\omega \cdot \mathbf{v})\mathbf{I}_3 - \mathbf{v}\omega^T \right],
\end{aligned} \tag{14.49}$$

$$\begin{aligned}
\frac{\partial (A\mathbf{a})}{\partial \omega} = {}& -\frac{\theta \Delta t - \sin(\theta \Delta t)}{\theta^3} \tilde{\mathbf{a}} + \frac{3\sin(\theta \Delta t) - \theta \Delta t(2 + \cos(\theta \Delta t))}{\theta^5} (\tilde{\omega}\mathbf{a})\omega^T \\
& + \frac{4(1 - \cos(\theta \Delta t)) - (\theta \Delta t)^2 - \theta \Delta t \sin(\theta \Delta t)}{\theta^6} (\tilde{\omega}(\tilde{\omega}\mathbf{a}))\omega^T \\
& + \frac{(\theta \Delta t)^2 - 2(1 - \cos(\theta \Delta t))}{2\theta^4} \left[-\widetilde{\tilde{\omega}\mathbf{a}} + (\omega \cdot \mathbf{a})\mathbf{I}_3 - \mathbf{a}\omega^T \right].
\end{aligned} \tag{14.50}$$

where $\omega \cdot \mathbf{u}$ means the inner product of the two vectors ω and $\mathbf{u}$.

When a token matches a segment in the current frame, we use the above formalism to update its kinematic parameters. The same process is applied to each token.

14.5 Linearized Kinematic Model

In this section, we propose a linearized approximation solution of (14.7).

14.5.1 Linear Approximation

A point $\mathbf{p}_t$ is supposed to undergo a motion with constant angular acceleration and constant translational acceleration. The following state equations can then be written down:

$$\begin{aligned}
\boldsymbol{\omega}_t &= \boldsymbol{\omega} + \boldsymbol{\mu}(t - t_0) , \\
\mathbf{v}_t &= \mathbf{v} + \mathbf{a}(t - t_0) ,
\end{aligned} \tag{14.51}$$

where $\boldsymbol{\omega}$ denotes the angular velocity at $t = t_0$, $\boldsymbol{\mu}$ denotes the constant angular acceleration, $\mathbf{v}$ denotes the translational velocity at $t = t_0$, and $\mathbf{a}$ denotes the constant of the translational acceleration. In the following, we sometimes replace $(t - t_0)$ by Δt.

If Δt is small, or if there is no great movement of $\mathbf{p}_t$ during Δt, we can approximate $\dot{\mathbf{p}}_t$ by

$$\frac{\mathbf{p}_t - \mathbf{p}_0}{\Delta t} ,$$

where $\mathbf{p}_0$ is the value of $\mathbf{p}_t$ at $t = t_0$. We can then rewrite (14.7) as

$$[\mathbf{I}_3 - (\tilde{\boldsymbol{\omega}} + \tilde{\boldsymbol{\mu}}\Delta t)\Delta t]\mathbf{p}_t = \mathbf{p}_0 + (\mathbf{v} + \mathbf{a}\Delta t)\Delta t . \tag{14.52}$$

Since $(\tilde{\boldsymbol{\omega}} + \tilde{\boldsymbol{\mu}}\Delta t)\Delta t$ is assumed small, we have

$$[\mathbf{I}_3 - (\tilde{\boldsymbol{\omega}} + \tilde{\boldsymbol{\mu}}\Delta t)\Delta t]^{-1} = [\mathbf{I}_3 + (\tilde{\boldsymbol{\omega}} + \tilde{\boldsymbol{\mu}}\Delta t)\Delta t] , \tag{14.53}$$

and Equation (14.52) can be rewritten as

$$\mathbf{p}_t = [\mathbf{I}_3 + (\tilde{\boldsymbol{\omega}} + \tilde{\boldsymbol{\mu}}\Delta t)\Delta t][\mathbf{p}_0 + (\mathbf{v} + \mathbf{a}\Delta t)\Delta t] . \tag{14.54}$$

The above equation is the linearized kinematic model.

14.5.2 State Transition Equation

Let the angular velocity at time t_i be $\boldsymbol{\omega}_i$, the angular acceleration $\boldsymbol{\mu}_i$, the translational velocity $\mathbf{v}_i$ and the translational acceleration $\mathbf{a}_i$. Define the state vector as

$$\mathbf{s}_i = [\boldsymbol{\omega}_i^T \ \boldsymbol{\mu}_i^T \ \mathbf{v}_i^T \ \mathbf{a}_i^T]^T . \tag{14.55}$$

We then have the state transition equation as follows:

$$\mathbf{s}_i = H\mathbf{s}_{i-1} + \mathbf{n}_{i-1} , \tag{14.56}$$

where

$$H = \begin{bmatrix} \mathbf{I}_3 & \mathbf{I}_3\Delta t & 0 & 0 \\ 0 & \mathbf{I}_3 & 0 & 0 \\ 0 & 0 & \mathbf{I}_3 & \mathbf{I}_3\Delta t \\ 0 & 0 & 0 & \mathbf{I}_3 \end{bmatrix} .$$

We then replace $\frac{\partial \mathbf{h}_i}{\partial \mathbf{s}_i}$ in Algorithm 2.2 by H, since the transition function is linear. The $\mathbf{n}_{i-1}$ in (14.40) is the random disturbance, with

$$E[\mathbf{n}_{i-1}] = 0 \quad \text{and} \quad \Lambda_{\mathbf{n}_{i-1}} = Q_{i-1} .$$

The discussion on the use of a random noise term in Sect. 14.4.1 is applied here, too.

14.5.3 Measurement Equations

As described in Sect. 4.2, a segment S is represented by ψ and $\mathbf{m}$ and their covariance matrix. Suppose the token being considered has been matched with S_1 at t_{i-1} and is now matched with S_2 at t_i. We define the same measurement vector as in (14.41).

Using the linearized kinematic model, we have the following equation:

$$\mathbf{m}_2 = W\mathbf{m}_1 + W\mathbf{d}\Delta t , \tag{14.57}$$

where

$$\begin{aligned} W &= \mathbf{I}_3 + (\tilde{\omega} + \tilde{\mu}\Delta t)\Delta t , \\ \mathbf{d} &= \mathbf{v} + \mathbf{a}\Delta t . \end{aligned} \tag{14.58}$$

Let $\mathbf{u}_1$ be the unit direction vector of segment S_1 and $\mathbf{u}_2$ that of segment S_2. We have the following relation:

$$\mathbf{u}_2 = W\mathbf{u}_1 .$$

We then have the following measurement equation

$$\mathbf{f}(\mathbf{x}, \mathbf{s}) = \left[\begin{array}{c} \mathbf{g}(W\,\mathbf{h}(\psi_1)) - \psi_2 \\ W\mathbf{m}_1 + W\mathbf{d}\Delta t - \mathbf{m}_2 \end{array} \right] = \mathbf{0} . \tag{14.59}$$

See (14.44) for the meanings of $\mathbf{g}(.)$ and $\mathbf{h}(.)$. This is a 5-dimensional vector equation. In the following, the first two elements in $\mathbf{f}(\mathbf{x}, \mathbf{s})$ is denoted by $\mathbf{f}_1$ and the last three elements by $\mathbf{f}_2$.

In order to apply the EKF algorithm, it is necessary to compute the derivatives of $\mathbf{f}(\mathbf{x}, \mathbf{s})$ with respect to $\mathbf{s}$ and $\mathbf{x}$. In fact,

$$\frac{\partial \mathbf{f}}{\partial \mathbf{s}} = \left[\begin{array}{cccc} \frac{\partial \mathbf{f}_1}{\partial \omega} & \frac{\partial \mathbf{f}_1}{\partial \mu} & 0 & 0 \\ \frac{\partial \mathbf{f}_2}{\partial \omega} & \frac{\partial \mathbf{f}_2}{\partial \mu} & W\Delta t & W\Delta t^2 \end{array} \right] , \tag{14.60}$$

$$\frac{\partial f}{\partial \mathbf{x}} = \left[\begin{array}{cccc} \frac{\partial \mathbf{f}_1}{\partial \psi_1} & 0 & -\mathbf{I}_2 & 0 \\ 0 & W & 0 & -\mathbf{I}_3 \end{array} \right] , \tag{14.61}$$

where

$$\frac{\partial \mathbf{f}_1}{\partial \omega} = -\frac{\partial \mathbf{g}}{\partial \mathbf{u}_1'} \widetilde{\mathbf{u}}_1 \Delta t ,$$

$$\frac{\partial \mathbf{f}_1}{\partial \boldsymbol{\mu}} = -\frac{\partial \mathbf{g}}{\partial \mathbf{u}_1'}\widetilde{\mathbf{u}_1}\Delta t^2 = \frac{\partial \mathbf{f}_1}{\partial \boldsymbol{\omega}}\Delta t,$$

$$\frac{\partial \mathbf{f}_2}{\partial \boldsymbol{\omega}} = -\widetilde{\mathbf{m}_1}\Delta t - \tilde{\mathbf{d}}\Delta t,$$

$$\frac{\partial \mathbf{f}_2}{\partial \boldsymbol{\mu}} = -\widetilde{\mathbf{m}_1}\Delta t^2 - \tilde{\mathbf{d}}\Delta t^2 = \frac{\partial \mathbf{f}_2}{\partial \boldsymbol{\omega}}\Delta t,$$

$$\frac{\partial \mathbf{f}_1}{\partial \boldsymbol{\psi}_1} = \frac{\partial \mathbf{g}}{\partial \mathbf{u}_1'}W\frac{\partial \mathbf{h}}{\partial \boldsymbol{\psi}_1},$$

with $\mathbf{u}_1 = \mathbf{h}(\boldsymbol{\psi}_1)$, $\mathbf{u}_1' = W\mathbf{u}_1$ and $\mathbf{d}$ as defined earlier. $\partial \mathbf{g}/\partial \mathbf{u}_1'$ is given by (4.24). $\partial \mathbf{h}/\partial \boldsymbol{\psi}_1$ is given by (4.29).

When a token matches a segment in the current frame, we use the above formalism to update its kinematic parameters.

14.5.4 Discussions

The linearized kinematic model is simple and less computation is required than using the accurate model. We have implemented it together with the matching and grouping techniques described in the following chapter. Several experiments have been carried out, and using the linearized model suffers from several problems:

1. The linearization is only effective for a short period. The estimate given by using the linearized model diverges from the true value after a long period if no noise term is added to the model. To track a long sequence, we must add a (relatively large) noise term in the kinematic model.
2. The uncertainty in kinematic parameters decreases much slower in filtering over time than in the case of using the accurate models described in the previous sections. The estimate using the linearized model is also less precise.

In [14.11], *Gennery* used also a (simpler) linearized model to track known 3D objects. He claimed that in order to obtain greater accuracy, two additional effects have been included in the implemented program:

- First, the influence of the uncertainties in the previous orientation and in the angular velocity on the current orientation by the rotation that has occurred during the elapsed time interval.
- Second, the influence on position and orientation of the random acceleration during the elapsed time interval.

Although no more details have been given in that paper, adding those two effects is in fact to compensate for the inadequacies of the linearized model.

Here, the difference between using the accurate model and using the linearized model is similar to that between using the *extended Kalman filter* and using the *Linearized Kalman Filter* (LKF). *Maybeck* [14.10] gave in volume 2 of his book (pp. 46–48) an example of their difference: the EKF can track an object whose trajectory is an ellipse while the LKF cannot.

No more results about the linearized kinematic model will be presented in this monograph.

14.6 Summary

In this chapter, we have described object motions using classical kinematics, expressed in the form of a differential equation. We have integrated the equation under some assumptions on motion. Closed-form solution exists if angular velocity is constant and if translational velocity is given by a polynomial. In particular, we modelize motion with constant angular and translational velocities, and motion with constant angular velocity and constant translational acceleration. We believe this model can well approximate any smooth motion as long as the interval between frames is sufficiently small. We have not found the closed-form solutions for more complex motions. We have formulated the motion tracking problem as a kinematic parameter estimation problem. The state vector consists of the kinematic parameters, and the measurement vector is composed of the line segment parameters observed at two consecutive instants. Since the relation between the state vector and the measurement vector is not linear, the extended Kalman filter is applied. This formalism is used to predict the position of a token at the next instant and to update the kinematic parameters once a new observation is given. The linearized kinematic model has also been formulated. From several experiments we have carried out, it is apparent that using the linearized model suffers from inaccuracy of motion estimation.

15. Implementation Details and Experimental Results

We have proposed in Chap. 13 a framework to deal with the motion tracking problem in long sequences of stereo frames. In Chap. 14, we have modeled object motions using classical kinematics and derived closed-form solutions for several special but useful motions. Estimating motion kinematics has been formulated as a parameter estimation problem using the extended Kalman filter. As described in Chap. 13, we intend to solve the motion estimation and the token tracking at the same time. In this chapter, we develop the token tracking technique with which we shall integrate the formulation developed in Chap. 14.

Besides tracking each individual token, we want to group tokens into objects. The grouping criterion is the similarity in motion kinematics of tokens: tokens belonging to a single *rigid* object must have the same motion parameters. If a set of tokens have the same motion parameters, we consider them as a single object. Thus two different objects will be merged as a single one if they undergo the same motion in our algorithm. If we want to separate them, some other criteria such as spatial neighborhood should be added. The grouping process has several important consequences:

- An individual token is of little significance. The grouping process allows us to organize the disordered tokens into more coherent descriptions which can facilitate the tasks of higher-level analysis and decision.
- The estimated kinematic parameters for each individual token usually have big uncertainty. By grouping, the uncertainty in the kinematic parameters of an object will be much reduced.

From the above discussion, the motion tracking problem can be considered as an integration problem, and our motion tracking algorithm consists of an integration in two separate directions: both temporal and spatial (see Fig. 15.1). The temporal integration takes into account all information *from different instants* to estimate the kinematic parameters for each token and to help track the token from frame to frame. The spatial integration tries, *at each instant*, to divide tokens into several coherent groups and to fuse information from all tokens in each group.

Finally, we shall provide the experimental results of our motion tracking algorithm using both synthetic and real data.

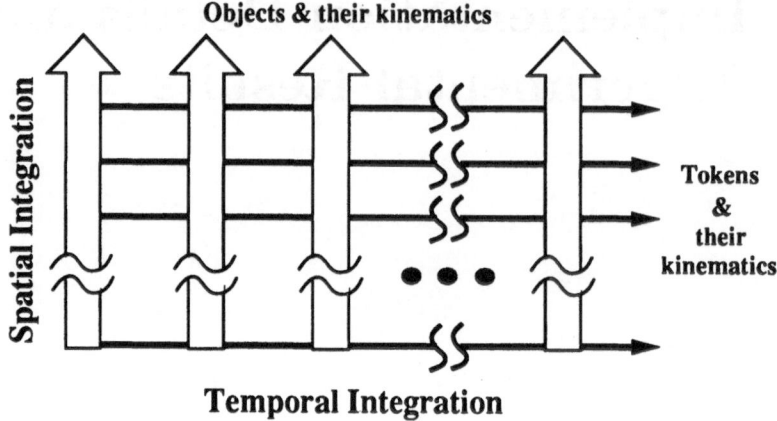

Fig. 15.1. Spatiotemporal integration in motion tracking

15.1 Matching Segments

In this section, we describe how to match a token being tracked with a segment in the current frame. The matching technique is based on the Mahalanobis distance, which can be considered as an Euclidean distance in parameter space weighted by uncertainty.

15.1.1 Prediction of a Token

Let $\mathbf{z} = [\boldsymbol{\psi}^T,\ \mathbf{m}^T]^T$ be the parameter vector of the token being tracked. The token kinematic parameters are $[\boldsymbol{\omega}^T,\ \mathbf{v}^T,\ \mathbf{a}^T]^T$. We can use them to *predict* its parameter vector $\widehat{\mathbf{z}} = [\widehat{\boldsymbol{\psi}}^T,\ \widehat{\mathbf{m}}^T]^T$ at the next time instant. From Theorem 14.2, $\widehat{\mathbf{z}}$ is given by

$$\begin{cases} \widehat{\boldsymbol{\psi}} &= \mathbf{g}(W\mathbf{h}(\boldsymbol{\psi}))\,, \\ \widehat{\mathbf{m}} &= W\mathbf{m} + V\mathbf{v} + A\mathbf{a}\,, \end{cases} \tag{15.1}$$

where the functions $\mathbf{g}$ and $\mathbf{h}$ are defined as in (14.44) in Sect. 14.4. Due to noise from multiple sources, it is very unlikely that a segment can be found with exactly the parameters $\widehat{\mathbf{z}} = [\widehat{\boldsymbol{\psi}}^T,\ \widehat{\mathbf{m}}^T]^T$ and we have to design a matching strategy.

15.1.2 Matching Criterion

Let $\{\cdots,\ [\boldsymbol{\psi}_i^T,\ \mathbf{m}_i^T],\ \cdots\}$ be the set of observed segments in the scene and $[\widehat{\boldsymbol{\psi}}^T,\ \widehat{\mathbf{m}}^T]^T$ be the expected segment. All segments have their measures of uncertainty attached (covariance matrices): $\{\cdots,\ \Lambda_i,\ \cdots\}$ and Λ_{token}. Λ_{token} is the covariance matrix of the predicted parameters $\widehat{\mathbf{z}} = [\widehat{\boldsymbol{\psi}}^T,\ \widehat{\mathbf{m}}^T]^T$ of the token being tracked, whose computation is given below. The *Mahalanobis distance* between the expected segment and each segment in the current frame is then given by

$$d_i^M = \mathbf{r}_i^T \Lambda_{\mathbf{r}_i}^{-1} \mathbf{r}_i \; , \tag{15.2}$$

where

$$\mathbf{r}_i \;\; = \;\; \mathbf{z}_i - \hat{\mathbf{z}} = \left[\begin{array}{c} \psi_i - \hat{\psi} \\ \mathbf{m}_i - \hat{\mathbf{m}} \end{array} \right] \; ,$$

$$\Lambda_{\mathbf{r}_i} \;\; = \;\; \Lambda_i + \Lambda_{\text{token}} \; .$$

The variable d_i^M is a scalar random variable following a χ^2 distribution with 5 degrees of freedom.

By looking up the χ^2 distribution table, we can choose an appropriate threshold ϵ on the Mahalanobis distance. For example, we can take $\epsilon = 11.07$ which corresponds to a probability of 95% to have the distance d_i^M less than ϵ if the match is correct. Thus segments in the current frame can be considered as *plausible matches*, if they verify the inequality:

$$d_i^M < \epsilon \; . \tag{15.3}$$

Using the above technique, a token may have multiple matches in the current frame. This problem has been described in Sect. 13.3.

Before computing the Mahalanobis distance (15.2), we must take care of the discontinuity of ϕ when a segment is nearly parallel to the plane $y = 0$ (see Sect. 4.2). The idea is the following. If a segment is represented by $\psi = [\phi, \theta]^T$, it is also represented by $[\phi - 2\pi, \theta]^T$. Therefore, when comparing the representations of two segments S and S', we perform the following tests and actions. If $\phi < \pi/2$ and $\hat{\phi} > 3\pi/2$, then set $\hat{\phi}$ to be $\hat{\phi} - 2\pi$; else if $\phi > 3\pi/2$ and $\hat{\phi} < \pi/2$, then set ϕ to be $\phi - 2\pi$; else do nothing. Notice that adding a constant to a random variable does not affect its covariance matrix.

Now we return to the problem of computing Λ_{token}. Given the original parameters of the token $[\psi^T, \mathbf{m}^T]^T$ and their covariance matrix $\Lambda_{(\psi,\mathbf{m})}$, and given the kinematic parameters $[\omega^T, \mathbf{v}^T, \mathbf{a}^T]^T$ and their covariance matrix $\Lambda_{(\omega,\mathbf{v},\mathbf{a})}$. Under the first order approximation, the covariance matrix of the expected segment is given, following Theorem 2.3, by

$$\Lambda_{\text{token}} = J_{(\psi,\mathbf{m})} \Lambda_{(\psi,\mathbf{m})} J_{(\psi,\mathbf{m})}^T + J_{(\omega,\mathbf{v},\mathbf{a})} \Lambda_{(\omega,\mathbf{v},\mathbf{a})} J_{(\omega,\mathbf{v},\mathbf{a})}^T \; . \tag{15.4}$$

From (15.1), the Jacobian matrix with respect to $[\psi^T, \mathbf{m}^T]^T$ is

$$J_{(\psi,\mathbf{m})} = \left[\begin{array}{cc} \dfrac{\partial \mathbf{g}}{\partial \mathbf{u}'} W \dfrac{\partial \mathbf{h}}{\partial \psi} & 0 \\ 0 & W \end{array} \right] \; , \tag{15.5}$$

and the Jacobian matrix with respect to $[\omega^T, \mathbf{v}^T, \mathbf{a}^T]^T$ is

$$J_{(\omega,\mathbf{v},\mathbf{a})} = \left[\begin{array}{cccc} \dfrac{\partial \mathbf{g}}{\partial \mathbf{u}'} \dfrac{\partial (W\mathbf{u})}{\partial \omega} & & 0 & 0 \\ \dfrac{\partial (W\mathbf{m})}{\partial \omega} + \dfrac{\partial (V\mathbf{v})}{\partial \omega} + \dfrac{\partial (A\mathbf{a})}{\partial \omega} & & V & A \end{array} \right] \; , \tag{15.6}$$

where $\mathbf{u} = \mathbf{h}(\boldsymbol{\psi})$ and $\mathbf{u}' = W\mathbf{u}$. All derivatives in the above equations are computed as in Sect. 14.4.

15.1.3 Reducing the Complexity by Bucketing Techniques

Although the complexity of matching one token is linear in the number of segments present in the current frame, the matching process may be slow, especially when there is a large number of segments. This is because the computation of the Mahalanobis distance is relatively expensive (it involves the inversion of a 5×5 matrix). If we can compute the distances d_i^M of the expected segment to only a subset of segments which are near the expected one, we can considerably speed up the matching process; this can be achieved by the use of *bucketing techniques.* which are now standard in Computational Geometry [15.1].

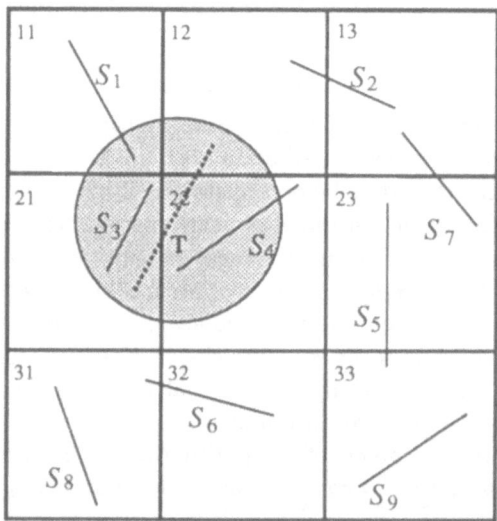

Fig. 15.2. Bucketing technique

We can apply the bucketing techniques either to 3D space or to 2D space. Bucketing in the image plane of one camera is preferred because it is cheaper. The image plane is partitioned into m^2 square windows (buckets) W_{ij} (in our implementation, $m = 16$). To each window W_{ij} we attach the list of segments $\{S_k\}$ intersecting W_{ij}. The key idea of bucketing is that on the average the number of segments intersecting a bucket is much smaller, and in practice constant, than the total number of segments in the frame (see for example [15.2] for details). Given a predicted token to be matched, we first compute the buckets which the disk defined by the predicted segment intersects. The disk is defined as follows: its center coincides with the midpoint of the predicted segment and its diameter equals its length plus a number corresponding to the projected uncertainty of its midpoint. Again, the idea is that, on the average, this disk will intersect a small number of buckets, except for a token which just appeared. Since we initialize such a token with a big uncertainty in motion, its corresponding disk is quite big and may intersect many buckets. The set of potential matching candidates is then the union of the lists of segments intersecting these buckets. This set is now considerably reduced, on the average. Figure 15.2 illustrates the bucketing technique. In the figure, $\mathbf{T}$ is the expected segment. Thanks to the bucketing technique, it finds 4 potential matching candidates $\{S_1, S_2, S_3, S_4\}$ instead of 9 in total. After computing the matching criterion described in the above section, there remain only two candidates S_3 and S_4. Note that the computation of buckets can be performed very quickly by an algorithm whose complexity is linear in the number of segments in the current frame.

15.2 Support of Existence

In this section we describe in detail how the beam search strategy sketched in Sect. 13.3 makes tracking much more robust by allowing multiple matches.

Indeed, in practice, a token being tracked may find several matches in the current frame. The most common strategy is to choose the nearest segment as in [15.3,4] and to discard the other possibilities.

Our implementation is based on the work of *Bar-Shalom* and *Fortmann* [15.5] and is much less sensitive to false matches. The idea is to keep open the possibility of accepting several or no matches for any given token. But, if tokens never disappear we may rapidly reach a computational explosion. To avoid this we compute for each token a number that we call its *support of existence* which measures the adequateness of the token with the measurements: if the token has not found any correspondences in a long time then it is bound either to be the result of a false match that happened in the past or to have disappeared from the scene.

We use the notations of Sect. 15.1 and denote the sequence of measurements corresponding to the token being tracked up to time t_k as $Z^k \triangleq \{\mathbf{z}(t_1), \cdots, \mathbf{z}(t_k)\}$ in which $\mathbf{z}(t_i) = [\boldsymbol{\psi}^T(t_i), \mathbf{m}^T(t_i)]^T$ is the parameter vector of the segment observed at time t_i. Denote the event that Z^k yields a correct token, i.e., that its components $\mathbf{z}(t_i)$ were produced by the same segment moving in space, by $e \triangleq \{Z^k \text{ yields a correct token}\}$. The likelihood function of this sequence yielding a correct token is the joint Probability Density Function (or PDF):

$$L^k(e) = p(Z^k|e) = p[\mathbf{z}(t_1), \cdots, \mathbf{z}(t_k)|e] . \tag{15.7}$$

From the definition of a conditional PDF, $L^k(e)$ can be written by recurrence as

$$
\begin{aligned}
L^k(e) &= p[Z^{k-1}, \mathbf{z}(t_k)|e] = p[\mathbf{z}(t_k)|Z^{k-1}, e] \, p[Z^{k-1}|e] = \cdots \\
&= \prod_{i=1}^{k} p[\mathbf{z}(t_i)|Z^{i-1}, e] ,
\end{aligned}
\tag{15.8}
$$

where Z^0 represents the prior information.

As in Sect. 15.1, we denote the measurement residual as $\mathbf{r}$, i.e., $\mathbf{r}_i = \mathbf{z}(t_i) - \hat{\mathbf{z}}(t_i)$. Then $p(\mathbf{r}_i) = N[\mathbf{r}_i; \mathbf{0}, \Lambda_{\mathbf{r}_i}]$ with $\Lambda_{\mathbf{r}_i} = \Lambda_i + \Lambda_{\text{token}}$. We use $N[\mathbf{x}; \bar{\mathbf{x}}, \Lambda]$ to denote the Gaussian density function of the random variable $\mathbf{x}$ with mean $\bar{\mathbf{x}}$ and covariance Λ. We now make the admittedly strong assumption that the $\mathbf{r}_i$'s are Gaussian and uncorrelated. We thus write:

$$p[\mathbf{z}(t_i)|Z^{i-1}, e] = N[\mathbf{r}_i; \mathbf{0}, \Lambda_{\mathbf{r}_i}] . \tag{15.9}$$

It follows that under the previous assumption:

$$L^k(e) = \left[\prod_{i=1}^{k} |2\pi\Lambda_{\mathbf{r}_i}|^{-1/2}\right] \exp\left[-\frac{1}{2}\sum_{i=1}^{k} \mathbf{r}_i^T \Lambda_{\mathbf{r}_i}^{-1} \mathbf{r}_i\right] .$$

Note that $\mathbf{r}_i^T \Lambda_{\mathbf{r}_i}^{-1} \mathbf{r}_i = d_i^M$ (see (15.2)). The modified log-likelihood function, corresponding to the exponent of $L^k(e)$, is defined as

$$l_k \stackrel{\Delta}{=} -2\ln\left[L^k(e)/\prod_{i=1}^{k} |2\pi\Lambda_{\mathbf{r}_i}|^{-1/2}\right] = \sum_{i=1}^{k} d_i^M$$

and can be computed recursively as follows:

$$l_k = l_{k-1} + d_k^M .$$

The last term has a χ^2 distribution with $n_r = 5$ degrees of freedom. Since the $\mathbf{r}_k$ are assumed to be independent, l_k has a χ^2 distribution with kn_r degrees of freedom.

The statistical test for deciding that Z^k yields a correct token is that the log-likelihood function satisfies

$$l_k \leq \kappa , \tag{15.10}$$

where the threshold κ is obtained from the χ^2 table with kn_r degrees of freedom by setting $P(\kappa, kn_r) = \alpha$, where α is typically equal to 95%. We call l_k the *support of existence* of the token, because when l_k is small it means that there is a very good accordance between the token and the measurements and thereby there is a strong support for the existence of the token. On the other hand, when l_k is large, it implies that the fit of the token to the measurements is not very good, and thereby there is only a poor support for the existence of the token.

In practice, the test (15.10) cannot be used for long sequences because the likelihood function is dominated by old measurements and responds very slowly to recent ones. In order to limit the "memory" of system, we can multiply the likelihood function at each step by a discount factor $c < 1$. This results in the fading-memory likelihood function:

$$l_k = cl_{k-1} + d_k^M = \sum_{i=1}^{k} c^{k-i} d_i^M .$$

The effective memory of l_k is now $(1-c)^{-1}$, and in steady state l_k is approximately a χ^2 random variable with $n_r(1+c)/(1-c)$ degrees of freedom, mean $n_r/(1-c)$, and variance $2n_r/(1-c^2)$. See [15.5] for the proof. In our implementation, $c = 0.75$.

In the above discussion, we assume implicitly that a match is detected at each sampling time. As described earlier, match detection may fail from time to time for a number of reasons. In these cases, it means that:

$$d_k^M \geq \epsilon \,,$$

as described in Sect. 15.1. Thus, if at time t_k no match is found, the fit between the prediction and the observation is not very good. But note that even in that case we may still have $l_k \leq \kappa$ and the processing of the token will continue. This allows us to cope with problems such as occlusion, disappearance and absence. Of course, if the Mahalanobis distances stay over the threshold ϵ at too many consecutive time instants, i.e., if the token does not find any good match in the scene too often, then l_k will go beyond the threshold κ, and the token will be discarded, as expected. In practice we set $d_k^M = \alpha\epsilon$ where $\alpha = 1.2$ in our implementation if match detection fails.

Our concept of support differs significantly from the techniques reported in the literature for discarding false matches in that it takes into account not only the number of times a token has not been present, but also the number of times it has been present in the past and how well the measurements agreed with the prediction. See Sect. 13.6 for a discussion.

15.3   Algorithm of the Token Tracking Process

The tracking is performed in parallel for each token in a "matching-update-prediction" loop (see Sect. 13.3 and Fig. 13.3), which is summarized in Algorithm 15.1.

15.4   Grouping Tokens into Objects

In the previous sections, we have described how to track each token and estimate its kinematic parameters in parallel. In this section, we address the following problem:

> How can the supervisor group tokens into objects based on their kinematic parameters ?

We assume that moving objects in the scene are rigid. From Chap. 14, we know that tokens belonging to a single rigid object must have the same kinematic parameters with respect to a *common* point. In our algorithm, the kinematic

Algorithm 15.1: Token Tracking Procedure

- Bucket the current scene tokens /* Sect.15.1.3 */
- **for** each token **T** in the tracking team
  - → Find the scene tokens near **T** using the bucketing technique: $\{S\}$
  - → Find matches of **T** in $\{S\}$ satisfying (15.3): $\{\mathcal{M}\}$
  - → **if** the length of $\{\mathcal{M}\} = 0$ /* no match */
  - **then** ↪ Update the support of existence of **T** as described
            in the end of Sect.15.2
    - ↪ **if** the support > threshold **then** deactivate **T** **endif**
  - **else if** the length of $\{\mathcal{M}\} = 1$ /* unique match */
  - **then** ↪ Update the support of existence of **T** /* Sect.15.2 */
    - ↪ Update the kinematic parameters of **T** /* Sect.14.4 */
    - ↪ Replace the token parameters of **T** by those of $\mathcal{M}$
    - ↪ **if** the support > threshold **then** deactivate **T** **endif**
  - **else if** the length of $\{\mathcal{M}\} > 1$ /* multiple match */
  - **then** ↪ Choose the two best matches: $\mathcal{M}_1$ and $\mathcal{M}_2$
    - ↪ **if** $\mathcal{M}_1$ and $\mathcal{M}_2$ are collinear /* Sect.13.4 */
      - **then** ⇒ Merge $\mathcal{M}_1$ and $\mathcal{M}_2$ as a single one: $\mathcal{M}$
        - ⇒ Do the same thing as in the unique-match
              case for **T** and $\mathcal{M}$
      - **else** ⇒ Activate a new token **T′** /* splitting */
        - ⇒ Do the same thing as in the unique-match
              case for **T** and $\mathcal{M}_1$
        - ⇒ Do the same thing as in the unique-match
              case for **T′** and $\mathcal{M}_2$
    - ↪ **endif**
  - ← **endif**
- ⇑ **endfor**
- Initialize each unmatched scene token as a new tracking team member
- Predict the occurrence of each token in the tracking team
      for the next instant using (15.1)

parameters of all tokens are computed with respect to the same point — the origin of the system of reference. We can then define an object as follows: *an object is a set of tokens with the same kinematic parameters.* Under this definition, two different objects are grouped as a single one if they undergo the same motion. As described earlier, if we want to separate them, some other criteria such as spatial neighborhood should be used.

Clearly, what we want to solve is a *clustering problem*. A *cluster* is a subset of the set of active tokens in the tracking team. *Clustering* is to partition the set of active tokens into clusters such that each token lies in just one cluster and that all tokens in each cluster have the same kinematics. Thus, a cluster is just an object defined above.

Of course, the estimated kinematic parameters are uncertain, and have attached to them an uncertainty measure — their covariance matrix. One cannot expect to find two tokens having exactly the same kinematic parameters. The Mahalanobis distance is again used in this case to measure the discrepancy between kinematic parameters. Given two kinematic parameter vectors $\mathbf{s}_1$ and $\mathbf{s}_2$ and their covariance matrices $\Lambda_{\mathbf{s}_1}$ and $\Lambda_{\mathbf{s}_2}$, their Mahalanobis distance is given by

$$\delta^{\mathrm{M}} = [\mathbf{s}_1 - \mathbf{s}_2]^T (\Lambda_{\mathbf{s}_1} + \Lambda_{\mathbf{s}_2})^{-1} [\mathbf{s}_1 - \mathbf{s}_2] . \tag{15.11}$$

When $\mathbf{s} = [\boldsymbol{\omega}, \mathbf{v}, \mathbf{a}]^T$, δ^{M} is a random scalar following a χ^2 distribution with 9 degrees of freedom. By looking up the χ^2 distribution table, we can choose an appropriate threshold ε on the Mahalanobis distance. For example, we can take $\varepsilon = 16.92$ which corresponds to a probability of 95% that the distance δ^{M} is less than ε. Thus two tokens can be considered to belong to the same object, if the Mahalanobis distance of their kinematic parameters verifies the inequality:

$$\delta^{\mathrm{M}} < \varepsilon . \tag{15.12}$$

If two tokens are identified to belong to a single object by the above test, we can compute a better estimate $\mathbf{s}$ of the kinematic parameters for the object from those attached to each token by the Kalman filter or simply by the modified Kalman minimum-variance estimator (see Sect. 12.2):

$$\begin{aligned} \mathbf{s} &= \Lambda_{\mathbf{s}}^{-1}(\Lambda_{\mathbf{s}_1}^{-1}\mathbf{s}_1 + \Lambda_{\mathbf{s}_2}^{-1}\mathbf{s}_2) , \\ \Lambda_{\mathbf{s}} &= (\Lambda_{\mathbf{s}_1}^{-1} + \Lambda_{\mathbf{s}_2}^{-1})^{-1} , \end{aligned} \tag{15.13}$$

where $\Lambda_{\mathbf{s}}$ is the covariance matrix of the new estimate $\mathbf{s}$. This procedure can be performed incrementally when more tokens are identified to belong to the same object.

Many clustering techniques have been developed to solve a range of classification problems in diverse fields, for example, zoology, astronomy, medicine and even linguistics. *Hartigan* has described in his book [15.6] quite a number of algorithms. As our primary goal is not to develop an optimal clustering algorithm, but to demonstrate the overall performance of the framework we proposed in Chap. 13, we are content to use a simple algorithm (Algorithm 15.2).

Assume we have found, using Algorithm 15.2, n objects in the scene: O_b^i $(i = 1, \ldots, n)$. O_b^i contains m tokens: S_j^i $(j = 1, \ldots, m)$. Here, m changes from one object to another. Let $\mathbf{s}^i$ and $\Lambda_{\mathbf{s}^i}$ be the kinematic parameters associated with the object O_b^i and let $\mathbf{s}_j^i$ and $\Lambda_{\mathbf{s}_j^i}$ be those associated with the jth token in the ith object, i.e., S_j^i. Ideally, the Mahalanobis distance (15.11) between $\mathbf{s}_j^i$ and $\mathbf{s}^i$ for any j, denoted by ${}^i\delta_j^i$, is less than or equal to that between $\mathbf{s}_j^i$ and $\mathbf{s}^k$ $(k = 1, \ldots, n$ but $k \neq i)$, denoted by ${}^k\delta_j^i$. That is,

$${}^i\delta_j^i \leq {}^k\delta_j^i \quad (i, k = 1, \ldots, n, \; i \neq k, \; j = 1, \ldots, m) . \tag{15.14}$$

Algorithm 15.2: Token Grouping Procedure

- Set *list_tokens* = all active tokens
  /* *list_tokens* contains the tokens not yet identified */
- Set *list_objects* to NIL
  /* *list_objects* contains the objects recovered */
- **for** a segment S_1 in *list_tokens*
  → Set *cluster* to NIL   /* *cluster* contains
                   tokens identified for the current object */
  → Retain S_1 in *cluster* and delete it from *list_tokens*
  → Set s and Λ_s to the kinematic parameters of S_1
  → **for** a segment S_2 in *list_tokens*
    ↪ **if** the kinematic parameters of S_2 are similar to s and Λ_s
          in the sense of the criterion (15.12)
      ⇒ Retain S_2 in *cluster* and delete it from *list_tokens*
      ⇒ Update s and Λ_s by using (15.13)
    ↩ **endif**
  ← **endfor**
  → Retain *cluster* in *list_objects* and associate s and Λ_s with it
⇑ **endfor**

If the above requirement is met, then we can say that the clustering is optimal in the sense that the difference between the motions of any two tokens of one object is smaller than the difference between the motions of any two tokens not belonging to a single object. In other words, an object thus obtained represents the most coherent set of tokens in motion.

Unfortunately, the grouping result given by Algorithm 15.2 is not optimal in the sense of (15.14). This is due to the fact that the search for a compatible token is sequential in the algorithm and that the Mahalanobis distances between a relatively uncertain token and more than one object are possibly all less than the threshold given in (15.12). Therefore, it is possible that $\varepsilon > {}^i\delta_j^i > {}^k\delta_j^i$ for $k > i$[1]. The nonoptimality of the algorithm does not cause any problem if the objects in the scene move quite differently, as in the experiment with synthetical data described below. This does cause some minor problem if two objects in the scene undergo similar motions, especially in the beginning of the tracking process. In the beginning of the tracking process, the estimated kinematic parameters for each token have a relatively big uncertainty and a token is probably grouped into the object with a similar motion. As the tracking process continues, the estimated kinematic parameters become more and more precise and the nonoptimality of the algorithm has less and less effect.

---

[1]However, our algorithm guarantees that ${}^i\delta_j^k \geq \varepsilon > {}^k\delta_j^k$ for $k > i$, because, if not, the jth token in O_b^k would be grouped into O_b^i due to the sequential nature of the grouping algorithm.

We are satisfied with the result given by Algorithm 15.2. However, its nonoptimality could be remedied by adding a posterior checking: if a token in O_b^i is closer to O_b^k ($k > i$), then switch the token from O_b^i to O_b^k. The kinematic parameters associated with O_b^i and O_b^k should be recomputed.

15.5 Experimental Results

We have implemented the proposed algorithm on a SUN workstation using the C language. Our program can display and control interactively the motion tracking process. In this section, results with synthetic and real data are described to show the performance of our algorithm. The angular velocity unit is radians/unit-time, and the translational velocity unit is millimeters/unit-time, except when stated otherwise.

15.5.1 Synthetic Data

Our synthetic data include two objects: a box and a model car. Noise-free 3D data are first constructed for each object. The box has 16 segments and the model car has 17 segments. We then apply a motion with $s_1 = [0, 0.03, 0, 9.55, 0, 30.15]^T$ (without translational acceleration) to the first object (the box). We get 16 frames. The box moves with constant angular and translational velocities. Similarly, we apply a motion with $s_2 = [0, 0, 0, -40, -25, 45]^T$ (without translational acceleration) to the second object (the model car). We also get 16 frames. The model car moves with pure translation (constant velocity). By combining the two sequences of 3D frames, we obtain a sequence of noise-free 3D frames

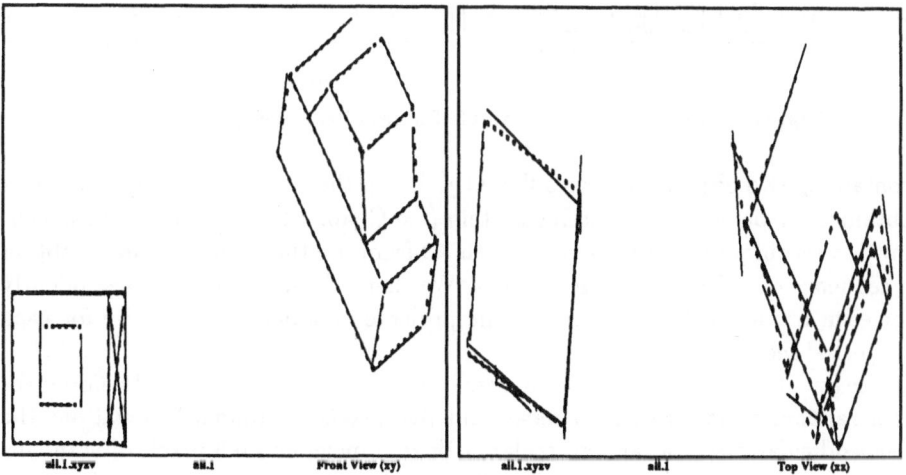

Fig. 15.3. Front and top views of the first frame of the generated sequence (in solid lines), compared with its original noise-free data (in dashed lines)

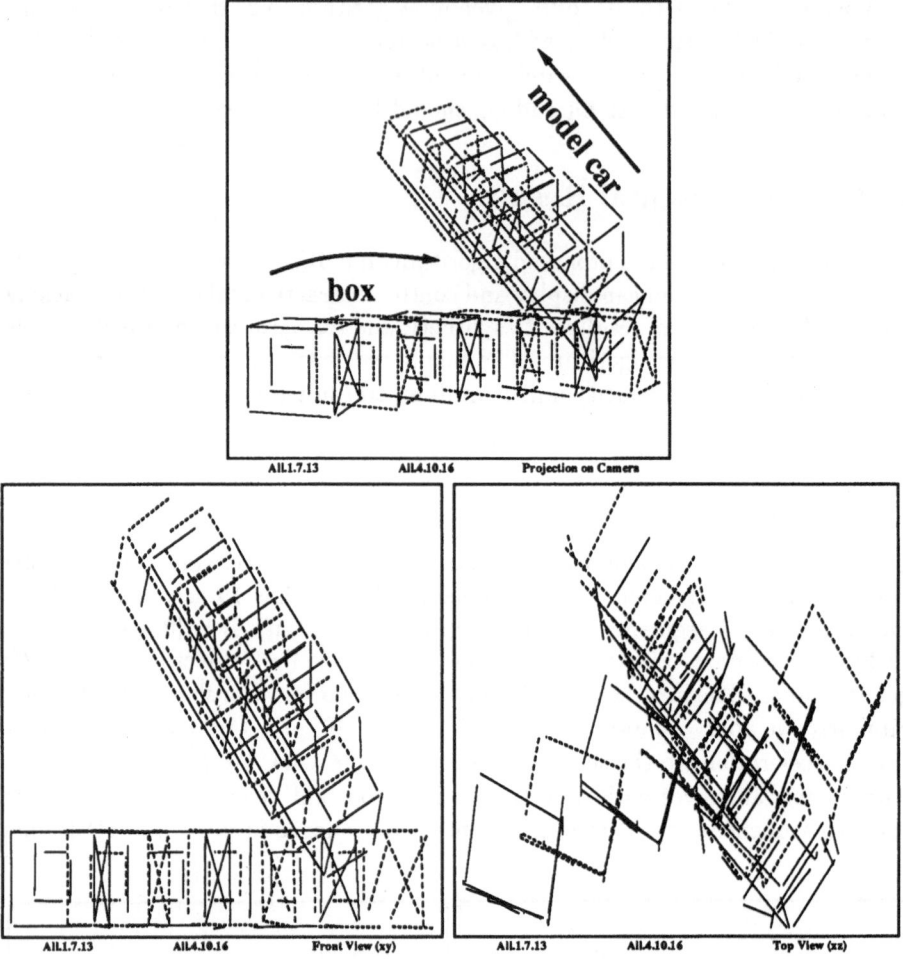

Fig. 15.4. Superposition of sample frames of the generated sequence

containing two objects moving differently. Using the real calibration parameters obtained for our trinocular stereo system (see Chap. 3 for the details of the calibration technique), we project each such 3D frame on the cameras, and we obtain a sequence of noise-free image triplets. We then add white noise independently for each image to the endpoints of the projected segments. The noise for each projected segment has two independent components: one component parallel to the segment and another perpendicular to the segment. The parallel component is a random scalar with zero mean and five pixels of standard deviation; the perpendicular one is a random scalar with zero mean and 0.5 pixels of standard deviation. Such noises are added to both endpoints of a segment. This partly models the precision of our edge detection and polygonal approximation process: more uncertainty is present along the segment than in its normal direction. At

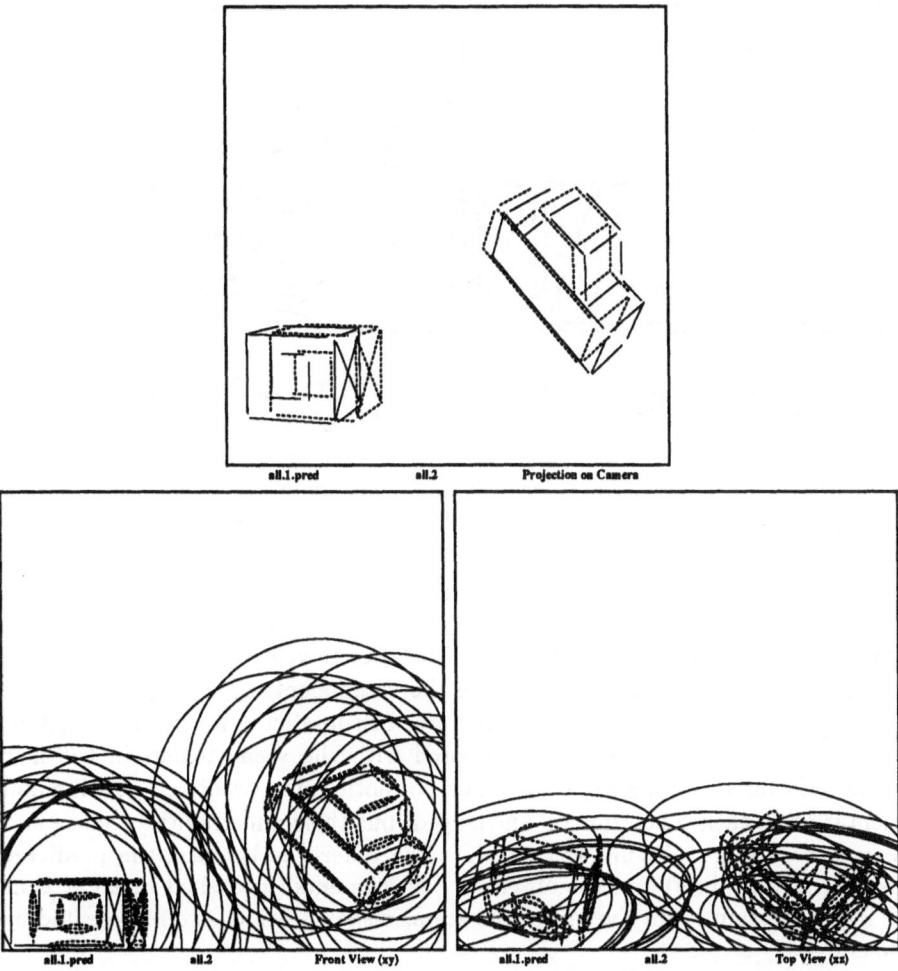

Fig. 15.5. The predicted segments (in solid lines) and observed segments (in dashed lines) at time t_2 and their covariance ellipses

this point, we generate a sequence of noisy image triplets. Each noisy triplet is then supplied to the trinocular stereo system (see Chap. 3 for the details), and a 3D frame is finally reconstructed. We thus obtain a sequence of noisy 3D frames. Figure 15.3 shows the first frame reconstructed (in solid lines), compared with its original noise-free data (in dashed lines). We can observe there exists a relatively big difference in depth between them, as illustrated by the top view in Fig. 15.3.

Figure 15.4 illustrates the sample sequence. The 1st, 7th and 13th frames are in solid lines. The 4th, 10th and 16th frames are in dashed lines. The box moves from left to right; the model car moves from low right to high left.

Each segment in the first frame is initialized as a token in the tracking team.

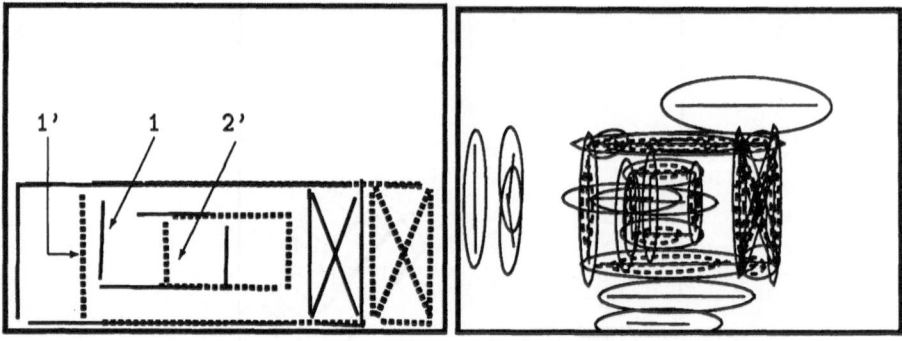

Fig. 15.6. The close up of the front view of the box at t_2

Fig. 15.7. The close up of the front view of the box at t_3

Since the motion tracking algorithm is recursive, some *a priori* information on the kinematics is required. A reasonable assumption may be that objects do not move, as the interframe motion is expected to be small. The kinematic parameters are thus all initialized to zero, but with a fairly large uncertainty: the standard deviation for each angular velocity component is 0.0873 radians/unit-time; the standard deviation for the translational velocity component is 100 millimeters/unit-time. The variances of the components of acceleration are set to zero (no acceleration). These tokens are then predicted for the next instant t_2 and the predicted tokens are compared with the scene tokens in the new frame. Of course, since we have assumed no motion, the predicted position and orientation of each token remain unchanged, but their uncertainties change and become very large. Figure 15.5 displays the difference between the predicted tokens (in solid lines) and the observed line segments (in dashed lines) at time t_2. The uncertainties in the midpoints are also shown, which are represented by the ellipses.

Each token then searches the uncertainty region of its prediction for a match based on the criterion described in Sect. 15.1. As expected, multiple matches occur for most of tokens. Techniques based only on the "best" match usually fail at this stage, since the nearest segment is not always the correct match. See the close up of the box in Fig. 15.6. Consider for example the predicted token 1. Its "best" match is segment 1', but its correct match is segment 2'. We retain the two best matches if a token has multiple matches. The token updates its kinematic parameters using its best match. The supervisor initializes a new token by combining the token and its second best match which is used to estimate its kinematic parameters. Figure 15.8 shows the superposition of the predicted segments for the instant t_3 and the segments observed at t_3. As can be observed, more active tokens (in solid lines) exist at this moment, but the uncertainty for most tokens is significantly reduced. Most of the tokens due to false matching in the preceding instant cannot find matches at this instant. Each of the correct tokens finds only one match. The superposition between them is already very good (see the close up in Fig. 15.7).

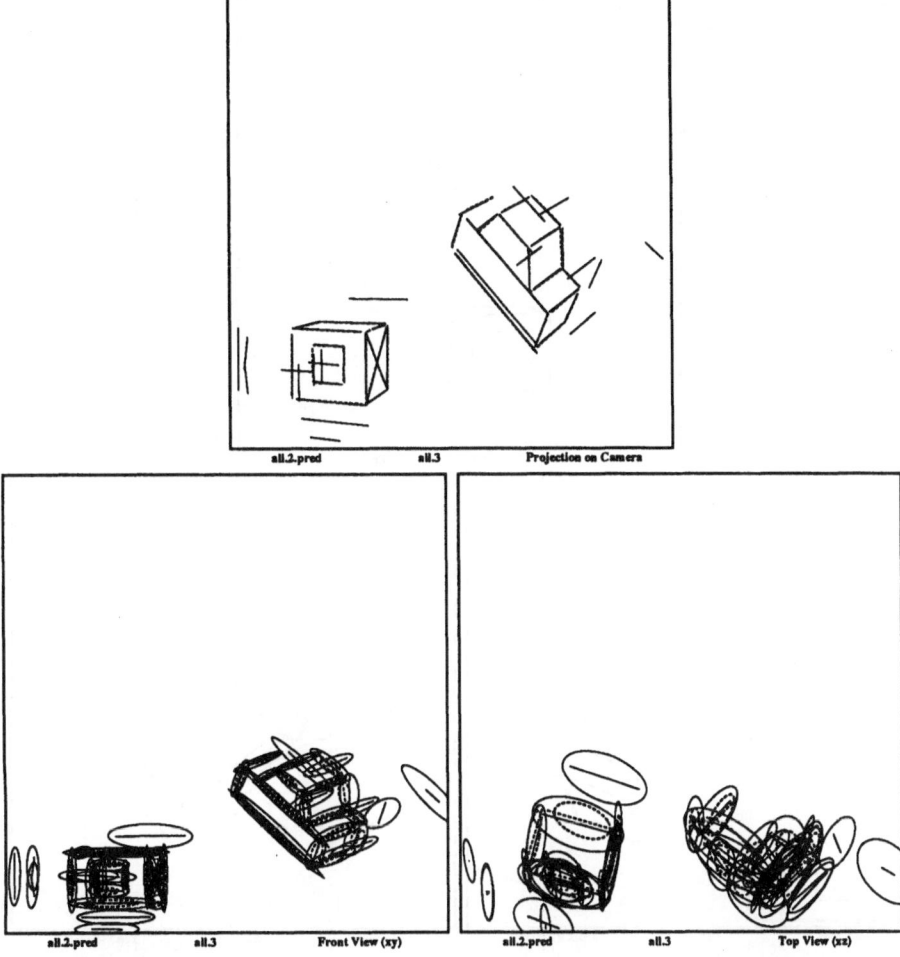

Fig. 15.8. The predicted segments (in solid lines) and observed segments (in dashed lines) at time t_3 and their covariance ellipses

We continue the tracking in the same manner. After several instants, incorrect tokens lose their supports of existence and are deactivated by the supervisor. For example, at time t_6, there remains only one incorrect token (see Fig. 15.9). Figure 15.9 displays also the observed segments at time t_6 (in dashed lines). As one can notice, the precision of the predicted positions is good. Figure 15.10 shows the evolution of the number of active tokens and the number of matches at each instant. As can be observed, incorrect tokens are active only during 3 or 4 views. At instant t_{16}, there is a segment moving out of the view field of the cameras.

The supervisor successfully segments the scene into objects based on the information in the kinematic parameters. Only at t_3, t_4 and t_5, the supervisor

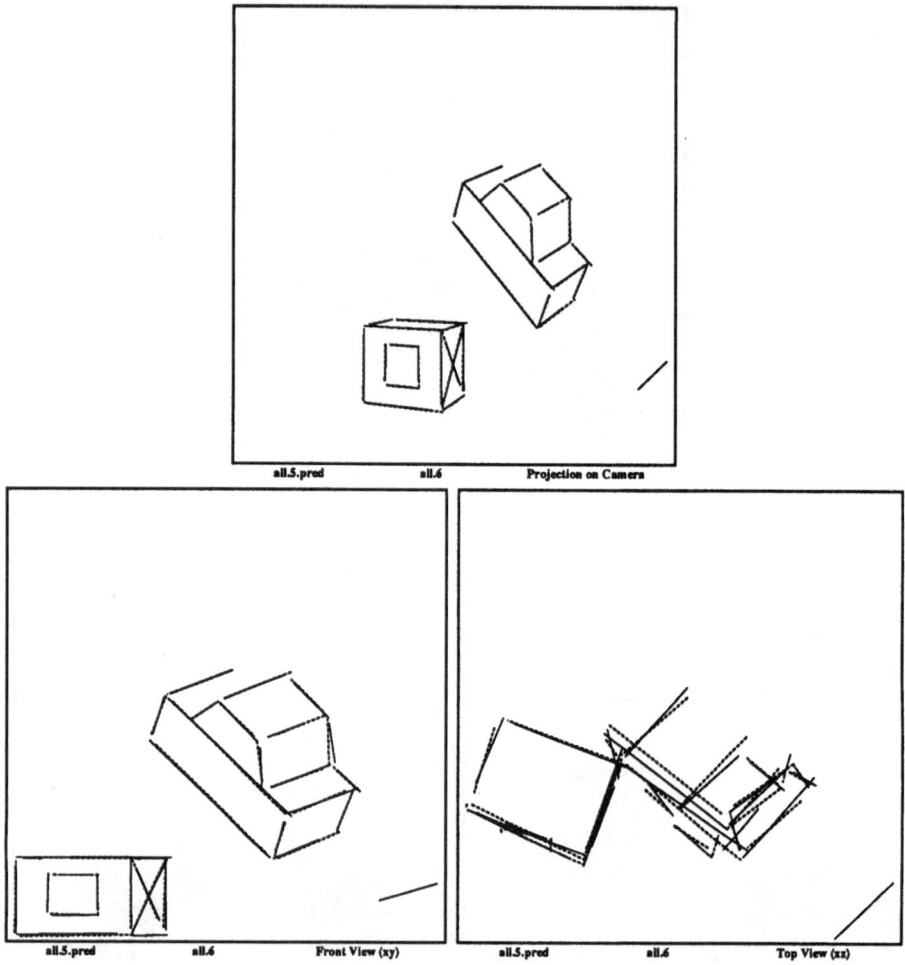

Fig. 15.9. The predicted segments and observed segments at time t_6

finds one more token for each object, since there exist many tokens, several of which have large uncertainties. The results of motion estimation given by the supervisor for each object are displayed in Figs. 15.11 to 15.14. Figures 15.11 and 15.12 show the *absolute* errors in the angular and translational velocities of the box. Figures 15.13 and 15.14 show the *absolute* errors in the angular and translational velocities of the model car. The error in the angular velocity is given in *milliradians*/unit-time and that in translational velocity is given in *millimeters*/unit-time. As can be observed, all errors converge to zero while more frames are processed. After a few initial (four or five) frames, the accuracy is of the order of milliradians/unit-time for the angular velocity and of millimeters/unit-time for the translational velocity. One can remark that the er-

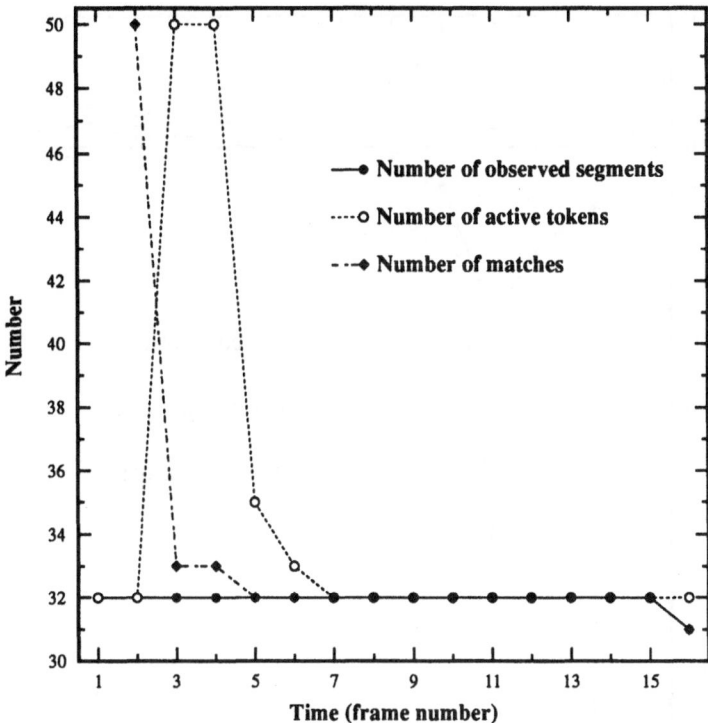

Fig. 15.10. Evolution of the numbers of observed segments, active tokens and matches

rors for the model car are relatively bigger than those for the box. This is because several segments of the model car are very noisy (see Fig. 15.3). This can be explained by the 3D reconstruction process. If a segment in one image is almost parallel to the epipolar line of its corresponding segment in another image, the precision of the reconstructed 3D segment is very poor, especially in its orientation. This bears upon in the estimation of the motion parameters: the error in the angular velocity is relatively big, and this in turn affects the precision in the translational velocity. The coupling between angular and translational velocities has also been noticed by other researchers [15.7]. It is difficult to distinguish the effect of a small rotation from that of a small translation. After processing over time, this ambiguity is of course decreasing. The final estimation for the box is
$\hat{s}_1 = [-3.932\text{e}{-04}, 2.923\text{e}{-02}, -9.592\text{e}{-05}, 1.174\text{e}{+01}, -1.182\text{e}{+00}, 2.984\text{e}{+01}]^T$,
and that for the model car is
$\hat{s}_2 = [-3.680\text{e}{-05}, -1.735\text{e}{-03}, 2.736\text{e}{-04}, -3.495\text{e}{+01}, -2.516\text{e}{+01}, 4.422\text{e}{+01}]^T$.
The relative error of the angular velocity of the box is 2.9% and that of the translational velocity is 7.93%. The angle between the true and estimated axes of rotation is 0.8 degrees, and that between the true and estimated translation vectors is 4.4 degrees. The relative error of the translational velocity of the car is 7.84%. The angle between the true and estimated translation vectors is 3.4 degrees.

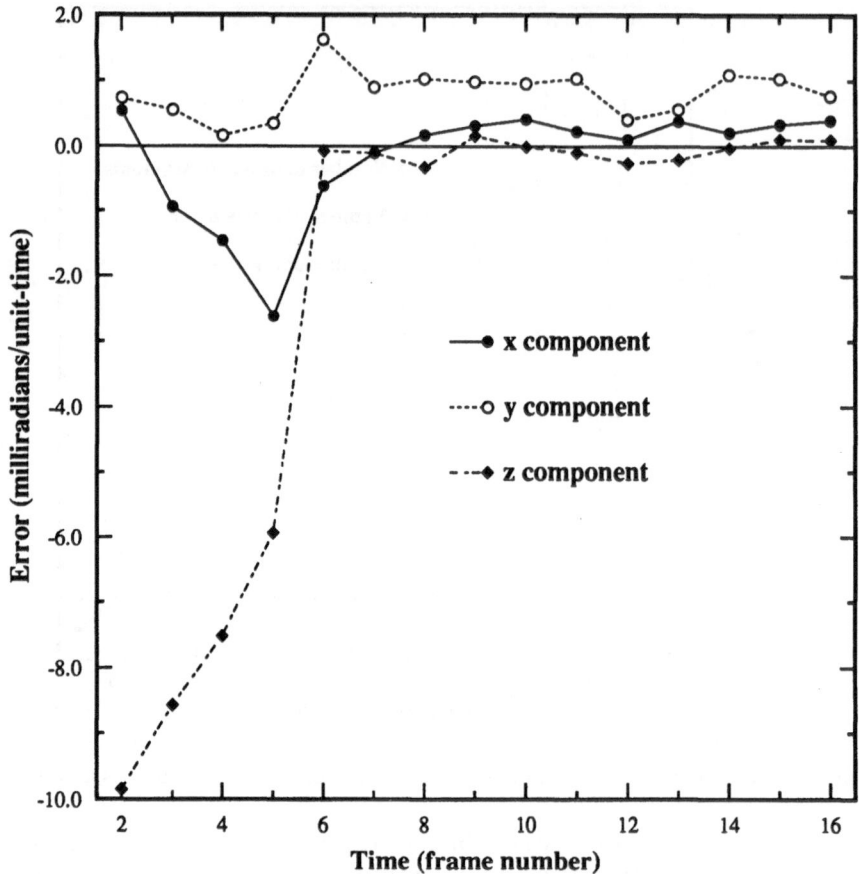

Fig. 15.11. Error evolution in the angular velocity of the box

In the following, we describe the performance of our algorithm. The program is run on a SUN 3/60 workstation. The average user time for the prediction of each token is about 0.05 seconds. The average user time for the matching and update of each token is about 0.15 seconds. That is, if we implement the algorithm on a parallel machine with the same performance as SUN 3/60, the time required to process one frame to another is 0.2 seconds. The average user time required to group tokens into objects is about 2.44 seconds.

To examine whether our algorithm can cope with the occlusion problem, we remove one arbitrary segment from each object in the 9th frame and in the 3 consecutive frames. That is to say, the occlusion occurs over four frames. At instants t_9, t_{10}, t_{11} and t_{12}, their corresponding tokens cannot find any matches, but they keep tracking by hypothetical extension as described in Sects. 13.3 and 15.2 since they have still the support of existence. As expected, these tokens

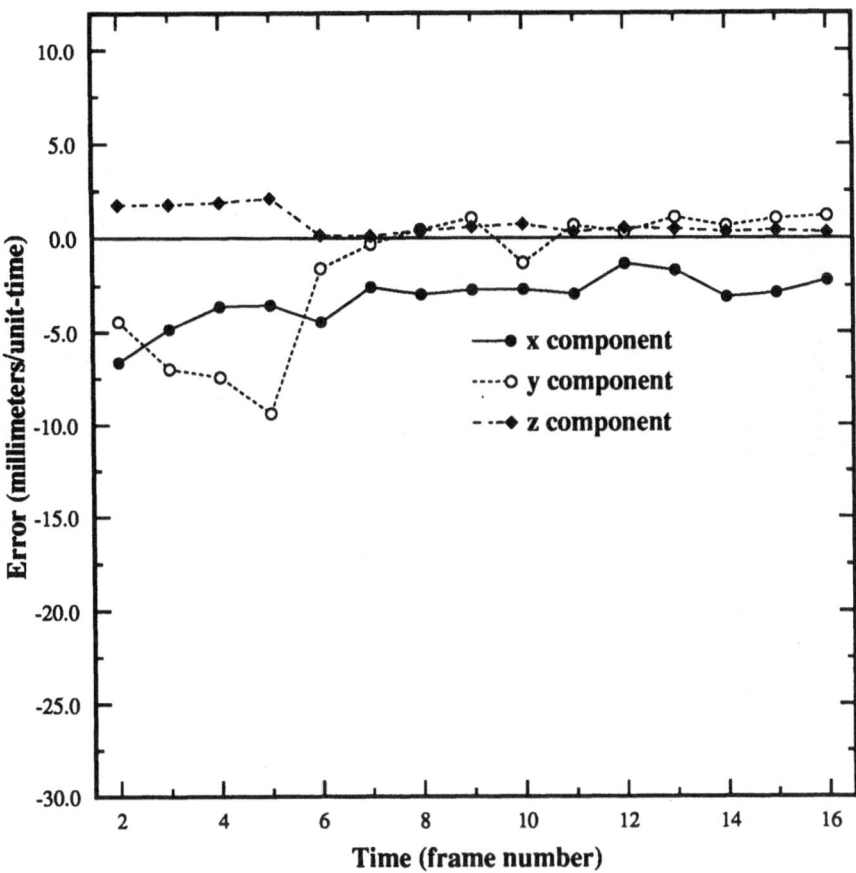

Fig. 15.12. Error evolution in the translational velocity of the box

match correctly the segments as they reappear at time t_{13}. The changes in the estimations of the object kinematic parameters due to occlusion are small.

15.5.2   Real Data with Controlled Motion

The data are acquired as follows: the trinocular stereo rig is in front of a rotating table (at about 2.5 meters). The rotating table has two degrees of freedom: vertical translation and rotation around the vertical. The motion is controlled manually. Boxes are put on the table and will be considered by the program as an object. The table undergoes a general motion, combination of a rotation and a translation parallel to the rotation axis. It rotates of 3 degrees (clockwise, if viewed from the top), and at the same time, translates from bottom to top of 50 millimeters between adjacent frames. Thus, the motion corresponds exactly to the constant velocity model. There are in fact three "objects" in the scene (see

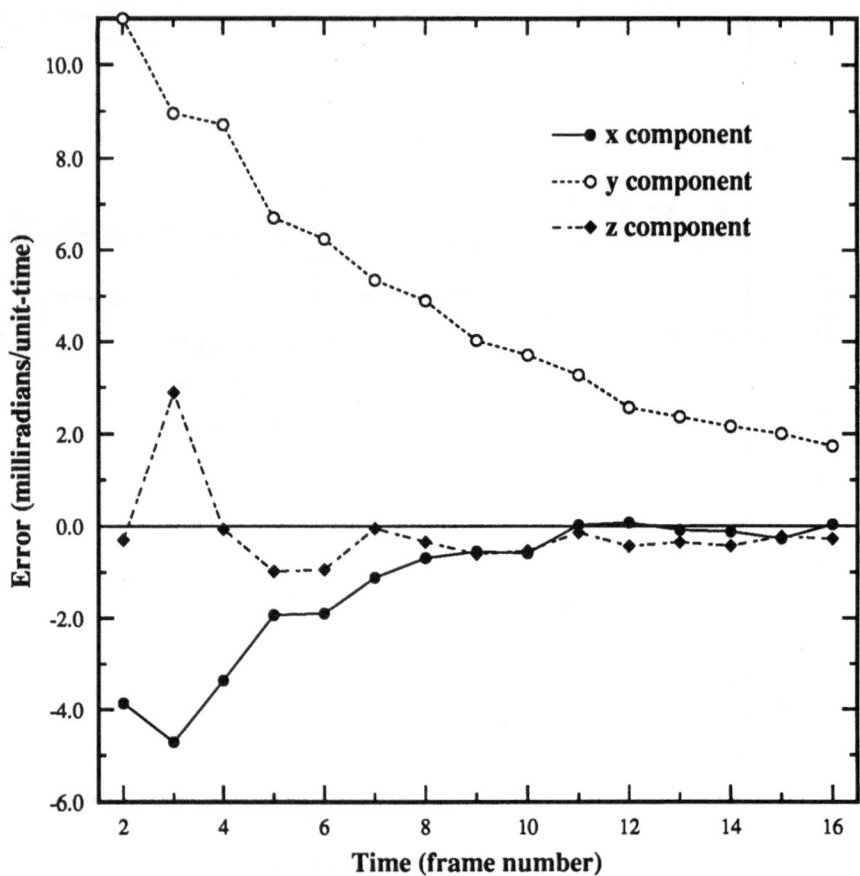

Fig. 15.13. Error evolution in the angular velocity of the model car

Fig. 15.15): the second is the static support of the table, and the third is what is below and attached to the rotating table, and undergoes the same translation but no rotation.

Ten frames have been acquired for this experiment. Each frame contains about 130 3D line segments. Figure 15.16 displays the images taken by the first camera at t_1 and t_{10}. To show the motion, the first and second frames are superimposed in Fig. 15.17 and the first and the last in Fig. 15.18, together with a pair of stereograms which allow the reader to perceive the depth through cross-eye fusion. We find that the data are very noisy even for the static object and that occlusion, appearance, disappearance and absence problems are very severe. We can also observe that the object in translation is not detected until the third frame is taken.

Each segment in the first frame is initialized as a token to be tracked. Since the motion tracking algorithm is recursive, some *a priori* information on the kine-

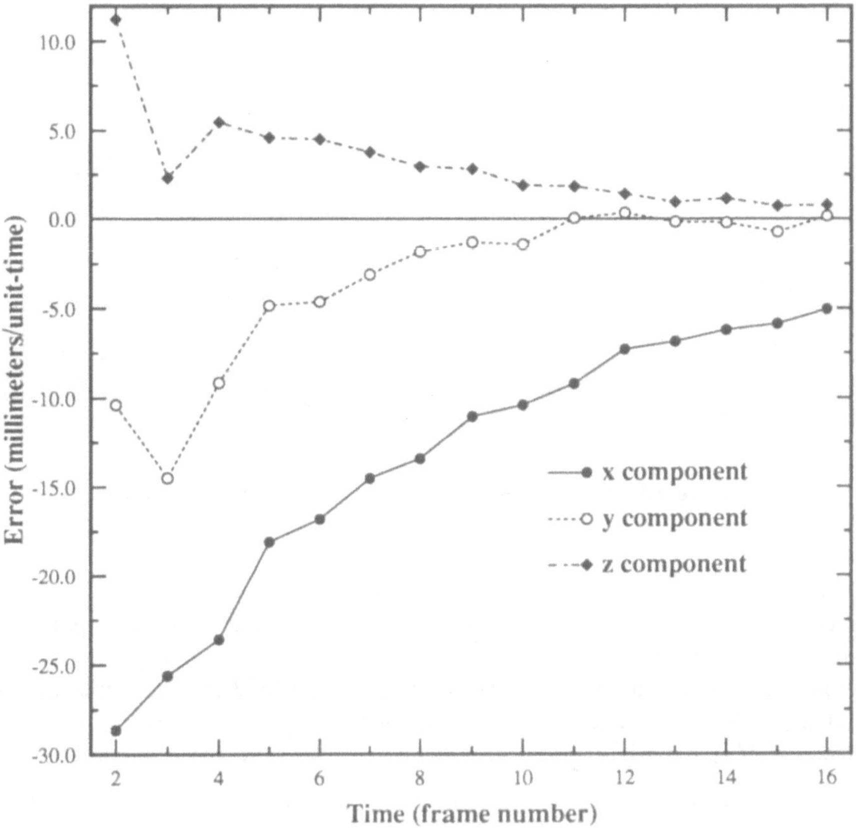

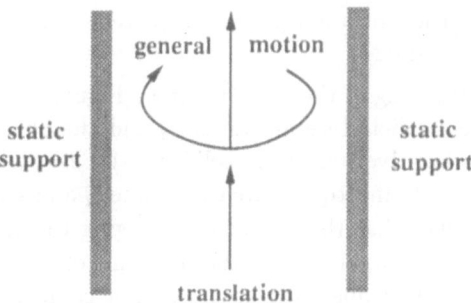

Fig. 15.14. Error evolution in the translational velocity of the car

Fig. 15.15. Description of the rotating table undergoing general motion

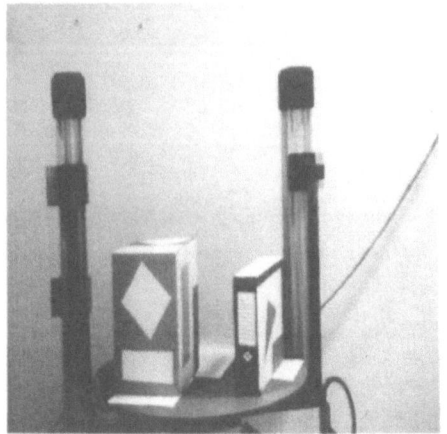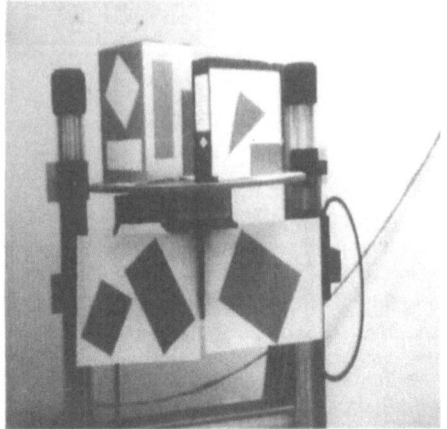

Fig. 15.16. The images taken by the first camera at t_1 and t_{10}

matics is required. A reasonable assumption may be that objects do not move, as the interframe motion is expected to be small. The kinematic parameters are thus all initialized to zero, but with fairly large uncertainty: the standard deviation for each angular velocity component is 0.0873 radians/unit-time, and that for each translational velocity component is 150 millimeters/unit-time. The variances of the acceleration components are set to zero (no acceleration).

Those tokens are then predicted for the next instant t_2 and the predicted tokens are compared with those in the new frame. Of course, since we have assumed no motion, the predicted position and orientation of each token remain unchanged, but their uncertainties change and become very large. Thus Figure 15.17 displays also the difference between the predicted tokens (in solid lines) and the observed line segments (in dashed lines) at time t_2. As expected, multiple matches occur for most of tokens, which are resolved in the same manner as in the previous experiment.

We continue the tracking in the same manner. Figure 15.19 shows the superposition between the predicted segments for t_3 and the observed segments. As can be observed, more active tokens (in solid lines) exist at this moment: some have been activated due to multiple matches at time t_2 and some just entered the field of view. We observe that the segments belonging to the object undergoing general motion coincide well, and so do the segments of the static support. The segments belonging to the translating object do not coincide well, because they just appeared in the field of view and no information about their kinematics is available. Figure 15.20 shows the superposition between the predicted segments for t_{10} and the observed segments. Almost all segments are well superimposed except those in the middle part. Those segments correspond to the outline of the rotating table which is an ellipse. Using line segments to approximate an ellipse

Fig. 15.17. The superposition of the first (in solid lines) and the second (in dashed lines) frames

is of course difficult, and this is clearly one of the limitations of our system. One can notice that most of the tokens due to false matching in the preceding instants have disappeared, because they have lost their supports of existence.

The supervisor successfully segments the scene into 3 groups of objects based on the information of the kinematics of each token. In the following, we show the result of the estimated kinematic parameters of the object undergoing general motion. If we express the motion in the stereo coordinate system, the kinematic parameter vector is $\mathbf{s} = [0.0, 5.236\mathrm{e}{-}02, 0.0, ?, 50, ?]^T$. The x and z components of the translational velocity are not known, since because of the calibration method we have used, the rotation axis (i.e., the axis of the cylinder) is not known in the stereo coordinate system. We know the y component of the translational

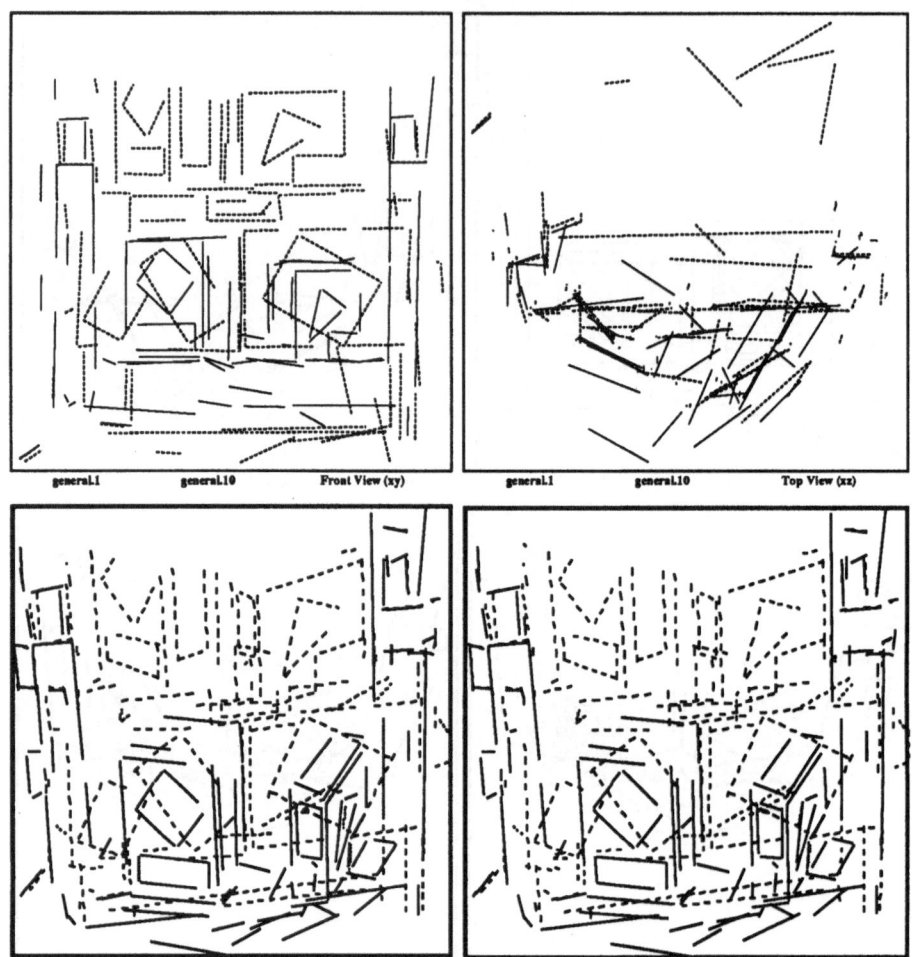

Fig. 15.18. The superposition of the first (in solid lines) and the last (in dashed lines) frames

velocity because the rotation axis is parallel to the y axis. The influence of the change of the rotation center on the translation has been discussed in Chap. 11. (In fact, we can use the technique described in Chap. 11 to recover the rotation axis of the rotating table.) Figure 15.21 shows the variation of the absolute errors in the estimation of the angular velocity. Figure 15.22 shows only the variation of the absolute error in the estimation of the y component of the translational velocity. The results are good: after a few (four or five) frames, the error in the angular velocity is less than 2.5 milliradians/unit-time (compared with 52.36), and the error in the translational velocity is less than 1.5 millimeters/unit-time (compared with 50). The final estimation for this object is

$$\hat{s} = [4.291e{-}04,\ 5.328e{-}02,\ -7.059e{-}04,\ -7.387e{+}00,\ -4.951e{+}01,\ 3.677e{+}01]^T.$$

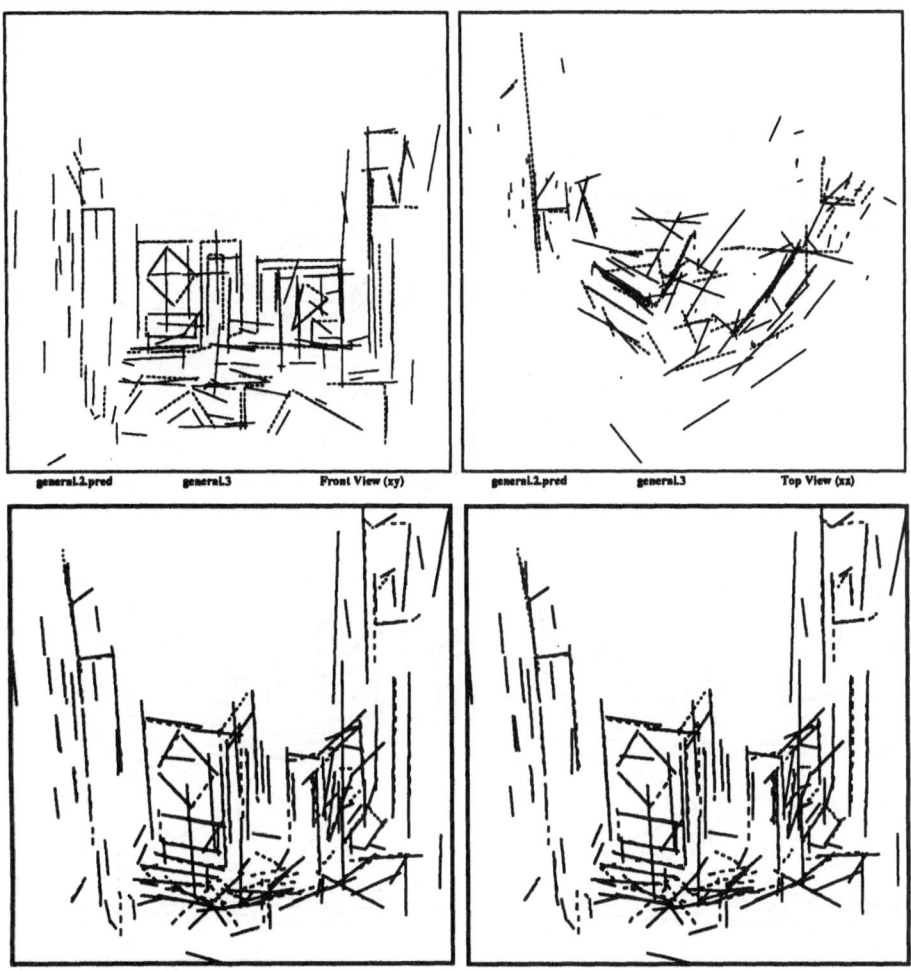

Fig. 15.19. The superposition of the predicted (in solid lines) and the observed (in dashed lines) segments at time t_3

The relative error of the angular velocity is 2.36%. The angle between the true and estimated axes of rotation is 0.9 degrees. The relative error of the y component of the translational velocity is 0.98%.

Similar results are observed for the other objects. The final motion estimate corresponding to the translating object is $[3.437e–04, 1.353e–03, –7.303e–04, –2.009e–01, –4.937e+01, 8.361e–01]^T$, which should be compared with $[0, 0, 0, 0, –50, 0]$. The relative error of the translational velocity is 2.13%. The final motion estimate corresponding to the static object is $[–2.321e–04, 5.009e–04, 2.981e–04, 2.681e–02, –3.399e–01, 4.799e–01]^T$.

We now show that the position of the axis of the rotating table in the stereo coordinate system can be computed from the estimated kinematics $\hat{\mathbf{s}}$ and the

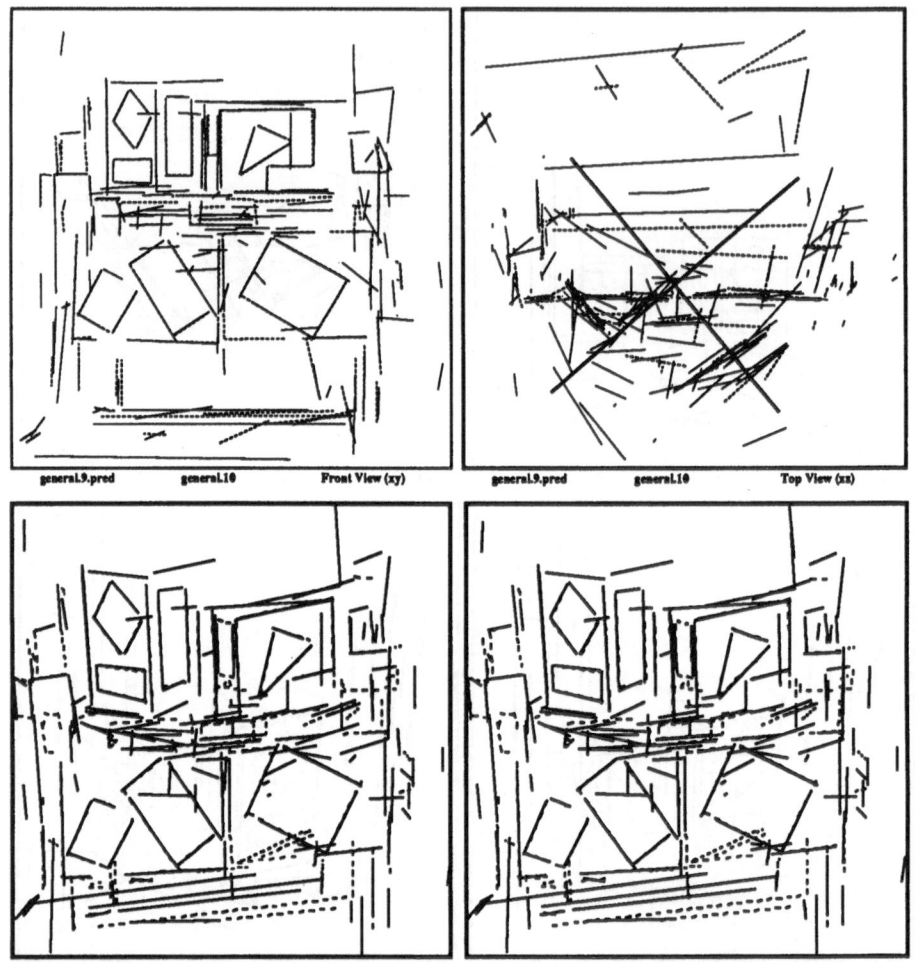

Fig. 15.20. The superposition of the predicted (in solid lines) and the observed (in dashed lines) segments at time t_{10}

known nature of the motion. For a point not on the axis of the rotating table, its trajectory is a helix. For a point on that axis, its trajectory is a straight line. Let a point on the axis be **p**. From (14.6), the velocity of **p** is given by

$$\mathbf{v_p} = \mathbf{v} + \boldsymbol{\omega} \wedge \mathbf{p} \ .$$

Since the rotation is about the axis of the table, the velocity $\mathbf{v_p}$ of **p** must be parallel to the rotation axis $\boldsymbol{\omega}$, i.e., $\boldsymbol{\omega} \wedge \mathbf{v_p} = 0$. We thus have

$$\boldsymbol{\omega} \wedge \mathbf{v} + \boldsymbol{\omega} \wedge (\boldsymbol{\omega} \wedge \mathbf{p}) = 0 \ . \tag{15.15}$$

Equation (15.15) gives only two independent scalar equations, because the three components of a cross-product are not linearly independent. It defines the axis

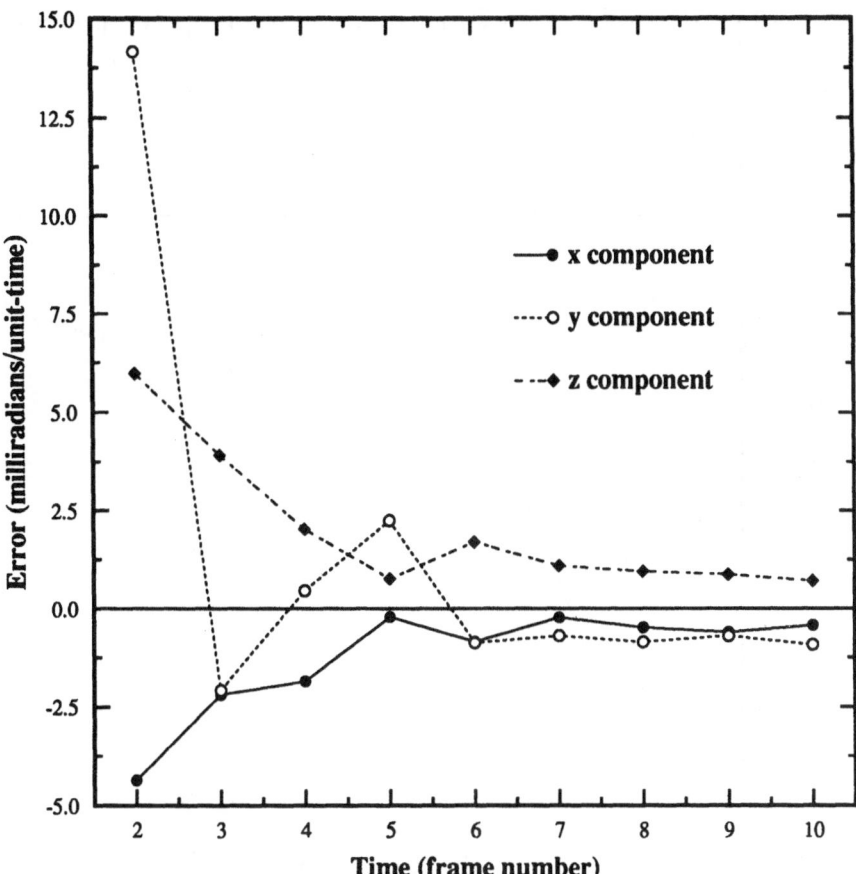

Fig. 15.21. Error evolution in the angular velocity of the rotating table undergoing general motion

of the table. In order to compute a point on it, we can choose the y component p_y of $\mathbf{p}$ equal to zero, and compute its x and z components: p_x and p_z. In this example, we get $p_x = 677.68$ and $p_z = 131.08$. That point is shown by the intersection of the two bold line segments in the top view of Fig. 15.20, and seems roughly correct.

15.5.3 Real Data with Uncontrolled Motion

The data described in this section are acquired as follows. The trinocular stereo rig is mounted on a mobile vehicle, which moves in the laboratory. Objects in the scene are about 2 to 7 meters away from the mobile vehicle. We have taken 15 stereo views while the vehicle was supposed to undergo a translation. Figure 15.23 displays the images taken by the first camera at t_1 and t_{15}. We

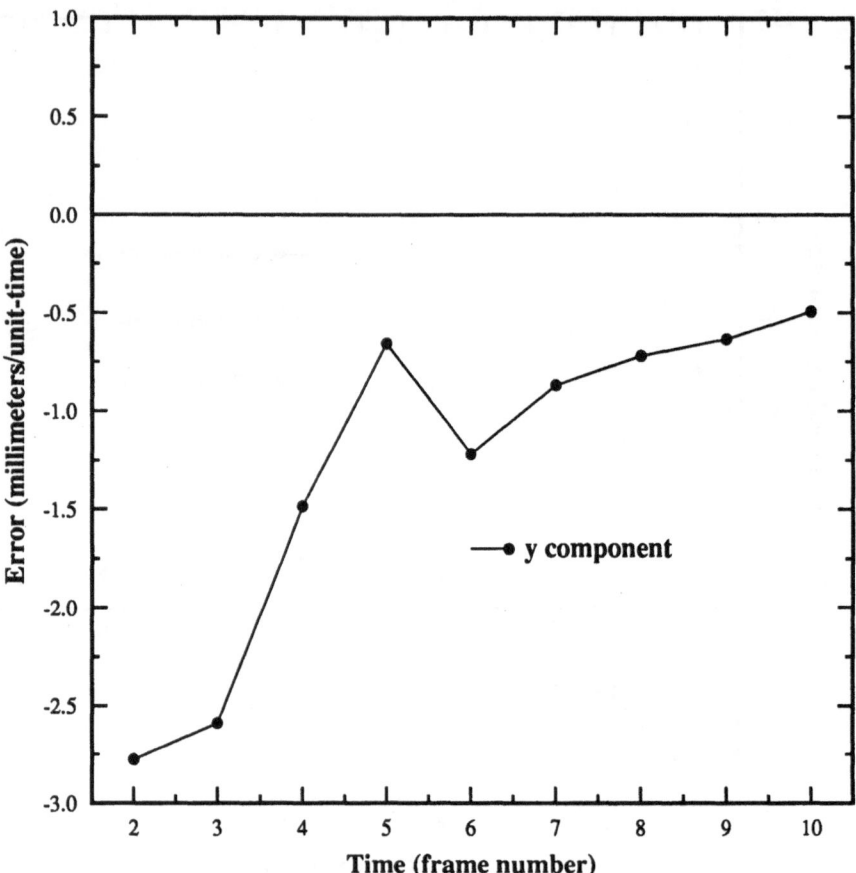

Fig. 15.22. Error evolution in the translational velocity of the rotating table undergoing general motion

have thus reconstructed 15 3D frames using our stereo system, each containing around 85 line segments.

Ideally, the motion of the vehicle is a pure translation. If described in our stereo coordinate system, the translation between successive views is about $[-25, 0, -98]^T$, i.e., a displacement of 100 millimeters. However, due to the precision of the mechanical system and slipping of the vehicle wheels, it is naive to believe that the vehicle motion is going to be exactly the requested one. To check this, we have applied the algorithm developed earlier for computing motion between two 3D frames (see Chap. 7 and [15.8]) to every successive 3D views. We have found that there is a random rotation of about one degree between successive views and a difference in translation of up to 30 millimeters in x or z direction. To give an idea, we show in Table 15.1 the x and z components of the computed motions (the sign is omitted). We can remark that the motion realized by the

Fig. 15.23. The images taken by the first camera at t_1 and t_{15}

mobile vehicle is not very smooth and does not satisfy either the constant velocity or the constant acceleration model. This is our ground truth data for checking the performances of the methods presented in this part of the monograph.

Table 15.1. The x and z components of the translations between successive views

views	1–2	2–3	3–4	4–5	5–6	6–7	7–8	8–9	9–10	10–11	11–12	12–13	13–14	14–15
x (mm)	30.8	34.8	18.7	24.7	21.0	29.5	33.1	26.2	7.1	30.7	9.8	19.9	21.1	25.4
z (mm)	104.1	85.7	91.1	99.7	107.7	108.7	89.4	95.0	126.8	97.3	104.8	104.3	103.7	93.8

We apply the techniques developed in the last and this chapters to the above data. As in the previous experiments, each segment in the first frame is initialized as a token to be tracked. The kinematic parameters are all initialized to zero, but with fairly large uncertainties: the standard deviation is 0.0873 radians/unit-time for each angular velocity component, and 150 millimeters/unit-time for each translational velocity component. The variances of the acceleration components are set to zero (no acceleration). The only difference is that the noise term $\mathbf{n}_{i-1}$ in the state equation (14.40) is not zero. We add at each step a noise with standard deviation 0.5 degrees/unit-time to the $\boldsymbol{\omega}$ components and a noise with standard deviation 2 millimeters/unit-time to the $\mathbf{v}$ components.

The position and orientation of those tokens are then predicted for the next instant t_2 and the predicted tokens are compared with those in the new frame. Of course, since no motion is assumed, the predicted position and orientation of each token remain unchanged, but their uncertainties change and become very large. Figure 15.24 displays a stereogram showing the difference between the predicted tokens (in solid lines) and the observed line segments (in dashed lines) at t_2.

Fig. 15.24. A stereogram showing the superposition of the predicted (in solid lines) and observed (in dashed lines) segments at t_2

Fig. 15.25. A stereogram showing the superposition of the predicted (in solid lines) and observed (in dashed lines) segments at t_3

The reader can perceive the 3D information by cross-eye fusion. As expected, multiple matches occur for most of tokens, which are treated in the same way as in the previous experiment.

Figure 15.25 displays a stereogram showing the superposition of the predicted (in solid lines) and observed (in dashed lines) segments at t_3. As can be observed from the figure, more active tokens (in solid lines) exist at this moment: some have been activated due to multiple matches at time t_2 and some just entered the field of view. We observe that most of the tokens have already a

Fig. 15.26. A stereogram showing the superposition of the predicted (in solid lines) and observed (in dashed lines) segments at t_{10}

Fig. 15.27. A stereogram showing the superposition of the predicted (in solid lines) and observed (in dashed lines) segments at t_{15}

good kinematics information since their predictions coincide well with their observations. Unfortunately, this is not always the case. For example, the motion between the ninth and tenth views is not coherent with the global motion (see Table 15.1). Figure 15.26 displays a stereogram showing the superposition of the predicted (in solid lines) and observed (in dashed lines) segments at t_{10}. We can observe a big difference between the prediction and the observation. After several views, such occasional incoherent motion will be compensated for by the algorithm. Figure 15.27 displays a stereogram showing the superposition of the predicted (in solid lines) and observed (in dashed lines) segments at t_{15}. Quite a good fitting between the prediction and observation can be observed. The final estimate of the angular velocity (in radians/unit-time) is $[-1.1\times10^{-5}, 2.6\times10^{-3}, -3.2\times10^{-4}]^T$. The final estimate of the translational velocity (in millimeters/unit-time) is $[-24.5, -1.25, -97.1]^T$.

The program runs on a SUN 4/60 workstation. The number of active tokens varies between 110 and 155, except in the first view (89 tokens). The average user time for predicting a token is about 6.7 milliseconds and that for matching and updating a token is about 44.6 milliseconds. That is, if we implement the algorithm on a parallel machine with the same performance as SUN 4/60, the time required to process two frames is a little more than 50 milliseconds. The average user time required to group tokens into objects is about 2.5 seconds. In the last view, 90 segments have been identified as belonging to the static environment.

15.6 Summary

We have presented in Chap. 13 a framework to deal with the problem of computing three-dimensional motion and segmenting objects according to their kinematics in a sequence of stereo frames. The motion tracking problem in a long sequence has been considered as a process of spatiotemporal integration, and our algorithm consists of two levels. The low level tracks tokens from frame to frame and estimates their kinematic parameters, which is the temporal integration. The processing at the low level is completely parallel for each token. Unlike the previous methods reported in the literature, our approach unifies the token-tracking and motion-estimation processes by integrating the motion kinematic model, instead of a model on the change in token parameters, with the tracking process. The high level groups tokens into objects based on kinematic parameters, controls the processing at the low level to cope with problems as occlusion, disappearance and appearance, and provides information to other components of the system. The grouping process can be considered as a spatial integration. The implementation details have been presented in this chapter together with the experimental results.

The tokens used in our algorithm are 3D line segments, although other tokens such as corner points can be considered without affecting our algorithm a great deal. The matching technique has been described. A token is first predicted at the next instant, and the matching is based on the measure of the Mahalanobis distance between the predicted token and observed segments. Multiple matches are allowed, and false matches are resolved later, thanks to the temporal continuity of motion. Techniques based only on the best match usually fail at this stage. This is especially important at the beginning of the motion tracking process, where no information about kinematics is known (i.e., the uncertainty of motion for each token is very large). A measure called the *support of existence* has been defined which measures the adequation of a token to the measurements.

We have proposed a technique based on a statistical test on the support of existence to decide whether to pursue the tracking of a token or to discard it from tracking.

We have described also how the supervisor groups tokens into objects based on their kinematic parameters. Tokens with similar kinematic parameters are considered as belonging to the same object. The similarity is again measured by the Mahalanobis distance. This stage is important, in order to organize the irrelevant data and to obtain higher-level representation of scenes.

Both synthetic data generated by real stereo parameters and real data acquired by our trinocular stereo system have been used. From these results, we have found that our algorithm is robust and efficient, and can cope with problems such as occlusion, appearance, disappearance and absence of tokens. Finally, note that the main part of our algorithm can be implemented in parallel, something we are doing at the moment, hoping to achieve a throughput of 5 Hz.

16. Conclusions and Perspectives

Motion analysis is a very important research area in Computer Vision and Robotics. Its applications include mobile robot navigation, scene segmentation, world model construction, dynamic surveillance and object tracking. Most research efforts in the past were on the motion analysis from sequences of monocular images. With the development of stereovision systems, motion from stereo becomes an attractive research field. We have described in this monograph the results of our research on the motion analysis from sequences of stereo frames. Stereo frames are obtained at different instants by a stereo rig mounted on a mobile robot which navigates in an unknown environment, possibly containing other moving rigid objects. The primitives reconstructed by our stereo system are 3D line segments.

16.1 Summary

An important feature of the work described in this book is that uncertainty is modeled as early as in the 2D images and manipulated during all subsequent stages of processing. This distinguishes our approach from most of those reported in the literature which usually do not tackle this issue. The necessity of modeling explicitly uncertainty is now widely recognized. Because of its power and simplicity, we have chosen a probabilistic approach. The extended Kalman filter has been applied to estimate motion parameters in order to take into account the different error distributions in the measurements. A comparative study has been done between different methods for estimating 3D motion from 3D line segment matches, which shows that we gain in robustness if taking uncertainty into account.

We have developed an approach based on the *hypothesize-and-verify* paradigm for registering two stereo frames and computing the 3D displacement between them. Rigidity constraints are used to generate hypotheses of feature correspondences between two frames. We have shown that the rigidity constraints we have formulated are complete for 3D line segments and that a unique rigid displacement can be computed from two pairings of segments satisfying those constraints. The uncertainty of measurements has been integrated into the formalism of the rigidity constraints. If two pairings of segments satisfy the rigidity constraint, they are retained as a hypothesis. An initial estimate of the displacement can then be computed for each hypothesis. This initial estimate is propagated to the whole frame in an attempt to match more segments. Each time a new match is obtained, the displacement estimate is updated. Finally, the best hypothesis is retained. This algorithm has been successfully tested with several hundreds of real stereo frames

We have generalized the algorithm to determine multiple-object motions. When the robot navigates in an environment in which there exist other moving rigid objects, our algorithm determines first the egomotion and then cancels it before recovering the object motions.

We have applied this algorithm to solve a simple model-based object recognition and localization problem. We consider the model as one stereo frame, and our algorithm is directly applicable, provided the model is represented in the same fashion as a stereo frame.

We have used this algorithm together with a *data fusion* module to incrementally build a global description of a robot room. An example has been presented which deals with the fusion of 35 stereo frames observed in different positions. Considerable improvement in the accuracy has been observed through the fusion process.

We have also used this algorithm to calibrate the stereo and odometric coordinate systems of our mobile robot. We have designed a closed-loop visual navigation demonstration: we first displace the robot with an arbitrary motion and then ask the robot to recover the displacement and to return to the original position. Many such demonstrations have been conducted, and good results have been obtained. For example, we have displaced in one demonstration the robot with about 10 degrees in rotation and 75 centimeters in translation. The error between the original and final position is 0.6 degrees in rotation and 2.5 centimeters in translation.

We have proposed a framework based on a kinematic model to track objects in a long sequence of stereo frames. The kinematic model includes constant angular velocity and polynomial translational velocity. The motion tracking problem in a long sequence has been considered as a process of spatiotemporal integration, and our algorithm consists of two levels. At the low level, each token is independently tracked from frame to frame. It is the temporal integration. At the high level, tokens with the same kinematic parameters are grouped as parts of objects and more precise kinematic parameters estimation becomes available because of this grouping. It is the spatial integration.

Unlike the previous methods reported in the literature, our approach unifies the token-tracking and motion-estimation processes by integrating the motion kinematics model into the tracking process (rather than a model on the changes in token parameters). Our approach can, at least partially, cope with several interesting problems occurring with long sequences, such as occlusion, appearance, disappearance and absence of tokens. Again, our implementation has been with 3D line segments as tokens. Several experiments have been carried out using both synthetic and real data. The results have shown that our algorithm is robust and efficient for many of the scenes we have analyzed.

16.2 Perspectives

Although we have proposed many methods to solve 3D motion estimation problems, tested them using a large number of real 3D frames reconstructed by a stereovision system, and obtained very interesting results, many problems remain open and need to be further investigated.

Stereovision is only one of the many possibilities to perceive 3D information of the environment. How to apply the methods described in this monograph to 3D data acquired by other sensing modalities is not known. Our intuition is that the extension should be simple provided that the 3D data contains many line segments and that the uncertainty is appropriately modeled.

All methods proposed in this monograph have been implemented using 3D line segments as primitives. We have described in Sect. 3.1 why we made such a decision. However, 3D line segments are rather limited in their descriptive power and are usually not sufficient to describe complex scenes, e.g., outdoor scenes. The methods described in this monograph can easily be extended to include stable points such as corners, but this is not sufficient, either. We have experienced the limitations of our approach in the experiments with the rock scenes (Sect. 7.5) and with the rotating table (Sect. 15.5). Preliminary ideas about how to cope with this problem have been recently reported in the literature [16.1–3]. However, it is not yet clear how to deal with curved objects in general, and considerable efforts should be devoted to this problem.

There are more than one way to use the 3D information contained in a sequence of stereo images. In this monograph, we have directly used 3D data reconstructed by a stereovision system. This has an advantage over using monocular sequences in that the former works in the right space — Euclidean space where it is easy to model motion kinematics and that the later works in the projected space. However, we have not made full use of the information contained in the stereo images. For example, a part of the scene may be visible to one camera but not to the other, and vice versa. This part of the information is lost in the 3D reconstruction process. In order to make full use of the information and potentially reduce the overall processing time (doing stereo matching and 3D reconstruction at each instant is computationally expensive), we can make the stereo and monocular motion processes cooperate [16.4–6]. This approach has recently attracted several researchers and one can expect to see interesting results from this direction in a near future.

Until recently, vision has been used mainly as a sensor for perceiving and interpreting an environment. Little has been done to use vision as a sensory feedback in *reactive* systems. However, it is indispensable for real applications to control the motion of a robot or to adjust the cameras parameters to adapt itself to the changing situation based on the information provided by the vision system.

Furthermore, a vision system can create useful information by being active (self-motion, focusing, cameras convergence, etc.) [16.7–9]. There are many important issues that need to be explored in this domain.

Although a "general-purpose" vision system is unlikely to become available in the near feature [16.10], Computer Vision remain an exciting and rapidly evolving research field. Its success will depend in the end on its ability to solve specific applications.

Appendix: Vector Manipulation and Differentiation

C olumn vectors are used in this monograph. The number of elements in a vector is its dimension. A n-dimensional vector may be thought of as a matrix of n rows and 1 column. The superscript ^{T} is used to denote the transpose of a vector or a matrix.

In the following, we use 3-dimensional vectors to explain vector operations. Let $\mathbf{a} = [a_1, \ a_2, \ a_3]^T$, $\mathbf{b} = [b_1, \ b_2, \ b_3]^T$, $\mathbf{c} = [c_1, \ c_2, \ c_3]^T$ and λ be an arbitrary scalar.

A.1 Manipulation of Vectors

- **vector norm:** (vector length or magnitude)

$$\|\mathbf{a}\| = \sqrt{a_1^2 + a_2^2 + a_3^2}$$

- **vector addition:**

$$\mathbf{a} + \mathbf{b} = [a_1 + b_1, \ a_2 + b_2, \ a_3 + b_3]^T$$

- **scalar multiplication:**

$$\lambda \mathbf{a} = [\lambda a_1, \ \lambda a_2, \ \lambda a_3]^T$$

- **dot product:** (inner product)

$$\mathbf{a} \cdot \mathbf{b} = a_1 b_1 + a_2 b_2 + a_3 b_3$$

- **cross product:**

$$\mathbf{a} \wedge \mathbf{b} = [a_2 b_3 - a_3 b_2, \ a_3 b_1 - a_1 b_3, \ a_1 b_2 - a_2 b_1]^T$$

We have

$$\|\mathbf{a} \wedge \mathbf{b}\|^2 = \|\mathbf{a}\|^2 \|\mathbf{b}\|^2 - (\mathbf{a} \cdot \mathbf{b})^2$$

- **tilde of a vector:**

$$\tilde{\mathbf{a}} = \begin{bmatrix} 0 & -a_3 & a_2 \\ a_3 & 0 & -a_1 \\ -a_2 & a_1 & 0 \end{bmatrix}$$

We have

$$\mathbf{a} \wedge \mathbf{b} = \tilde{\mathbf{a}}\mathbf{b}$$

- **triple product:**

$$< \mathbf{a}, \mathbf{b}, \mathbf{c} > = \mathbf{a} \cdot (\mathbf{b} \wedge \mathbf{c}) = (\mathbf{a} \wedge \mathbf{b}) \cdot \mathbf{c}$$

$$= \begin{vmatrix} a_1 & b_1 & c_1 \\ a_2 & b_2 & c_2 \\ a_3 & b_3 & c_3 \end{vmatrix}$$

- **other products:**

$$\mathbf{a} \wedge (\mathbf{b} \wedge \mathbf{c}) = (\mathbf{c} \cdot \mathbf{a})\mathbf{b} - (\mathbf{a} \cdot \mathbf{b})\mathbf{c}$$
$$(\mathbf{a} \wedge \mathbf{b}) \wedge \mathbf{c} = (\mathbf{c} \cdot \mathbf{a})\mathbf{b} - (\mathbf{b} \cdot \mathbf{c})\mathbf{a}$$

A.2 Differentiation of Vectors

- **derivative of a scalar with respect to a vector:**

$$\frac{d\lambda}{d\mathbf{a}} = [\frac{\partial\lambda}{\partial a_1}, \frac{\partial\lambda}{\partial a_2}, \frac{\partial\lambda}{\partial a_3}]^T$$

We have

$$\frac{d}{d\mathbf{a}}(\mathbf{a} \cdot \mathbf{b}) = \mathbf{b} \qquad \frac{d}{d\mathbf{b}}(\mathbf{a} \cdot \mathbf{b}) = \mathbf{a}$$
$$\frac{d}{d\mathbf{a}}\|\mathbf{a}\|^2 = 2\mathbf{a} \qquad \frac{d}{d\mathbf{a}}\|\mathbf{a}\| = \frac{\mathbf{a}}{\|\mathbf{a}\|}$$
$$\frac{d}{d\mathbf{a}}\mathbf{a}^T M\mathbf{b} = M\mathbf{b} \qquad \frac{d}{d\mathbf{b}}\mathbf{a}^T M\mathbf{b} = M^T\mathbf{a}$$

where M is a matrix.
- **derivative of a vector with respect to a vector:**

$$\frac{d\mathbf{b}}{d\mathbf{a}} = \begin{bmatrix} \frac{\partial b_1}{\partial a_1} & \frac{\partial b_1}{\partial a_2} & \frac{\partial b_1}{\partial a_3} \\ \frac{\partial b_2}{\partial a_1} & \frac{\partial b_2}{\partial a_2} & \frac{\partial b_2}{\partial a_3} \\ \frac{\partial b_3}{\partial a_1} & \frac{\partial b_3}{\partial a_2} & \frac{\partial b_3}{\partial a_3} \end{bmatrix}$$

We have

$$\frac{d}{d\mathbf{a}}M\mathbf{a} = M$$

$$\frac{d}{d\mathbf{b}}(\mathbf{a} \wedge \mathbf{b}) = \tilde{\mathbf{a}}$$

$$\frac{d}{d\mathbf{a}}(\mathbf{a} \wedge \mathbf{b}) = -\tilde{\mathbf{b}}$$

$$\frac{d}{d\mathbf{a}}(f(\mathbf{a})\mathbf{b}) = \mathbf{b}\left[\frac{df(\mathbf{a})}{d\mathbf{a}}\right]^T$$

$$\frac{d}{d\mathbf{a}}(\|\mathbf{a}\|\mathbf{b}) = \frac{\mathbf{b}\mathbf{a}^T}{\|\mathbf{a}\|}$$

$$\frac{d}{d\mathbf{a}}\frac{\mathbf{a}}{\|\mathbf{a}\|} = \frac{\mathbf{I}_3}{\|\mathbf{a}\|} - \frac{\mathbf{a}\mathbf{a}^T}{\|\mathbf{a}\|^3}$$

where M is a matrix and $f(\mathbf{a})$ is a scalar function of $\mathbf{a}$.

- **derivative of a scalar with respect to a matrix:**

Let $A = \begin{bmatrix} a_{11} & a_{12} & \cdots & a_{1n} \\ a_{21} & a_{22} & \cdots & a_{2n} \\ \vdots & \vdots & \ddots & \vdots \\ a_{m1} & a_{m2} & \cdots & a_{mn} \end{bmatrix}$, then the derivative of λ with respect to A

is defined by

$$\frac{d\lambda}{dA} = \begin{bmatrix} \frac{\partial\lambda}{\partial a_{11}} & \frac{\partial\lambda}{\partial a_{12}} & \cdots & \frac{\partial\lambda}{\partial a_{1n}} \\ \frac{\partial\lambda}{\partial a_{21}} & \frac{\partial\lambda}{\partial a_{22}} & \cdots & \frac{\partial\lambda}{\partial a_{2n}} \\ \vdots & \vdots & \ddots & \vdots \\ \frac{\partial\lambda}{\partial a_{m1}} & \frac{\partial\lambda}{\partial a_{m2}} & \cdots & \frac{\partial\lambda}{\partial a_{mn}} \end{bmatrix}.$$

Of particular interest are the derivative of the matrix trace and determinant. Recall the trace of a square matrix A $(m = n)$ is defined by

$$tr(A) = \sum_{i=1}^{n} a_{ii} .$$

We have

$$\frac{d}{dA}tr(A) = \mathbf{I}$$

$$\frac{d}{dA}tr(e^A) = e^{A^T}$$

$$\frac{d}{dA}tr(BAC) = B^T C^T$$

$$\frac{d}{dA}tr(ABA^T) = A(B + B^T)$$

$$\frac{d}{dA}|BAC| = |BAC|(A^{-1})^T$$

where A, B and C are all square matrices.

References

Chapter 1

1.1 M. D. Levine: *Vision in Man and Machine* (McGraw-Hill, New York 1985)

1.2 J. Hochberg: Machines should not see as people do, but must know how people see. Comput. Vision, Graphics Image Process. **37**, 221–237 (1987)

1.3 S. Tsuji, J. Zheng: Visual path planning by a mobile robot. Proc. Int'l Joint Conf. Artif. Intell., Milan, Italy (1987) pp. 1127–1130

1.4 Z. Zhang, O. Faugeras: Calibration of a mobile robot with application to visual navigation. Proc. IEEE Workshop Visual Motion, Irvine, CA (1989) pp. 306–313

1.5 H. Chen, T. Huang: Maximal matching of 3-D points for multiple-object motion estimation. Pattern Recog. **21**(2), 75–90 (1988)

1.6 H. Chen, T. Huang: Matching 3-D line segments with applications to multiple-object motion estimation. IEEE Trans. PAMI-**12**(10), 1002–1008 (1990)

1.7 Z. Zhang, O. Faugeras, N. Ayache: Analysis of a sequence of stereo scenes containing multiple moving objects using rigidity constraints. Proc. Second Int'l Conf. Comput. Vision, Tampa, FL (1988) pp. 177–186

1.8 N. Ayache, O. Faugeras: Building, registrating and fusing noisy visual maps. Proc. First Int'l Conf. Comput. Vision, London, UK (1987) pp. 73–82

1.9 M. Asada: Building a 3-D world model for a mobile robot from sensory data. Proc. Int'l Conf. Robotics Automation, Philadelphia, PA (1988) pp. 918–923

1.10 E. Dickmanns: 4D-dynamic scene analysis with integral spatio-temporal models. Proc. ISSR, Santa-Cruz, CA (1987) pp. 73–80

1.11 E. Dickmanns, V. Graefe: Applications of dynamic monocular machine vision. Machine Vision and Applications 1, 241–261 (1988)

1.12 E. Dubois, B. Prasada, M. Sabri: Image sequence coding, in *Image Sequence Analysis*, ed. by T. Huang, Springer Ser. Info. Sci., Vol. 5 (Springer, Berlin, Heidelberg 1981) Chap. 3, pp. 229–287

1.13 D. Marr, E. Hildreth: Theory of edge detection. Proc. Royal Society of London **207**, 187–217 (1980)

1.14 A. Rosenfeld, A. Kak: *Digital Picture Processing*, Vols. 1 & 2, 2nd edn. (Academic, New York 1982)

1.15 J. Canny: *Finding edges and lines in images*, A.I. Memo. 720, MIT. Artif. Intell. Lab (Cambridge, MA) (1983)

1.16 J. Canny: A computational approach to edge detection. IEEE Trans. PAMI-8(6), 679–698 (1986)

1.17 R. Deriche: Using canny's criteria to derive an optimal edge detector recursively implemented. Int'l J. Comput. Vision **2**, 167–187 (1987)

1.18 J. Shen, S. Castan: An optimal linear operator for edge detection. Proc. IEEE Conf. Comput. Vision Pattern Recog., Miami, FL (1986) pp. 109–114

1.19 S. Castan, J. Zhao, J. Shen: Optimal filter for edge detection methods and results. Proc. First European Conf. Comput. Vision, Antibes, France (1990) pp. 13–17

1.20 O. Faugeras et al.: Toward a flexible vision system, in *Robot Vision*, ed. by A. Pugh (IFS, UK 1983) Chap. 3, pp. 129–142

1.21 Y. Kim, J. Aggarwal: Finding range from stereo images. Proc. IEEE Conf. Comput. Vision Pattern Recog., San Francisco, CA (1985) pp. 189–294

1.22 F. Lustman: Vision stéréoscopique et perception du mouvement en vision artificielle. Dissertation, University of Paris XI (Orsay, Paris, France) (1987)

1.23 N. Ayache: *Artificial Vision for Mobile Robots: Stereo Vision and Multisensory Perception* (MIT Press, Cambridge, MA 1991)

1.24 J. Aggarwal, L. Davis, W. Martin: Correspondence processes in dynamic scene analysis. Proc. IEEE **69**(6), 562–572 (1981)

1.25 S. Barnard, W. Thompson: Disparity analysis of images. IEEE Trans. PAMI-**2**, 333–340 (1980)

1.26 N. Ayache, B. Faverjon: Efficient registration of stereo images by matching graph descriptions of edge segments. Int'l J. Comput. Vision **1**(2), 107–131 (1987)

1.27 M. Jenkin, J. Tsotsos: Applying temporal constraints to the dynamic stereo problem. Comput. Vision, Graphics Image Process. **24**, 16–32 (1986)

1.28 Y. Kim, J. Aggarwal: Determining object motion in a sequence of stereo images. IEEE J. RA-**3**(6), 599–614 (1987)

1.29 S. Ullman: *The Interpretation of Visual Motion* (MIT Press, Cambridge, MA 1979)

1.30 R. Tsai, T. Huang: Estimating 3-D motion parameters of a rigid planar patch, i. IEEE Trans. ASSP-**29**(6), 1147–1152 (1981)

1.31 H. Longuet-Higgins: A computer algorithm for reconstructing a scene from two projections. Nature **293**, 133–135 (1981)

1.32 R. Tsai, T. Huang, W. Zhu: Estimating three-dimensional motion parameters of a rigid planar patch, ii: singular value decomposition. IEEE Trans. ASSP-**30**(4), 525–534 (1982)

1.33 R. Tsai, T. Huang: Uniqueness and estimation of three-dimensional motion parameters of rigid objects with curved surface. IEEE Trans. PAMI-**6**(1), 13–26 (1984)

1.34 X. Zhuang, R. Haralick: Two view motion analysis. Proc. IEEE Conf. Comput. Vision Pattern Recog., San Francisco, CA (1985) pp. 686–690

1.35 O. Faugeras, F. Lustman, G. Toscani: Motion and structure from motion from point and line matches. Proc. First Int'l Conf. Comput. Vision, London, UK (1987) pp. 25–34

1.36 R. R. Kumar, A. Tirumalai, R. Jain: A non-linear optimization algorithm for the estimation of structure and motion parameters. Proc. IEEE Conf. Comput. Vision Pattern Recog., San Diego, CA (1989) pp. 136–143

1.37 T. Viéville, O. Faugeras: Feed-forward recovery of motion and structure from a sequence of 2D-lines matches. Proc. Third Int'l Conf. Comput. Vision, Osaka, Japan (1990) pp. 517–520

1.38   S. Blostein, T. Huang: Estimation 3-D motion from range data. Proc. First Conf. Artif. Intell. Applications, Denver, CO (1984) pp. 246–250

1.39   T. Huang: Motion analysis, in *AI Encyclopedia* (Wiley, New York 1986) pp. 620–632

1.40   O. Faugeras, M. Hebert: The representation, recognition, and locating of 3D shapes from range data. Int'l J. Robotics Res. **5**(3), 27–52 (1986)

1.41   K. Arun, T. Huang, S. Blostein: Least-squares fitting of two 3-D point sets. IEEE Trans. PAMI-**9**(5), 698–700 (1987)

1.42   N. Navab, R. Deriche, O. Faugeras: Recovering 3D motion and structure from stereo and 2D token tracking cooperation. Proc. Third Int'l Conf. Comput. Vision, Osaka, Japan (1990) pp. 513–517

1.43   B. Horn, B. Schunk: Determining optical flow. Artif. Intell. **20**, 199–228 (1981)

1.44   B. Horn: *Robot Vision* (MIT Press, Cambridge, MA and McGraw–Hill, New York 1986)

1.45   H. Nagel: Displacement vectors derived from second-order intensity variations in image sequences. Comput. Vision, Graphics Image Process. **21**, 85–117 (1983)

1.46   E. Hildreth: *The Measurement of Visual Motion* (MIT Press, Cambridge, MA 1983)

1.47   A. Bruss, B. Horn: Passive navigation. Comput. Vision, Graphics Image Process. **21**, 3–20 (1983)

1.48   P. Bouthemy: A maximum likelihood framework for determining moving edges. IEEE Trans. PAMI-**11**(5), 499–511 (1989)

1.49   G. Adiv: Inherent ambiguities in recovering 3D motion and structure from a noisy flow field. Proc. IEEE Conf. Comput. Vision Pattern Recog., San Francisco, CA (1985) pp. 70–77

1.50   S. Maybank: A theoretical study of optical flow. Ph.D. thesis, Birkbeck College, University of London (London, UK) (1987)

1.51   M. Subbarao: Solution and uniqueness of image flow equations for rigid curved surfaces in motion. Proc. First Int'l Conf. Comput. Vision, London, UK (1987) pp. 687–692

1.52   A. Waxman, S. Sinha: Dynamic stereo: passive ranging to moving objects from relative image flows. IEEE Trans. PAMI-**8**(4), 406–412 (1986)

1.53   A. Waxman, J. Duncan: Binocular image flows: steps toward stereo-motion fusion. IEEE Trans. PAMI-**8**(6), 715–729 (1986)

1.54   A. Verri, T. Poggio: Against quantitative optical flow. Proc. First Int'l Conf. Comput. Vision, London, UK (1987) pp. 171–180

1.55   S. Liou, R. Jain: Motion detection in spatio-temporal space. Comput. Vision, Graphics Image Process. **45**, 227–250 (1989)

1.56   H. Nagel: From image sequences towards conceptual descriptions. Image and Vision Computing **6**(2), 59–74 (1988)

1.57   R. Nelson, J. Aloimonos: Using flow field divergence for obstacle avoidance in visual navigation. IEEE Trans. PAMI-**11**(10), 1102–1106 (1989)

1.58   E. François, P. Bouthemy: The derivation of qualitative information in motion analysis. Proc. First European Conf. Comput. Vision, Antibes, France (1990) pp. 226–230

1.59 B. Horn, J. Harris: Rigid body motion from range image sequences. CVGIP: Image Understanding **53**(1), 1–13 (1991)

1.60 T. Broida, R. Chellappa: Kinematics and structure of a rigid object from a sequence of noisy images. Proc. IEEE Workshop on Motion: Representation and Analysis, Charleston, SC (1986) pp. 95–100

1.61 T. Broida, R. Chellappa: Kinematics and structure of a rigid object from a sequence of noisy images: a batch approach. Proc. IEEE Conf. Comput. Vision Pattern Recog., Miami, FL (1986) pp. 176–182

1.62 T. Broida, R. Chellappa: Experiments and uniqueness results on object structure and kinematics from a sequences of monocular images. Proc. IEEE Workshop Visual Motion, Irvine, CA (1989) pp. 21–30

1.63 J. Weng, T. Huang, N. Ahuja: 3-D motion estimation, understanding, and prediction from noisy image sequences. IEEE Trans. PAMI-**9**(3), 370–389 (1987)

1.64 G. Young, R. Chellappa: 3–D motion estimation using a sequence of noisy stereo images: models, estimation, and uniqueness results. IEEE Trans. PAMI-**12**(8), 735–759 (1990)

1.65 J. Heel: Dynamic motion vision. Proc. DARPA Image Understanding Workshop, Palo Alto, CA (1989)

1.66 H. Nagel: Image sequences - ten (octal) years- from phenomenology towards a theoretical foundation. Proc. 8th Int'l Conf. Pattern Recog., Paris, France (1986) pp. 1174–1185

1.67 J. Aggarwal, N. Nandhakumar: On the computation of motion from sequences of images — a review. Proc. IEEE **76**(8), 917–935 (1988)

1.68 W. Grimson, T. Lozano-Perez: Model-based recognition and localization from sparse range or tactile data. Int'l J. Robotics Res. **5**(3), 3–34 (1984)

1.69 T. Huang, R. Tsai: Image sequence analysis: motion estimation, in *Image Sequence Analysis*, ed. by T. Huang, Springer Ser. Info. Sci., Vol. 5 (Springer, Berlin, Heidelberg 1981) Chap. 3, pp. 1–18

1.70 B. Yen, T. Huang: Determining 3-D motion/structure of a rigid body over 3 frames using straight line correspondences. Proc. IEEE Conf. Comput. Vision Pattern Recog., Washington, DC (1983) pp. 267–272

1.71 Y. Liu, T. Huang: Estimation of rigid body motion using straight line correspondences. Proc. Workshop on Motion: Representation and Analysis, Charleston, SC (1986) pp. 47–51

1.72 A. Mitiche, S. Seida, J. Aggarwal: Line based computation of structure and motion using angular invariance. Proc. Workshop on Motion: Representation and Analysis, Charleston, SC (1986) pp. 175–180

1.73 J. Aggarwal, Y. Wang: Analysis of a sequence of images using point and line correspondences. Proc. Int'l Conf. Robotics Automation, Raleigh, NC (1987) pp. 1275–1280

1.74 O. Faugeras, S. Maybank: Motion from point matches: multiplicity of solutions. Int'l J. Comput. Vision 4(3), 225–246 (1990)

1.75 O. Faugeras, G. Toscani: The calibration problem for stereo. Proc. IEEE Conf. Comput. Vision Pattern Recog., Miami, FL (1986) pp. 15–20

1.76 R. Tsai: An efficient and accurate camera calibration technique for 3D machine vision. Proc. IEEE Conf. Comput. Vision Pattern Recog., Miami, FL (1986) pp. 364–374

1.77   O. Faugeras, N. Ayache, B. Faverjon: Building visual maps by combining noisy stereo measurements. Proc. Int'l Conf. Robotics Automation, San Francisco, CA (1986) pp. 1433–1438

1.78   N. Ayache, O. D. Faugeras: Maintaining Representations of the Environment of a Mobile Robot. IEEE Trans. RA-5(6), 804–819 (1989)

Chapter 2

2.1   M. Brady: Problems in robotics, in *The Robotics Review I*, ed. by O. Khatib, J. Craig, T. Lozano-Pérez (MIT Press, Cambridge, MA 1989) pp. 1–25

2.2   T. Henderson, E. Shilcrat: Logical sensor systems. J. Robotic Systems 1(2), 169–193 (1984)

2.3   W. Grimson, T. Lozano-Perez: Model-based recognition and localization from sparse range or tactile data. Int'l J. Robotics Res. 5(3), 3–34 (1984)

2.4   H. Chen, T. Huang: Matching 3-D line segments with applications to multiple-object motion estimation. IEEE Trans. PAMI-12(10), 1002–1008 (1990)

2.5   P. Winston: *Artificial Intelligence* (Addison Wesley, Reading, MA 1984)

2.6   N. Ayache, O. Faugeras: Building, registrating and fusing noisy visual maps. Proc. First Int'l Conf. Comput. Vision, London, UK (1987) pp. 73–82

2.7   N. Ayache, O. D. Faugeras: Maintaining Representations of the Environment of a Mobile Robot. IEEE Trans. RA-5(6), 804–819 (1989)

2.8   H. Durrant-Whyte: Uncertain geometry in robotics. IEEE J. RA-4(1), 23–31 (1988)

2.9   H. Durrant-Whyte: Sensor models and multisensor integration. Int'l J. Robotics Res. 7(6), 97–113 (1988)

2.10   G. Hager: Active reduction of uncertainty in multi-sensor systems. Ph.D. thesis, University of Pennsylvania (Philadelphia, PA) (1988)

2.11   J. Porrill: Optimal combination and constraints for geometrical sensor data. Int'l J. Robotics Res. 7(6), 66–77 (1988)

2.12   Z. Zhang, O. Faugeras, N. Ayache: Analysis of a sequence of stereo scenes containing multiple moving objects using rigidity constraints. Proc. Second Int'l Conf. Comput. Vision, Tampa, FL (1988) pp. 177–186

2.13   O. Faugeras, N. Ayache, Z. Zhang: A preliminary investigation of the problem of determining ego- and object motions from stereo. Proc. 9th Int'l Conf. Pattern Recog., Rome, Italy (1988) pp. 242–246

2.14   A. Jazwinsky: *Stochastic Processes and Filtering Theory* (Academic, New York 1970)

2.15   P. Maybeck: *Stochastic Models, Estimation and Control*, Vol. 1 (Academic, New York 1979)

2.16   F. Lustman: Vision stéréoscopique et perception du mouvement en vision artificielle. Dissertation, University of Paris XI (Orsay, Paris, France) (1987)

2.17   N. Ayache: *Artificial Vision for Mobile Robots: Stereo Vision and Multisensory Perception* (MIT Press, Cambridge, MA 1991)

2.18   M. Kendall, R. Moran: *Geometric Probability* (Griffin Academic, New York 1963)

2.19   R. Bolle, D. Cooper: On optimally combining pieces of information with application to estimating 3-D complex-object position from range data. IEEE Trans. PAMI-8(5), 619–638 (1986)

2.20 O. Faugeras, N. Ayache, B. Faverjon: Building visual maps by combining noisy stereo measurements. Proc. Int'l Conf. Robotics Automation, San Francisco, CA (1986) pp. 1433–1438

2.21 H. Durrant-Whyte: *Integration, Coordination, and Control of Multi-sensor* (Kluwer Academic, Boston, MA 1988)

2.22 W. J. Rey: *Introduction to Robust and Quasi-Robust Statistical Methods* (Springer, Berlin, Heidelberg 1983)

2.23 R. Szeliski: Bayesian modeling of uncertainty in low-level vision. Int'l J. Comput. Vision **5**(3), 271–301 (1990)

2.24 D. G. Lowe: Review of "TINA: the Sheffield AIVRU vision system" by J. Porrill et al., in *The Robotics Review I*, ed. by O. Khatib, J. Craig, T. Lozano-Pérez (MIT Press, Cambridge, MA 1989) pp. 195–198

2.25 C. Chui, G. Chen: *Kalman Filtering with Real-Time Applications*, Springer Ser. Info. Sci., Vol. 17 (Springer, Berlin, Heidelberg 1987)

2.26 A. Papoulis: *Probability, Random Variables, and Stochastic Processes* (McGraw-Hill, New York 1965)

2.27 W. Förstner: Reliability analysis of parameter estimation in linear models with application to mensuration problems in computer vision. Comput. Vision, Graphics Image Process. **40**, 273–310 (1987)

2.28 S. Maybank: Filter based estimates of depth. Proc. British Machine Vision Conf., University of Oxford, London, UK (1990) pp. 349–354

2.29 P. Maybeck: *Stochastic Models, Estimation and Control*, Vol. 2 (Academic, New York 1982)

2.30 R. Kumar, A. Hanson: Analysis of different robust methods for pose refinement. Proc. Int'l Workshop Robust Comput. Vision, Seattle, WA (1990) pp. 167–182

Chapter 3

3.1   O. Faugeras et al.: Toward a flexible vision system, in *Robot Vision*, ed. by A. Pugh (IFS, UK 1983) Chap. 3, pp. 129–142

3.2   S. T. Barnard, M. A. Fishler: Computational stereo. ACM Computing Surveys **14**(4), 553–572 (1982)

3.3   W. Grimson: Computational experiments with a feature based stereo algorithm. IEEE Trans. PAMI-7(1), 17–34 (1985)

3.4   S. Pollard, J. Mayhew, J. Frisby: PMF: a stereo correspondence algorithm using a disparity gradient limit. Perception **14**, 449–470 (1985)

3.5   N. Ayache, B. Faverjon: Efficient registration of stereo images by matching graph descriptions of edge segments. Int'l J. Comput. Vision **1**(2), 107–131 (1987)

3.6   Y. Kim, J. Aggarwal: Finding range from stereo images. Proc. IEEE Conf. Comput. Vision Pattern Recog., San Francisco, CA (1985) pp. 189–294

3.7   F. Lustman: Vision stéréoscopique et perception du mouvement en vision artificielle. Dissertation, University of Paris XI (Orsay, Paris, France) (1987)

3.8   N. Ayache: *Artificial Vision for Mobile Robots: Stereo Vision and Multisensory Perception* (MIT Press, Cambridge, MA 1991)

3.9   J. Shen, S. Castan, J. Zhao: A new trinocular stereo vision method. Proc. 6th Scandinavian Conf. Image Analysis, Oulu, Finland (1989) pp. 88–95

3.10   R. Bolles, H. Baker, D. Marimont: Epipolar-plane image analysis: an approach to determining structure from motion. Int'l J. Comput. Vision 1(1), 7–55 (1987)

3.11   O. Faugeras et al.: Depth and motion analysis: the machine being developed within esprit project 940. Proc. IAPR Workshop Comput. Vision—Special Hardware and Industrial Applications, Tokyo, Japon (1988) pp. 35–44

3.12   M. Berthod: *Approximation polygonale de chaînes de contours*, Programmes C, INRIA (INRIA Sophia-Antipolis, F-06565 Valbonne cedex) (1986)

3.13   J. Canny: A computational approach to edge detection. IEEE Trans. PAMI-8(6), 679–698 (1986)

3.14   R. Deriche: Using canny's criteria to derive an optimal edge detector recursively implemented. Int'l J. Comput. Vision 2, 167–187 (1987)

3.15   O. Faugeras, G. Toscani: The calibration problem for stereo. Proc. IEEE Conf. Comput. Vision Pattern Recog., Miami, FL (1986) pp. 15–20

3.16   G. Toscani: Système de calibration optique et perception du mouvement en vision artificielle. Dissertation, University of Paris XI (Orsay, Paris, France) (1987)

3.17   R. Vaillant: Géométrie différentielle et vision par ordinateur: détection et reconstruction des contours d'occultation de la surface d'un objet non-polyédrique. Dissertation, University of Paris XI (Orsay, Paris, France) (1990)

3.18   R. Tsai: An efficient and accurate camera calibration technique for 3D machine vision. Proc. IEEE Conf. Comput. Vision Pattern Recog., Miami, FL (1986) pp. 364–374

3.19   R. Lenz, R. Tsai: Techniques for calibrating of the scale factor and image center for high accuracy 3D machine vision metrology. Proc. Int'l Conf. Robotics Automation, Raleigh, NC (1987) pp. 68–75

3.20   G. Giraudon: *Chaînage efficace contour*, Rapport de Recherche 605, INRIA (Sophia-Antipolis, France) (1987)

3.21   R. Vaillant, R. Deriche, O. Faugeras: 3D vision on the parallel machine capitan. Int'l Workshop on Industrial Application of Machine Intelligence and Vision, Japan (1989)

3.22   C. Hansen, N. Ayache, F. Lustman: Towards real-time trinocular stereo. Proc. Second Int'l Conf. Comput. Vision, Tampa, FL (1988) pp. 129–133

3.23   R. Deriche, R. Vaillant, O. Faugeras: From noisy edge points to 3D reconstruction of a scene: a robust approach and its uncertainty analysis. Proc. Scandinavian Conf. Image Analysis, Aalborg University, Denmark (1991)

Chapter 4

4.1   S. Ullman: *The Interpretation of Visual Motion* (MIT Press, Cambridge, MA 1979)

4.2   O. D. Faugeras: *Three-Dimensional Computer Vision* (MIT Press, Cambridge, MA 1991)

4.3   B. Horn: *Robot Vision* (MIT Press, Cambridge, MA and McGraw–Hill, New York 1986)

4.4   E. Pervin, J. Webb: Quaternions in computer vision and robotics. Proc. IEEE Conf. Comput. Vision Pattern Recog., Washington (1983) pp. 382–383

4.5   N. Ayache, O. D. Faugeras: Maintaining Representations of the Environment of a Mobile Robot. IEEE Trans. RA–5(6), 804–819 (1989)

4.6   H. Durrant-Whyte: Uncertain geometry in robotics. IEEE J. RA–4(1), 23–31 (1988)

4.7   L. Matthies, S. A. Shafer: Error modeling in stereo navigation. IEEE J. RA–3(3), 239–248 (1987)

4.8   F. Ramparany: Perception multisensorielle de la structure géométrique d'une scène. Dissertation, Institut National Polytechnique de Grenoble (Grenoble, France) (1989)

4.9   Z. Zhang, O. Faugeras, N. Ayache: Analysis of a sequence of stereo scenes containing multiple moving objects using rigidity constraints. Proc. Second Int'l Conf. Comput. Vision, Tampa, FL (1988) pp. 177–186

4.10 J. Crowley, P. Stelmaszyk: Measurement and integration of 3-D structures by tracking edge lines. Proc. First European Conf. Comput. Vision, Antibes, France (1990) pp. 269–280

4.11 Y. Kim, J. Aggarwal: Determining object motion in a sequence of stereo images. IEEE J. RA–3(6), 599–614 (1987)

4.12 N. Ayache, O. Faugeras: Building, registrating and fusing noisy visual maps. Proc. First Int'l Conf. Comput. Vision, London, UK (1987) pp. 73–82

4.13 O. Faugeras, N. Ayache, Z. Zhang: A preliminary investigation of the problem of determining ego- and object motions from stereo. Proc. 9th Int'l Conf. Pattern Recog., Rome, Italy (1988) pp. 242–246

4.14 N. Ayache, O. Faugeras, F. Lustman, Z. Zhang: Visual navigation of a mobile robot: recent steps. Proc. Int'l Symposium and Exposition on Robots, Sydney, Australia (1988) pp. 725–740

4.15 N. Ayache: *Artificial Vision for Mobile Robots: Stereo Vision and Multisensory Perception* (MIT Press, Cambridge, MA 1991)

4.16 K. Roberts: A new representation for a line. Proc. IEEE Conf. Comput. Vision Pattern Recog., Ann Arbor, Michigan (1988) pp. 635–640

4.17 M. Buffa, O. Faugeras, Z. Zhang: Obstacle avoidance and trajectory planning for an indoors mobile robot using stereo vision and Delaunay triangulation. Proc. Roundtable Discussion on Vision-Based Vehicle Guidance, Science University of Tokyo, Japan (1990) pp. 12–1–12–8

Chapter 5

5.1   R. Szeliski: Estimating motion from sparse range data without correspondence. Proc. Second Int'l Conf. Comput. Vision, Tampa, FL (1988) pp. 207–216

5.2   S. Blostein, T. Huang: Estimation 3-D motion from range data. Proc. First Conf. Artif. Intell. Applications, Denver, CO (1984) pp. 246–250

5.3   T. Huang: Motion analysis, in *AI Encyclopedia* (Wiley, New York 1986) pp. 620–632

5.4   O. Faugeras, M. Hebert: The representation, recognition, and locating of 3D shapes from range data. Int'l J. Robotics Res. **5**(3), 27–52 (1986)

5.5   K. Arun, T. Huang, S. Blostein: Least-squares fitting of two 3-D point sets. IEEE Trans. PAMI–**9**(5), 698–700 (1987)

5.6   R. Haralick et al.: Pose estimation from corresponding point data. IEEE Trans. SMC–**19**(6), 1426–1446 (1989)

5.7   Y. Kim, J. Aggarwal: Determining object motion in a sequence of stereo images. IEEE J. RA–**3**(6), 599–614 (1987)

5.8   N. Ayache, O. Faugeras: Building, registrating and fusing noisy visual maps. Proc. First Int'l Conf. Comput. Vision, London, UK (1987) pp. 73–82

5.9   *NAg, FORTRAN Library Manual*, Numerical Algorithms Group, mark 14 edn. (1990)

5.10   Z. Zhang, O. Faugeras, N. Ayache: Analysis of a sequence of stereo scenes containing multiple moving objects using rigidity constraints. Proc. Second Int'l Conf. Comput. Vision, Tampa, FL (1988) pp. 177–186

Chapter 6

6.1   Z. Lin, H. Lee, T. Huang: Finding 3-D point correspondence in motion estimation. Proc. 8th Int'l Conf. Pattern Recog., Paris, France (1986) pp. 303–305

6.2   K. Sugihara: An $n \log n$ algorithm for determining the congruity of polyhedra. J. Comput. System Science **29**, 36–47 (1984)

6.3   O. Faugeras, M. Hebert: A 3-D recognition and positioning algorithm using geometrical constraints between primitive surfaces. Proc. Int'l Joint Conf. Artif. Intell., Karlshrue, Germany (1983) pp. 996–1002

6.4   W. Grimson, T. Lozano-Perez: Model-based recognition and localization from sparse range or tactile data. Int'l J. Robotics Res. **5**(3), 3–34 (1984)

6.5   O. Faugeras, M. Hebert: The representation, recognition, and locating of 3D shapes from range data. Int'l J. Robotics Res. **5**(3), 27–52 (1986)

6.6   W. E. L. Grimson: *Object Recognition by Computer: The Role of Geometric Constraints* (MIT Press, Cambridge, MA 1990)

6.7   S. Pollard, J. Porrill, J. Mayhew, J. Frisby: Matching geometrical descriptions in three-space. Image and Vision Computing **5**(2), 73–78 (1987)

6.8   H. Chen, T. Huang: Matching 3-D line segments with applications to multiple-object motion estimation. IEEE Trans. PAMI–**12**(10), 1002–1008 (1990)

Chapter 7

7.1   R. Bolles, R. Cain: Recognizing and locating partially visible objects, the local-feature-focus method. Int'l J. Robotics Res. **1**(3), 57–82 (1982)

7.2   O. Faugeras, M. Hebert: A 3-D recognition and positioning algorithm using geometrical constraints between primitive surfaces. Proc. Int'l Joint Conf. Artif. Intell., Karlshrue, Germany (1983) pp. 996–1002

7.3   P. Horaud, R. Bolles: 3DPO's strategy for matching three-dimensional objects in range data. Proc. Int'l Conf. Robotics Automation, Atlanta, GA (1984) pp. 78–85

7.4 O. Faugeras, M. Hebert: The representation, recognition, and locating of 3D shapes from range data. Int'l J. Robotics Res. **5**(3), 27–52 (1986)

7.5 N. Ayache, O. Faugeras: HYPER: a new approach for the recognition and positioning of two-dimensional objects. IEEE Trans. PAMI–**8**(1), 44–54 (1986)

7.6 S. Pollard, J. Porrill, J. Mayhew, J. Frisby: Matching geometrical descriptions in three-space. Image and Vision Computing **5**(2), 73–78 (1987)

7.7 M. Dhome, M. Richetin, J. Lapresté, G. Rives: Determination of the attitude of 3-D objects from a single perspective view. IEEE Trans. PAMI–**11**(12), 1265–1278 (1989)

Chapter 8

8.1 R. Bolles, R. Cain: Recognizing and locating partially visible objects, the local-feature-focus method. Int'l J. Robotics Res. **1**(3), 57–82 (1982)

8.2 P. Horaud, R. Bolles: 3DPO's strategy for matching three-dimensional objects in range data. Proc. Int'l Conf. Robotics Automation, Atlanta, GA (1984) pp. 78–85

8.3 R. Bolles, P. Horaud: 3DPO: a three-dimensional part orientation system. Int'l J. Robotics Res. **5**(3), 3–26 (1986)

8.4 N. Ayache, O. Faugeras: HYPER: a new approach for the recognition and positioning of two-dimensional objects. IEEE Trans. PAMI–**8**(1), 44–54 (1986)

8.5 D. Lowe: *Perceptual Organization and Visual Recognition* (Kluwer Academic, Boston, MA 1985)

8.6 W. Grimson, T. Lozano-Perez: Model-based recognition and localization from sparse range or tactile data. Int'l J. Robotics Res. **5**(3), 3–34 (1984)

8.7 O. Faugeras, M. Hebert: The representation, recognition, and locating of 3D shapes from range data. Int'l J. Robotics Res. **5**(3), 27–52 (1986)

8.8 O. Faugeras, F. Lustman: Let us suppose that the world is piecewise planar, in *Proc. Int'l Symposium Robotics Res.*, ed. by O. D. Faugeras, G. Giralt (MIT Press, Cambridge, MA 1986) pp. 33–40

8.9 O. D. Faugeras, F. Lustman: Motion and structure from motion in a piecewise planar environment. Int'l J. Pattern Recog. Artif. Intell. **2**(3), 485–508 (1988)

8.10 M. Thonnat: Semantic interpretation of 3-D stereo data: finding the main structures. Int'l J. Pattern Recog. Artif. Intell. **2**(3), 509–525 (1988)

8.11 P. Grossmann: *Building Planar Surfaces From Raw Data*, Technical Report R4.1.2, ESPRIT Project P940 (1987)

8.12 P. Grossmann: From 3D line segments to objects and spaces. Proc. IEEE Conf. Comput. Vision Pattern Recog., San Diego, CA (1989) pp. 216–221

8.13 O. D. Faugeras: *Three-Dimensional Computer Vision* (MIT Press, Cambridge, MA 1991)

Chapter 9

9.1 G. Adiv: Determining three-dimensional motion and structure from optical flow generated by several moving objects. IEEE Trans. PAMI–**7**, 384–401 (1985)

9.2   A. Verri, T. Poggio: Against quantitative optical flow. Proc. First Int'l Conf. Comput. Vision, London, UK (1987) pp. 171–180

9.3   D. Marr: *Vision* (W.H. Freeman, San Francisco, CA 1982)

9.4   R. Nelson, J. Aloimonos: Using flow field divergence for obstacle avoidance in visual navigation. IEEE Trans. PAMI-11(10), 1102–1106 (1989)

9.5   H. Nagel: From image sequences towards conceptual descriptions. Image and Vision Computing 6(2), 59–74 (1988)

9.6   D. Koller, N. Heinze, H. Nagel: Algorithmic characterization of vehicle trajectories from image sequences by motion verbs. Proc. IEEE Conf. Comput. Vision Pattern Recog., Maui, Hawaii (1991) pp. 90–95

9.7   H. Chen, T. Huang: Maximal matching of 3-D points for multiple-object motion estimation. Pattern Recog. 21(2), 75–90 (1988)

9.8   H. Chen, T. Huang: Matching 3-D line segments with applications to multiple-object motion estimation. IEEE Trans. PAMI-12(10), 1002–1008 (1990)

Chapter 10

10.1   P. Besl, R. Jain: Three-dimensional object recognition. ACM Computing Surveys 17(1), 75–145 (1985)

10.2   R. Chin, C. Dyer: Model-based recognition in robot vision. ACM Computing Surveys 18(1), 67–108 (1986)

10.3   W. E. L. Grimson: *Object Recognition by Computer: The Role of Geometric Constraints* (MIT Press, Cambridge, MA 1990)

10.4   O. D. Faugeras: *Three-Dimensional Computer Vision* (MIT Press, Cambridge, MA 1991)

10.5   J. Porrill: Optimal combination and constraints for geometrical sensor data. Int'l J. Robotics Res. 7(6), 66–77 (1988)

10.6   W. Grimson, T. Lozano-Perez: Model-based recognition and localization from sparse range or tactile data. Int'l J. Robotics Res. 5(3), 3–34 (1984)

10.7   N. Ayache, O. Faugeras: HYPER: a new approach for the recognition and positioning of two-dimensional objects. IEEE Trans. PAMI-8(1), 44–54 (1986)

10.8   O. Faugeras, M. Hebert: The representation, recognition, and locating of 3D shapes from range data. Int'l J. Robotics Res. 5(3), 27–52 (1986)

10.9   D. Huttenlocher, S. Ullman: Recognizing solid objects by alignment with an image. Int'l J. Comput. Vision 5(2), 195–212 (1990)

10.10   D. Lowe: *Perceptual Organization and Visual Recognition* (Kluwer Academic, Boston, MA 1985)

10.11   D. Lowe: The viewpoint consistency constraint. Int'l J. Comput. Vision (1), 57–72 (1987)

10.12   R. C. Bolles, M. A. Fischler: A RANSAC-based approach to model fitting and its application to finding cylinders in range data. Proc. Int'l Joint Conf. Artif. Intell., Vancouver, Canada (1981) pp. 637–643

10.13   R. Bolles, R. Cain: Recognizing and locating partially visible objects, the local-feature-focus method. Int'l J. Robotics Res. 1(3), 57–82 (1982)

10.14   P. Horaud, R. Bolles: 3DPO's strategy for matching three-dimensional objects in range data. Proc. Int'l Conf. Robotics Automation, Atlanta, GA (1984) pp. 78–85

10.15   R. Bolles, P. Horaud: 3DPO: a three-dimensional part orientation system. Int'l J. Robotics Res. **5**(3), 3–26 (1986)

10.16   Y. Lamdan, H. J. Wolfson: Geometric hashing: a general and efficient model-based recognition scheme. Proc. Second Int'l Conf. Comput. Vision, Tampa, FL (1988) pp. 238–249

10.17   M. Dhome, M. Richetin, G. Rives: Model-based recognition and location of local patterns in polygonal contours via hypothesis accumulation, in *Pattern Recognition in Practice II*, ed. by E. S. Gelsema, L. N. Kanal (North-Holland, Amsterdam 1986) pp. 211–218

10.18   G. Stockman: Object recognition and localization via pose clustering. Comput. Vision, Graphics Image Process. (40), 361–387 (1987)

10.19   C. Hansen, T. C. Henderson: CAGD-based computer vision. IEEE Trans. PAMI-**11**(11), 1181–1193 (1989)

10.20   K. Ikeuchi: Generating an interpretation tree from a CAD model for 3D-object recognition in bin-picking tasks. Int'l J. Comput. Vision **1**(2), 145–165 (1987)

10.21   K. Ikeuchi, T. Kanade: Applying sensor models to automatic generation of object recognition programs. Proc. Second Int'l Conf. Comput. Vision, Tampa, FL (1988) pp. 228–237

10.22   A. K. Jain, R. Hoffman: Evidence-based recognition of 3-D objects. IEEE Trans. PAMI-**10**(6), 783–802 (1988)

Chapter 11

11.1   O. Faugeras, G. Toscani: The calibration problem for stereo. Proc. IEEE Conf. Comput. Vision Pattern Recog., Miami, FL (1986) pp. 15–20

11.2   R. Tsai: An efficient and accurate camera calibration technique for 3D machine vision. Proc. IEEE Conf. Comput. Vision Pattern Recog., Miami, FL (1986) pp. 364–374

11.3   R. Tsai, R. Lenz: A new technique for fully autonomous and efficient 3D robotics hand-eye calibration. Proc. Int'l Symposium Robotics Res., Santa Cruz, CA (1987)

11.4   M. Brady: Artificial intelligence and robotics. Artif. Intell. **26**, 79–121 (1985)

11.5   D. Kriegman, E. Triendl, T. Binford: A mobile robot: sensing, planning and locomotion. Proc. Int'l Conf. Robotics Automation, Raleigh, NC (1987) pp. 402–408

11.6   F. Lustman: Vision stéréoscopique et perception du mouvement en vision artificielle. Dissertation, University of Paris XI (Orsay, Paris, France) (1987)

11.7   N. Ayache: *Artificial Vision for Mobile Robots: Stereo Vision and Multisensory Perception* (MIT Press, Cambridge, MA 1991)

11.8   J. Weng, T. Huang, N. Ahuja: 3-D motion estimation, understanding, and prediction from noisy image sequences. IEEE Trans. PAMI-**9**(3), 370–389 (1987)

11.9   P. Tournassoud: Planification de trajectoires en robotique: complexité et approche pratique. Dissertation, University of Paris XI (Orsay, Paris, France) (1988)

11.10   M. Buffa, O. Faugeras, Z. Zhang: Obstacle avoidance and trajectory planning for an indoors mobile robot using stereo vision and Delaunay triangulation. Proc. Roundtable Discussion on Vision-Based Vehicle Guidance, Science University of Tokyo, Japan (1990) pp. 12-1–12-8

11.11   G. Toscani: Système de calibration optique et perception du mouvement en vision artificielle. Dissertation, University of Paris XI (Orsay, Paris, France) (1987)

Chapter 12

12.1   P. Tournassoud: Planification de trajectoires en robotique: complexité et approche pratique. Dissertation, University of Paris XI (Orsay, Paris, France) (1988)

12.2   T. Lozano-Pérez, M. Wesley: An algorithm for planning collision-free paths among polyhedral obstacles. Communications of ACM 560–570 (1979)

12.3   S. Tsuji, J. Zheng: Visual path planning by a mobile robot. Proc. Int'l Joint Conf. Artif. Intell., Milan, Italy (1987) pp. 1127–1130

12.4   O. Faugeras, N. Ayache, B. Faverjon: Building visual maps by combining noisy stereo measurements. Proc. Int'l Conf. Robotics Automation, San Francisco, CA (1986) pp. 1433–1438

12.5   N. Ayache, O. Faugeras: Building, registrating and fusing noisy visual maps. Proc. First Int'l Conf. Comput. Vision, London, UK (1987) pp. 73–82

12.6   G. Hager: Active reduction of uncertainty in multi-sensor systems. Ph.D. thesis, University of Pennsylvania (Philadelphia, PA) (1988)

12.7   J. Porrill: Optimal combination and constraints for geometrical sensor data. Int'l J. Robotics Res. **7**(6), 66–77 (1988)

12.8   H. Durrant-Whyte: Sensor models and multisensor integration. Int'l J. Robotics Res. **7**(6), 97–113 (1988)

12.9   M. Thonnat: Semantic interpretation of 3-D stereo data: finding the main structures. Int'l J. Pattern Recog. Artif. Intell. **2**(3), 509–525 (1988)

12.10   P. Grossmann: From 3D line segments to objects and spaces. Proc. IEEE Conf. Comput. Vision Pattern Recog., San Diego, CA (1989) pp. 216–221

12.11   J. Jezouin, N. Ayache: 3D structure from a monocular sequence of images. Proc. Third Int'l Conf. Comput. Vision, Osaka, Japan (1990) pp. 441–445

12.12   L. Matthies, A. Elfes: Integration of sonar and stereo range data using a grid-based representation. Proc. Int'l Conf. Robotics Automation, Philadelphia, PA (1988) pp. 727–733

12.13   M. Herman: Representation and incremental construction of a three-dimensional scene model, in *Techniques for 3-D Machine Perception*, ed. by A. Rosenfeld (Elsevier, North-Holland 1986) pp. 149–183

12.14   M. Herman, T. Kanade: Incremental reconstruction of 3D scenes from multiple, complex images. Artif. Intell. **30**, 289–341 (1986)

12.15   F. Lustman: Vision stéréoscopique et perception du mouvement en vision artificlelle. Dissertation, University of Paris XI (Orsay, Paris, France) (1987)

12.16   N. Ayache: *Artificial Vision for Mobile Robots: Stereo Vision and Multisensory Perception* (MIT Press, Cambridge, MA 1991)

12.17 M. Buffa, O. Faugeras, Z. Zhang: Obstacle avoidance and trajectory planning for an indoors mobile robot using stereo vision and Delaunay triangulation. Proc. Roundtable Discussion on Vision-Based Vehicle Guidance, Science University of Tokyo, Japan (1990) pp. 12-1–12-8

12.18 J. Robles: *Planification de trajectoires et évitement d'obstacles pour un robot mobile équipé de capteurs à ultrasons*, Rapport de DEA, University of Paris XI (Orsay, Paris, France) (1988)

12.19 Z. Zhang, O. D. Faugeras: A 3D world model builder with a mobile robot. Int'l J. Robotics Res. (1992)

12.20 R. Smith, P. Cheeseman: On the representation and estimation of spatial uncertainty. Int'l J. Robotics Res. **5**(4), 56–68 (1987)

12.21 P. Moutarlier, R. Chatila: Stochastic multisensory data fusion for mobile robot location and environment modelling. Proc. Int'l Symposium Robotics Res., Tokyo, Japan (1989) pp. 207–216

12.22 P. Grandjean, A. R. de Saint Vincent: *3-D Modeling of Indoor Scenes by Fusion of Noisy Range and Stereo Data*, Technical Report 89068, LAAS (Toulouse, France) (1989)

Chapter 13

13.1 T. Broida, R. Chellappa: Kinematics and structure of a rigid object from a sequence of noisy images: a batch approach. Proc. IEEE Conf. Comput. Vision Pattern Recog., Miami, FL (1986) pp. 176–182

13.2 T. Broida, R. Chellappa: Experiments and uniqueness results on object structure and kinematics from a sequences of monocular images. Proc. IEEE Workshop Visual Motion, Irvine, CA (1989) pp. 21–30

13.3 G. Young, R. Chellappa: 3–D motion estimation using a sequence of noisy stereo images: models, estimation, and uniqueness results. IEEE Trans. PAMI-**12**(8), 735–759 (1990)

13.4 M. Jenkin, J. Tsotsos: Applying temporal constraints to the dynamic stereo problem. Comput. Vision, Graphics Image Process. **24**, 16–32 (1986)

13.5 S. Sethi, R. Jain: Finding trajectories of feature points in a monocular image sequence. IEEE Trans. PAMI-**9**(1), 56–73 (1987)

13.6 J. Crowley, P. Stelmaszyk, C. Discours: Measuring image flow by tracking edge-lines. Proc. Second Int'l Conf. Comput. Vision, Tampa, FL (1988) pp. 658–664

13.7 J. Gambotto: Tracking points and line segments in image sequences. Proc. IEEE Workshop Visual Motion, Irvine, CA (1989) pp. 38–45

13.8 G. Gordon: On the tracking of featureless objects with occlusion. Proc. IEEE Workshop Visual Motion, Irvine, CA (1989) pp. 13–20

13.9 V. Hwang: Tracking feature points in time-varying images using an opportunistic selection approach. Pattern Recog. **22**(3), 247–256 (1989)

13.10 J. Weng, T. Huang, N. Ahuja: 3-D motion estimation, understanding, and prediction from noisy image sequences. IEEE Trans. PAMI-**9**(3), 370–389 (1987)

13.11 T. Broida, S. Chandrashekhar, R. Chellappa: Recursive 3-D motion estimation from a monocular image sequence. IEEE Trans. AES–**26**(4), 639–656 (1990)

13.12 G. Toscani, R. Deriche, O. Faugeras: 3D motion estimation using a token tracker. Proc. IAPR Workshop Comput. Vision: Special Hardware and Industrial Applications, Tokyo, Japan (1988) pp. 257–261

13.13 O. Faugeras, G. Toscani: The calibration problem for stereo. Proc. IEEE Conf. Comput. Vision Pattern Recog., Miami, FL (1986) pp. 15–20

13.14 R. Deriche, O. Faugeras: Tracking line segments. Proc. First European Conf. Comput. Vision, Antibes, France (1990) pp. 259–268

13.15 Y. Bar-Shalom, T. Fortmann: *Tracking and Data Association* (Academic, New York 1988)

13.16 B. Lowerre, R. Reddy: The harpy speech understanding system, in *Trends in Speech Recognition*, ed. by W. Lea (Prentice-Hall, NJ 1980)

13.17 J. Roach, J. Aggarwal: Computer tracking of objects moving in space. IEEE Trans. PAMI–**1**(2), 127–135 (1979)

13.18 D. Gennery: Tracking known three-dimensional objects. Proc. AAAI, Pittsburgh, PA (1982) pp. 13–17

Chapter 14

14.1 T. Broida, R. Chellappa: Kinematics and structure of a rigid object from a sequence of noisy images: a batch approach. Proc. IEEE Conf. Comput. Vision Pattern Recog., Miami, FL (1986) pp. 176–182

14.2 T. Broida, R. Chellappa: Experiments and uniqueness results on object structure and kinematics from a sequences of monocular images. Proc. IEEE Workshop Visual Motion, Irvine, CA (1989) pp. 21–30

14.3 J. Weng, T. Huang, N. Ahuja: 3-D motion estimation, understanding, and prediction from noisy image sequences. IEEE Trans. PAMI–**9**(3), 370–389 (1987)

14.4 G. Young, R. Chellappa: 3–D motion estimation using a sequence of noisy stereo images: models, estimation, and uniqueness results. IEEE Trans. PAMI–**12**(8), 735–759 (1990)

14.5 T. Broida, S. Chandrashekhar, R. Chellappa: Recursive 3-D motion estimation from a monocular image sequence. IEEE Trans. AES–**26**(4), 639–656 (1990)

14.6 J. Webb, J. Aggarwal: Structure from motion of rigid and jointed objects. Artif. Intell. **19**, 107–130 (1982)

14.7 H. Goldstein: *Classical Mechanics* (Addison Wesley, Reading, MA 1969)

14.8 W. L. Brogan: *Modern Control Theory* (Quantum Publishers, New York 1974)

14.9 Z. Zhang, O. Faugeras: Tracking 3D line segments: new developments. Proc. 5th Int'l Conf. Advanced Robotics, Pisa, Italy (1991) pp. 1365–1370

14.10 P. Maybeck: *Stochastic Models, Estimation and Control*, Vol. 2 (Academic, New York 1982)

14.11 D. Gennery: Tracking known three-dimensional objects. Proc. AAAI, Pittsburgh, PA (1982) pp. 13–17

Chapter 15

15.1   F. Preparata, M. Shamos: *Computational Geometry, An Introduction* (Springer, Berlin, Heidelberg 1986)
15.2   O. D. Faugeras, E. Lebras-Mehlman, J. Boissonnat: Representing Stereo data with the Delaunay Triangulation. Artif. Intell. **44**(1–2) (1990)
15.3   J. Crowley, P. Stelmaszyk, C. Discours: Measuring image flow by tracking edge-lines. Proc. Second Int'l Conf. Comput. Vision, Tampa, FL (1988) pp. 658–664
15.4   R. Deriche, O. Faugeras: Tracking line segments. Proc. First European Conf. Comput. Vision, Antibes, France (1990) pp. 259–268
15.5   Y. Bar-Shalom, T. Fortmann: *Tracking and Data Association* (Academic, New York 1988)
15.6   J. A. Hartigan: *Clustering Algorithms* (Wiley, New York 1974)
15.7   D. Murray, D. Castelow, B. Buxton: From an image sequence to a recognized polyhedral object. Image and Vision Computing **6**(2), 107–120 (1988)
15.8   Z. Zhang, O. Faugeras, N. Ayache: Analysis of a sequence of stereo scenes containing multiple moving objects using rigidity constraints. Proc. Second Int'l Conf. Comput. Vision, Tampa, FL (1988) pp. 177–186

Chapter 16

16.1   H. J. Wolfson: On curve matching. IEEE Trans. PAMI-**12**(5), 483–489 (1990)
16.2   E. Kishon, T. Hastie: 3D curve matching using splines. Proc. First European Conf. Comput. Vision, Antibes, France (1990) pp. 589–591
16.3   L. Robert, O. Faugeras: Curve-based stereo: figural continuity and curvature. Proc. IEEE Conf. Comput. Vision Pattern Recog., Maui, Hawaii (1991) pp. 57–62
16.4   E. Grosso, G. Sandini, M. Tistarelli: 3-D object reconstruction using stereo and motion. IEEE Trans. SMC-**19**(6), 1465–1476 (1990)
16.5   N. A. Thacker, Y. Zheng, R. Balckbourn: Using a combined stereo/temporal matcher to determine ego-motion. Proc. British Machine Vision Conf., University of Oxford, London, UK (1990) pp. 121–126
16.6   N. Navab, Z. Zhang, O. Faugeras: Tracking, motion and stereo: a robust and dynamic cooperation. Proc. Scandinavian Conf. Image Analysis, Aalborg University, Denmark (1991) pp. 98–105
16.7   J. Aloimonos, I. Weiss, A. Bandopadhay: Active vision. Int'l J. Comput. Vision **1**(4), 333–356 (1987)
16.8   A. L. Abbott, N. Ahuja: Surface reconstruction by dynamic integration of focus, camera vergence, and stereo. Proc. Second Int'l Conf. Comput. Vision, Tampa, FL (1988) pp. 532–543
16.9   G. Sandini, M. Tistarelli: Robust obstacle detection using optical flow. Proc. Int'l Workshop Robust Comput. Vision, Seattle, WA (1990) pp. 396–411
16.10 B. Horn: *Robot Vision* (MIT Press, Cambridge, MA and McGraw–Hill, New York 1986)

Subject Index